M. I. Page.

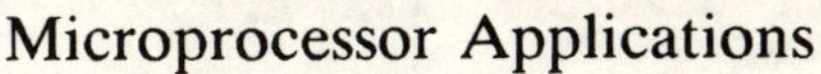

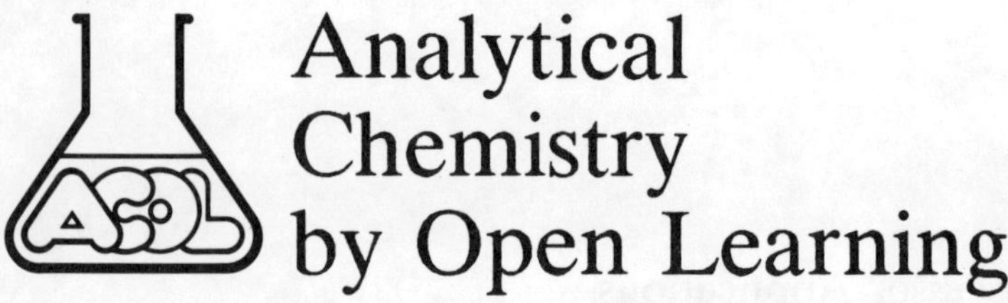

Titles in Series:

Samples and Standards
Sample Pretreatment
Classical Methods
Measurement, Statistics and Computation
Using Literature
Instrumentation
Chromatographic Separations
Gas Chromatography
High Performance Liquid Chromatography
Electrophoresis
Thin Layer Chromatography
Visible and Ultraviolet Spectroscopy
Fluorescence and Phosphorescence Spectroscopy
Infra Red Spectroscopy
Atomic Absorption and Emission Spectroscopy
Nuclear Magnetic Resonance Spectroscopy
X-Ray Methods
Mass Spectrometry
Scanning Electron Microscopy and X-Ray Microanalysis
Principles of Electroanalytical Methods
Potentiometry and Ion Selective Electrodes
Polarography and Other Voltammetric Methods
Radiochemical Methods
Clinical Specimens
Diagnostic Enzymology
Quantitative Bioassay
Assessment and Control of Biochemical Methods
Thermal Methods
Microprocessor Applications

Microprocessor Applications

Analytical Chemistry by Open Learning

Authors:
DONALD STEVENSON
Paisley College, UK

KEITH MILLER
Wolverhampton Polytechnic, UK

Editor:
ANTHONY C. NORRIS

on behalf of ACOL

Published on behalf of ACOL, Thames Polytechnic, London
by
JOHN WILEY & SONS
Chichester · New York · Brisbane · Toronto · Singapore

Library of Congress Cataloging in Publication Data:

Stevenson, Donald.
Analytical chemistry by open learning.
Microprocessor applications.

Bibliography: p.
1. Chemistry, Analytic—Data processing—Programmed instruction. 2. Microcomputers—Programming—Programmed instruction. I. Miller, K. J. (Keith John) II. Norris, A. C. (Anthony Charles) III. ACOL (Firm : London, England) IV. Title. V. Series: Analytical chemistry by open learning.
QD75.4.E4S74 1987 542'.8 86-28238
ISBN 0 471 91402 9
ISBN 0 471 91403 7 (pbk.)

British Library Cataloguing in Publication Data:

Stevenson, Donald
Microprocessor applications.—(Analytical chemistry)
1. Chemistry, Analytic–Automation
2. Microprocessors
I. Title II. Miller, Keith III. Norris, Anthony IV. Analytical Chemistry by Open Learning. *(Project)* V. Series
543'.0028'5416 QD75.4.A8

ISBN 0 471 91402 9
ISBN 0 471 91403 7 Pbk

Printed and bound in Great Britain

Analytical Chemistry

This series of texts is a result of an initiative by the Committee of Heads of Polytechnic Chemistry Departments in the United Kingdom. A project team based at Thames Polytechnic using funds available from the Manpower Services Commission 'Open Tech' Project has organised and managed the development of the material suitable for use by 'Distance Learners'. The contents of the various units have been identified, planned and written almost exclusively by groups of polytechnic staff, who are both expert in the subject area and are currently teaching in analytical chemistry.

The texts are for those interested in the basics of analytical chemistry and instrumental techniques who wish to study in a more flexible way than traditional institute attendance or to augment such attendance. A series of these units may be used by those undertaking courses leading to BTEC (levels IV and V), Royal Society of Chemistry (Certificates of Applied Chemistry) or other qualifications. The level is thus that of Senior Technician.

It is emphasised however that whilst the theoretical aspects of analytical chemistry can be studied in this way there is no substitute for the laboratory to learn the associated practical skills. In the U.K. there are nominated Polytechnics, Colleges and other Institutions who offer tutorial and practical support to achieve the practical objectives identified within each text. It is expected that many institutions worldwide will also provide such support.

The project will continue at Thames Polytechnic to support these 'Open Learning Texts', to continually refresh and update the material and to extend its coverage.

Further information about nominated support centres, the material or open learning techniques may be obtained from the project office at Thames Polytechnic, ACOL, Wellington St., Woolwich, London, SE18 6PF.

How to Use an Open Learning Text

Open learning texts are designed as a convenient and flexible way of studying for people who, for a variety of reasons cannot use conventional education courses. You will learn from this text the principles of one subject in Analytical Chemistry, but only by putting this knowledge into practice, under professional supervision, will you gain a full understanding of the analytical techniques described.

To achieve the full benefit from an open learning text you need to plan your place and time of study.

- Find the most suitable place to study where you can work without disturbance.

- If you have a tutor supervising your study discuss with him, or her, the date by which you should have completed this text.

- Some people study perfectly well in irregular bursts, however most students find that setting aside a certain number of hours each day is the most satisfactory method. It is for you to decide which pattern of study suits you best.

- If you decide to study for several hours at once, take short breaks of five or ten minutes every half hour or so. You will find that this method maintains a higher overall level of concentration.

Before you begin a detailed reading of the text, familiarise yourself with the general layout of the material. Have a look at the course contents list at the front of the book and flip through the pages to get a general impression of the way the subject is dealt with. You will find that there is space on the pages to make comments alongside the text as you study—your own notes for highlighting points that you feel are particularly important. Indicate in the margin the points you would like to discuss further with a tutor or fellow student. When you come to revise, these personal study notes will be very useful.

Π When you find a paragraph in the text marked with a symbol such as is shown here, this is where you get involved. At this point you are directed to do things: draw graphs, answer questions, perform calculations, etc. Do make an attempt at these activities. If necessary cover the succeeding response with a piece of paper until you are ready to read on. This is an opportunity for you to learn by participating in the subject and although the text continues by discussing your response, there is no better way to learn than by working things out for yourself.

We have introduced self assessment questions (SAQ) at appropriate places in the text. These SAQs provide for you a way of finding out if you understand what you have just been studying. There is space on the page for your answer and for any comments you want to add after reading the author's response. You will find the author's response to each SAQ at the end of the text. Compare what you have written with the response provided and read the discussion and advice.

At intervals in the text you will find a Summary and List of Objectives. The Summary will emphasise the important points covered by the material you have just read and the Objectives will give you a checklist of tasks you should then be able to achieve.

You can revise the Unit, perhaps for a formal examination, by re-reading the Summary and the Objectives, and by working through some of the SAQs. This should quickly alert you to areas of the text that need further study.

At the end of the book you will find for reference lists of commonly used scientific symbols and values, units of measurement and also a periodic table.

Contents

Study Guide

The last decade has seen a rapid growth in the use of computers in the analytical chemical laboratory. Most analytical instruments are now manufactured with an on-board microcomputer system for control purposes or for automatic processing of raw data. Additionally general purpose microcomputers, dedicated to analytical applications by specially written computer programs, are finding increased use. Such systems generally provide the analytical chemist with some flexibility about the way the collected data is to be manipulated. In some cases the software system provided allows a certain amount of programming by the analyst using a specialist language appropriate to the analytical task. Analytical chemists can also use general purpose microcomputers, interfaced to analytical instruments, to develop their own software, although care must be taken not to underestimate the development time required to produce a fully working and well tested system.

Because of the above developments in laboratory computing there is an increasing need for analytical chemists to become computer literate. Even an elementary knowledge of computer programming, computer interfacing, and an appreciation of the scope and limitations of computerisation, would allow the analytical chemist to use computerised equipment more confidently. A more advanced knowledge, although not necessarily appropriate for everyone, would increase the likelihood that existing hardware and software are fully exploited and allow new opportunities in analytical computing to be recognised.

The aim of this Unit is not to attempt to turn analytical chemical technicians into electronic engineers or computer programmers. Our aim is to provide a sufficiently good grounding in both computing and computer interfacing so that the analyst is confident in using computerised equipment and, if necessary, can overcome the 'jargon barrier' to discuss problems or developments with computer specialists and electronic engineers. Those who cope readily with the material of this Unit should find it a useful base for further study of computer applications in analytical chemistry.

Parts 1 to 3 provide an introduction to computing, programming and interfacing for the analytical chemist. Parts 1 and 2 are relatively free standing, but you must understand the contents of both of them before proceeding to Part 3. To satisfactorily complete Part 2 you will need access to a microcomputer with BASIC as a programming language.

Parts 4 to 6 are devoted to applications of the material covered in Parts 1 to 3 to analytical chemistry problems.

Bibliography

For an alternative presentation of much of the contents of Part 1 you may wish to consult the books listed below.

E. Morgan, *Laboratory Computing*, Sigma-Technical Press, Distributed by J. Wiley & Sons, Chichester, 1984.

C. G. Morgan, *The Micro in the Laboratory*, Sigma, Chichester, 1984.

You may also find it useful to refer to the introductory parts of the texts cited in the Overview for Part 3, Microcomputer Interfacing. All of these texts deal with number systems and have some general comments about microcomputers relevant to the content of Part 1.

For an alternative approach to programming (Part 2):

D. M. Monro, *Introduction to Computing with BASIC*, Arnold, 1974.

D. M. Monro, *Basic BASIC*, 2nd. Edn, Wiley, 1985.

C. Prigmore, *30 Hour BASIC*, National Extension College Correspondence Texts, Course M27, 1981.

A good general introduction to computers, computing, data processing and information technology (including some programming) is:

D. R. Sullivan, G. Lewis and C. R. Cook, *Using Computers Today*, Houghton Mifflen, 1986.

As general background reading you are recommended to study the following articles in the literature:

D. Malcolme-Lawes, *Microcomputers in the Chemical Laboratory*, *Chem. Brit.*, **20** (5) 1984, 425.

P. J. Farago, *Peek, Poke and Run*, *Chem. Brit.*, **18** (1), 1982, 40.

A. Hinchliffe, *Microcomputers in Chemistry Teaching*, *Educ. in Chem.* **20**, 1983, 44.

G. S. Owen, *Choosing an Appropriate Computer Language*, *J. Chem. Educ.*, **61** (2), 1984, 440.

R. E. Dessey, *Chemistry and the Microcomputer Revolution*, *J. Chem. Educ.* **59** (4), 1982, 321.

The following books and articles on data processing will be useful reference material in support of Parts 4, 5 and 6.

A. C. Norris, *Computational Chemistry*, Wiley, 1981.

William S. Dorn and Daniel D. McCracken, *Numerical Methods with Fortran IV Case Studies*, Wiley, 1972.

D. Binkley and R. J. Dessy, *J. Chem. Educ.*, **56**, 148, 1979.

A. Savitsky and M. J. E. Golay, *Anal. Chem.*, **36**, 1627, 1964. (Some corrections to the Savitsky and Golay tables are published in Steiner J. *et al*, *Anal. Chem.*, **44**, 1906, 1972.)

1. Microprocessors and Computing Concepts

Overview

On completing Sections 1.1 and 1.2 of this unit you should be able to relate binary numbers to their decimal and hexadecimal equivalents, and convert numbers from one system to another. You should be able to cope with the material of Section 1 without any previous knowledge of binary numbers.

The content of Section 1.2 requires a knowledge of the material in Section 1.1 to allow you to develop an understanding of the function of the principal components of a microcomputer system. Section 1.2 should introduce you to much of the jargon used with microprocessors. You should also become aware of the variety of microcomputers used in modern analytical laboratories and be able to relate the characteristics of the computer system to the needs of the analysis.

Section 1.3 provides an introduction to the different approaches to computer programming. From this section you should be able to distinguish between machine code, assembler programs and programs written in high-level languages. You should also start to appreciate the scope and limitations of each method of programming.

There are no pre-requisites for Part 1 but it must be studied before Part 3 and later units. Those who wish to study only BASIC programming can skip Part 1 and proceed directly to Part 2.

1.1. COMPUTERS AND NUMBER SYSTEMS

1.1.1. Binary Numbers

Laboratory applications of computers range from small systems dedicated to relatively simple apparatus such as a balance or a pH meter to quite sophisticated systems used with nuclear magnetic resonance spectrometers. Computers may be used to collect and store numeric data from instruments, to undertake calculations and store and retrieve information which may involve text.

All digital computers must reduce the above types of information, as well as the programs which specify what has to be done, to numeric codes expressed in binary. In the familiar decimal system we have ten digits 0, 1 ... 8, 9 but in binary we have only two digits namely 0 and 1. A binary digit is called a bit (Binary digIT). To express a number greater than 9 in decimal requires more than one decimal digit. In the same way to express a number greater than one in binary needs more than one bit. In fact to reach a number as high as 255 we need 8 bits, which is called a byte.

In writing a binary number, as with the decimal system, the least significant digit is written on the right and the most significant one is placed on the left, as shown below for an 8-bit number in which the individual bits have been labelled d0 (least significant bit) to d7.

	most significant bit							least significant bit
	d7	d6	d5	d4	d3	d2	d1	d0
weight of bit	2^7	2^6	2^5	2^4	2^3	2^2	2^1	2^0
	128	64	32	16	8	4	2	1

Thus to write the number 13 using 8 bits we would need to include the bits with weight 1, 4 and 8 as shown below.

0 0 0 0 1 1 0 1

The decimal equivalent of 13 is thus the sum of the weights of the bits set to 1 in the binary representation:

$$\text{decimal } 13 = 1 \times 2^3 + 1 \times 2^2 + 1 \times 2^0$$

SAQ 1.1a

What are the decimal numbers which correspond to each of the following bytes?

(*i*) 01101101

(*ii*) 10000000

(*iii*) 11111111

Often we need to work out the binary pattern which corresponds to a given decimal number. One way to do this is to find the weight of the bit which is smaller or equal to the given decimal value. This weight is then subtracted from the decimal number noting that the corresponding bit will be set to 1 in the answer. The process is repeated on the residue, and so on until no residue remains.

Suppose we wish to find the binary number which corresponds to 122. Referring as usual to the individual bits in a byte as d0, d1, ... d7, the weight of bit d7 is too high to contribute so in our answer d7 = 0. The weight of d6 is 64, so d6 = 1, and after subtracting 64 we obtain 58. From this we can subtract 32, so d5 = 1, giving 26. The weight of d4 is 16 so d4 = 1, and subtraction of 16 gives a residue of 10. Repeating this process gives the final bits as d3 = 1, d2 = 0, d1 = 1 and d0 = 0. The binary number equivalent 122 is therefore 01111010.

An alternative method that you may prefer is to repeatedly divide the decimal number by 2. The first division will give a remainder of 0 or 1 and this is the value of bit 0. Repeated division, noting the remainder each time generates the binary number starting at the least significant bit. Following this method, let us obtain the binary equivalent of 35.

35/2 = 17 + remainder 1

17/2 = 8 + remainder 1

8/2 = 4 + remainder 0

4/2 = 2 + remainder 0

2/2 = 1 + remainder 0

1/2 = 0 + remainder 1

The 8-bit binary equivalent of 35 is therefore 000100011.

SAQ 1.1b

Write the 8-bit binary numbers which correspond to each of the following decimal numbers.

(*i*) 38
(*ii*) 240
(*iii*) 15

The binary numbers above correspond to positive integers and are particularly important for passing binary data between computers and devices connected to them. All data stored in a digital computer are represented in binary, including negative and fractional numbers, although we shall only be concerned with binary numbers corresponding to positive integers or zero.

1.1.2. Hexadecimal Numbers

Binary numbers are quite difficult to remember and are rather tedious to write. They are easier to handle in groups of four which corresponds to a 'nibble' (a nibble is half a byte!). The four bits of a nibble can hold whole numbers in the range 0 to 15. This corresponds to the range of numbers used by the hexadecimal number system. In decimal we have 10 digits, but in hexadecimal there are 16 represented by the following, with the decimal values given directly below:

Hexa-decimal	0	1	2	3	4	5	6	7	8	9	A	B	C	D	E	F
Decimal	0	1	2	3	4	5	6	7	8	9	10	11	12	13	14	15

Notice that the symbols used for the hexadecimal digits are identical to those for decimal except in the range ten to fifteen where no corresponding single digits exist.

To represent an 8-bit binary number in terms of hexadecimal we first divide up the byte into least significant and most significant nibbles. Each nibble can then be equated to a single hexadecimal digit and so the original byte can be conveniently written as two hexadecimal digits. The right-hand digit is the least significant one and has a weight of 1, and the left-hand digit has a weight of 16.

Π What are the hexadecimal digits which represent the binary number 10101110?

Firstly, write the binary number as most significant and least significant nibbles:

1010	1110
most sig.	least sig.

The least significant nibble is equivalent to the hexadecimal digit E. Similarly the most significant nibble corresponds to the hexadecimal digit A. Putting the two hexadecimal digits together in order of significance gives AE as the equivalent of the original binary number.

It is possible to confuse hexadecimal and binary numbers. For example the number 99 may refer to decimal or hexadecimal unless we adopt some convention. Several methods are in use in the literature but a common one is to precede hexadecimal numbers with an ampersand (&) so that 99 is a decimal number but &99 is hexadecimal. In principle we could confuse binary with decimal and hexadecimal (consider 11 for example) but in practice we shall always write binary in groups of 4, or 8 and occasionally 16-bits with leading zeros included, and confusion is unlikely to arise.

SAQ 1.1c

Write each of the following binary numbers in hexadecimal.

(*i*) 01110011

(*ii*) 11111111

(*iii*) 11100001

1.1.3. Binary-coded Decimal

Many instruments provide data for input to a computer which are in Binary-coded Decimal (BCD). Here each decimal digit is coded separately as a 4-bit binary number, a nibble. Later, we shall be concerned with microcomputer input and the byte will play a special role. At this stage we should therefore note that two BCD digits can be contained in a single byte. The example shown below illustrates the BCD coding of the decimal number 3271 in two bytes.

Byte 1

d7	d6	d5	d4	d3	d2	d1	d0
0	0	1	1	0	0	1	0
Thousands				Hundreds			
3				2			

Byte 0

d7	d6	d5	d4	d3	d2	d1	d0
0	1	1	1	0	0	0	1
Tens				Units			
7				1			

SAQ 1.1d

Which of the following nibbles represent valid binary-coded decimal (BCD) digits?

	Nibble	Valid
(*i*)	1000	Y/N
(*ii*)	1001	Y/N
(*iii*)	1010	Y/N
(*iv*)	1111	Y/N

1.1.4. Memory Addressing

Modern computers usually have considerable memory capacity for the storage of data and programs. Individual memory locations are numbered consecutively from 0 and the number of a given location is known as the memory address. To access any memory location the computer has to specify the address in terms of a binary number. Even with small computers the number of memory locations available in principle usually extends from 0 to 64K (where 1K = 1024 bytes) or 65535. The range of possible memory addresses must also extend to 65535 and this can only be achieved by use of 16-bit binary numbers.

Memory addresses are therefore expressed in terms of decimal or as 4-digit hexadecimal numbers. To convert from hexadecimal addresses to decimal use the idea of each digit having a particular weight as shown below for the address &FE11.

Hexadecimal digit	d3	d2	d1	d0
	F	E	1	1
Weight of digit	4096	256	16	1
	16^3	16^2	16^1	16^0

&FE11 = 15 × 4096 + 14 × 256 + 1 × 16 + 1 × 1

i.e. &FE11 = 65041

SAQ 1.1e

Match each of the memory addresses on the left with one of the alternatives (*a*)–(*g*) on the right.

(*i*) &F000	(*a*) 61440
	(*b*) &0400
(*ii*) 1023	(*c*) &00FF
	(*d*) &03FF
(*iii*) &00F0	(*e*) 65535
	(*f*) 240
(*iv*) Binary 1111111111111111	(*g*) 15

SAQ 1.1e

If you have managed to complete all the self assessment questions satisfactorily you are well placed to proceed with the next section of this package which introduces some fundamental aspects of computers relevant to chemistry.

1.2. MICROCOMPUTER FUNDAMENTALS

1.2.1. Elements of Microcomputer Systems

Although there is tremendous variety in microcomputers used in laboratories, certain features, represented schematically in Fig. 1.2a, are common to most systems.

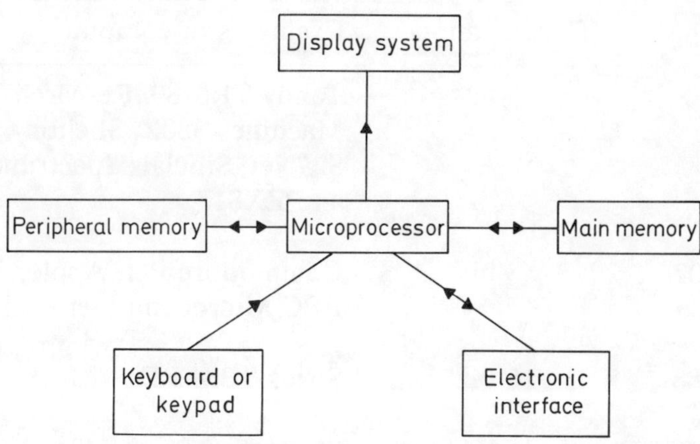

Fig. 1.2a. *Schematic diagram showing the principal components of a microcomputer system*

The arrows in the diagram show the possible directions of information flow. Notice that some links imply two-way communication but others do not. Each part of the diagram will be discussed in the remainder of this section.

1.2.2. The Microprocessor

This is the heart of any microcomputer system. It is an integrated circuit device which in spite of its complexity is relatively cheap because of the generality of application and the consequent economies of volume production. Before the advent of the microprocessor increasing sophistication of an integrated circuit device tended to imply greater specialisation in its use and therefore a more restricted market. The generality of a microprocessor can be dedicated to a specialist application by the provision of a program for the user. Some examples of currently available microprocessors are given in Fig. 1.2b.

Type	Classification	Examples of computer
Z80	8-bit	Tandy TRS-80, Research Machines 380Z, Shelton Sig/Net, Sinclair Spectrum and ZX81.
6502	8-bit	Commodore Pet, Apple, BBC Microcomputer.
8088	16-bit	Sirius, IBM PC.
8086	16-bit	Apricot, Olivetti M24.
68000	16-bit	Sage, Atari ST.

Fig. 1.2b. *Examples of microprocessors*

Microprocessors such as the Z80 or 6502 are classified as 8-bit microprocessors because the byte is the fundamental unit of data storage and processing. Such microprocessors have been available since the early 1970s and have been the basis of many popular microcomputer systems examples of which are included in Fig. 1.2b.

More recently 16-bit microprocessors have become available. Here the basic unit of data processing and storage is 16 bits although some of them transfer data 8 bits at a time, as in the case of the 8088. Other 16-bit microprocessor such as the 8086 or the 68000 use 16 bits for both data processing and data transfer. Research and development work in microprocessor technology continues at a pace and 32-bit microprocessors are now also available.

All microprocessors are designed to carry out basically two functions. One is to fetch instructions which are stored in the form of binary codes, and the other is to execute them in accordance with the manufacturer's design. Typical actions include moving 8-bit, or 16-bit, quantities from one storage location to another, simple arithmetic operations such as addition and subtraction (not multiplication or division), and the comparison of one byte of data with another.

SAQ 1.2a

By choosing the most appropriate phrase from the list below, complete the following sentence.

The microprocessor is .... part of a microcomputer system.

(*i*) the most important

(*ii*) the most expensive

(*iii*) a desirable optional

SAQ 1.2b

By ringing either T or F, indicate whether each of the following statements is true or false.

Microprocessors can undertake the following operations directly:

(*i*) multiplication or division (T / F)

(*ii*) addition or subtraction (T / F)

(*iii*) comparison of data bytes (T / F)

1.2.3. Main Memory

As we have just considered, the microprocessor fetches stored instructions and obeys them. The instructions are held as 8-bit quantities in solid-state memory chips (integrated circuit devices again) and, when taken as a consecutive sequence, the instructions define the program to be executed by the microprocessor.

Two types of memory are commonly used with microcomputers. One is called 'volatile memory' because once the power is switched off any stored information is lost. It is also called Random Access Memory (RAM), and whilst power is connected the contents of RAM locations can be changed. The user of the computer system normally places his program in RAM so that it is available for access by the microprocessor during execution.

The second type of memory is described as 'non-volatile' because the instructions contained in it are not lost when the power is switched off. ROM or Read-only-memory is an example of this type. As the name implies once information has been placed in ROM, for example by a manufacturer, it cannot be altered. However the information can be read by the microprocessor in its fetch-execute cycle and ROM is therefore used for programs which are needed for operation as soon as the power is switched on. A typical example of a program placed in ROM is one for scanning the keyboard of the computer in order to take in the instructions of the user.

Programmable-read-only-memory, PROM, can be programmed by the computer user using a special item of equipment known as a PROM programmer, but once programmed it cannot be changed. However some types of non-volatile memory can be reprogrammed. A type often used in prototype development is called EPROM which stands for Erasable-programmable-read-only-memory. An EPROM chip has a small window, as shown in Fig. 1.2c, through which ultra-violet light can be shone to erase the contents of the memory locations. A new program can then be placed in the EPROM chip by use of a PROM programmer.

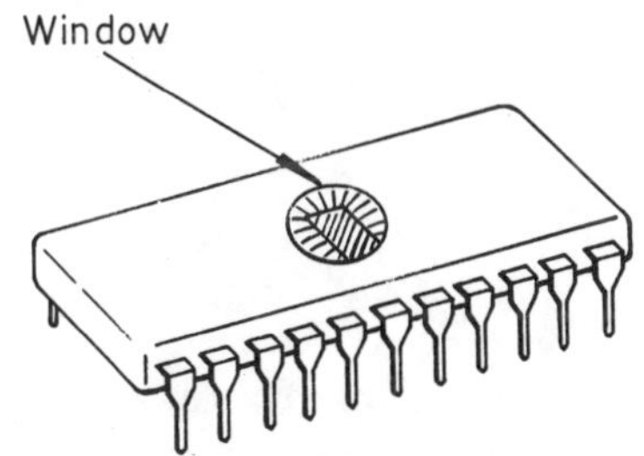

Fig. 1.2c. *A typical EPROM chip*

For volume production programs are placed in ROM rather than EPROM because a smaller unit cost can be achieved in spite of high initial costs. On the other hand EPROM is very useful for prototype read-only programs or when low numbers of units are to be produced.

SAQ 1.2c

Circle T or F to indicate whether each of the following is true or false.

(*i*) RAM can be used for the long-term storage of programs.

(T / F)

(*ii*) ROM is used for the storage of programs written by analytical chemists because the program is always available for use.

(T/ F)

(*iii*) EPROM can be programmed using an ordinary microcomputer in the same way as RAM.

(T / F)

(*iv*) The contents of RAM can be made non-volatile by the provision of supplementary power, in the form of a battery, which becomes active when the mains power is switched off.

(T / F)

(*v*) Prototype programs should never be committed to ROM.

(T / F)

(*vi*) All microcomputer systems use some programs contained in non-volatile memory but not all may require programs in RAM.

(T / F)

1.2.4. Peripheral Memory

The previous section concentrated on the microcomputer's main memory in which the program is stored ready for access and execution by the microcomputer. Clearly we need a convenient means of storing any programs we develop or data we collect from instruments. When programs are first developed they reside in RAM, along with any data collected. The problem is that this information will be lost when the mains power is removed, unless we first transfer it to a permanent storage medium which we shall refer to as peripheral memory.

The most common medium for permanent storage of data is magnetic disc or magnetic tape. Once the magnetic coating has been magnetised to indicate digital values of 0 or 1 (all information is stored in binary) a permanent magnetic copy of the program or data is obtained.

Magnetic tape used for audio recordings provide a cheap medium for peripheral memory. Using conventional tape recorders, the digital values of 0 and 1 are first converted to audio tones and then recorded on tape. Although inexpensive, this method can be unreliable since the ability to store and retrieve information depends on the recording and play-back levels as well as the setting of tone controls and the presence of background noise. Another disadvantage is that the transfer of information to or from tape is a slow process often taking several minutes to complete depending on the length of the program.

A preferred method is to use magnetic disc as a storage medium. This can take the form of floppy disc which, once formatted, contains a number of concentric tracks as shown in Fig. 1.2d.

The access to any track on the disc is relatively quick so that programs or data files can be loaded in a second or so. The capacity of the disc depends on the number of tracks available which varies from one computer system to another but in general one should have room for at least several hundred kilobytes information. The floppy discs are easily damaged and should always be kept in their protective sleeve when not in use. Also one should only label them

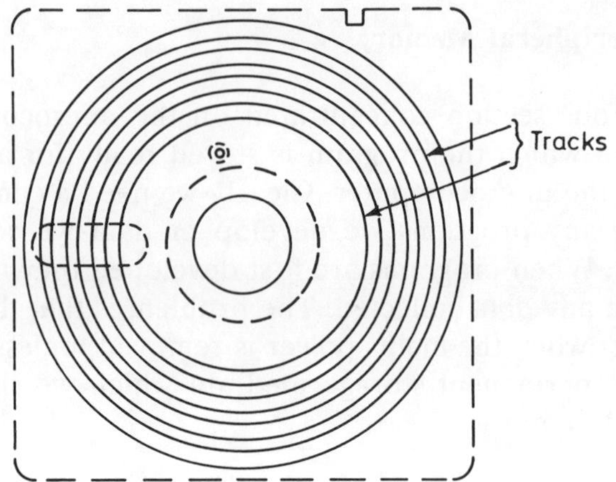

Fig. 1.2d. *A typical layout of tracks on a floppy disc*

using a felt-tipped pen. Finally make sure they are kept away from magnets, including the transformers used with instruments powered from the mains, or the stored data may be corrupted.

Magnetic storage of programs and data is also possible using so-called hard discs (a Winchester disc for example). These are completely sealed and are thus less prone to damage and are used with the more expensive microcomputer systems where a large storage of the order of tens of megabytes may be needed.

SAQ 1.2d

Circle T or F to indicate whether the statements below are true or false.

(*i*) During execution of the program the machine code instructions to be accessed by the microprocessor must be placed on either magnetic disc or tape.

(T / F)

(*ii*) All microcomputer systems must have facilities for either disc or tape storage.

(T / F)

⟶

SAQ 1.2d (cont.)

(*iii*) Tape storage is in general inferior to disc storage ONLY because of its lower reliability with respect to retrieval of information.

(T / F)

(*iv*) Magnetic materials can adversely affect the integrity of information stored on tape or disc.

(T / F)

1.2.5. Keyboard and Keypad

These provide the means for input of information (programs or data) to the computer. If you wish to program in a language like BASIC then you will need a full QWERTY keyboard of the type found on a typewriter. On the other hand if you are using a microcomputer system which is imbedded in an instrument such as an atomic absorption spectrophotometer then only a restricted keypad, as shown in Fig. 1.2e, will be required.

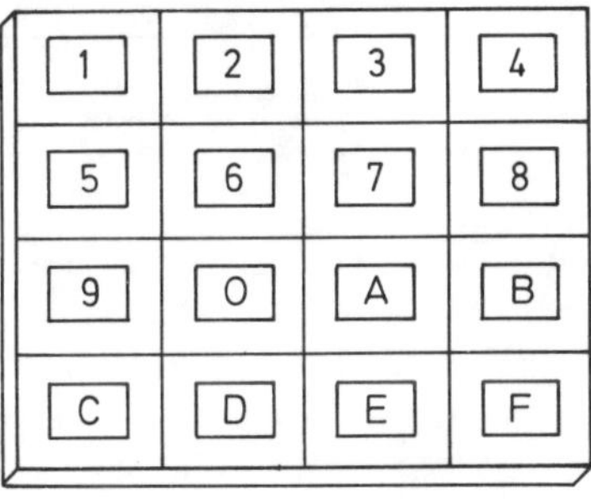

Fig. 1.2e. *A restricted keypad*

Virtually all the microcomputer systems that you come across in a laboratory environment will have either a full keyboard or a keypad for you to input information needed by the program to define its operation.

SAQ 1.2e

Assuming that you are using a microcomputer which has a keypad equipped with the digits 0–9, the letters A, B, C, D, E and F, and other specialist keys appropriate to the application in question, indicate by circling Y or N whether you could (Y) or could not (N) accomplish the following:

(*i*) Input a machine code program in hexadecimal.

(Y / N)

(*ii*) Specify numeric data such as the number of a sample.

(Y / N)

(*iii*) Program the microcomputer in BASIC.

(Y / N)

(*iv*) Specify the operating parameters of an instrument such as an infra-red spectrophotometer.

(Y / N)

1.2.6. Display System

This is the means of passing information from the microcomputer to the user. The type of display needed depends on the application. For example in infra-red spectroscopy a good graphic capability is required. Normally a microcomputer screen is divided into a number of small cells or picture elements (pixels). The larger the number of cells the greater will be the resolution which can be obtained allowing greater detail in plots to be displayed. As a general guide a screen with say 30 rows of 40 columns giving 1200 pixels would give rather low resolution, but one with say 400 × 500 pixels would provide very good resolution, certainly sufficient for an infra-red spectrum. Many computers reserve part of their main memory to store information currently displayed on the screen and the higher the resolution the greater the memory that is required.

Some applications may not require a graphics capability. For example the display of alphabetic and numeric characters is quite sufficient for word processing applications.

A very simple display is all that is required for many laboratory applications. Often only a single line of information has to be displayed at a time. This is often accomplished using a bank of seven-segment displays, or a liquid crystal display panel as shown in Fig. 1.2f.

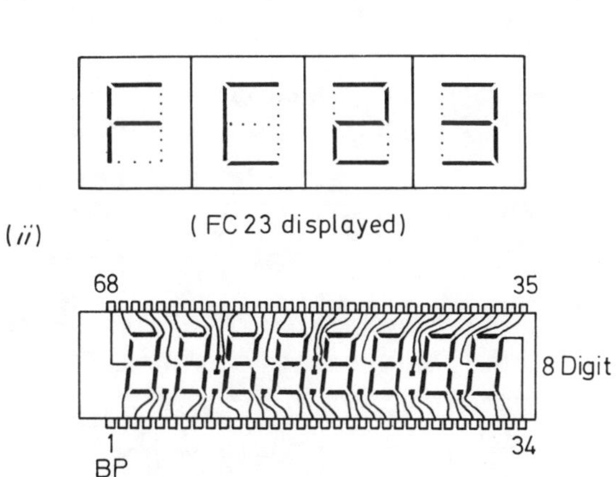

Fig. 1.2f. *Examples of (i) a bank of seven-segment displays, and (ii) a liquid crystal display*

The digital displays associated with for example pH meters or balances often use seven-segment displays. Portable computers sometimes have the facility to show one, or a few, lines of information using a liquid crystal display panel.

Almost all computer systems that we are likely to encounter in an analytical laboratory will have some type of visual display. If a permanent copy is required then a printer must be attached to the system. On sophisticated instruments, such as a modern recording spectrophotometer, alphanumeric information can be printed on the same chart as the spectrum.

SAQ 1.2f

A small microcomputer allows the programmer to select one of a number of modes of screen display with different resolutions. The highest resolution (640 × 256) has 640 pixels in a row and the screen is filled by 256 such rows. A medium resolution mode is available with 320 × 256 pixels. Both modes use main memory to store data on each pixel. For a monochrome display, which you can assume is in use in this example, each pixel is represented by 1 bit.

(*i*) How much memory has to be reserved when using

(*a*) high resolution,
(*b*) medium resolution?

(Express your answer in Kbytes (1K = 1024 bytes).

(*ii*) If the program takes 13K of memory and the computer has a total of 32K of memory available, could the high resolution graphics mode be used?

SAQ 1.2f

1.2.7. The Electronic Interface

This provides communication in terms of digital signals between the computer and external equipment including laboratory instrumentation. Microcomputers represent binary information in terms of two voltage levels, approximately 0 volts for digital 0 and about 5 volts for digital 1. All of the digital circuits in a microcomputer system use these voltages and they can be damaged if negative voltages or voltages appreciably greater than 5 volts are connected. The problem is that a lot of equipment used in laboratories works on 12, 24 volts DC or mains voltages (110 or 240 Volts AC). The electronics associated with a computer interface must be designed to isolate the computer from these potentially destructive voltages. Even when an external device operates on digital signals of 0 and 5 volts it is necessary to provide an electronic interface to provide the correct timing of input and output signals involving external equipment. Computer output signals are over in a very brief time (microseconds) and so we often need to design an interface which will catch and hold (ie 'latch') these signals to apply them over an extended time period for use with external devices.

Many electrical signals obtained from analytical instruments vary continuously with time and are not restricted to 0 or 5 volts. For example a signal fed to a chart recorder may vary between say 0 and 100 millivolts during a run. Such a continuously variable signal, or analogue signal, can be read by the computer via an appropriate interface called an analogue-to-digital converter, to be discussed in Part 3 of this unit. In some applications it may be necessary for the computer to produce an analogue signal as output. Again we would require a special interface, called a digital-to-analogue converter, which will also be discussed in Part 3.

SAQ 1.2g

(*i*) Is an interface necessary if we wish to read data into a computer from an instrument which provides digital signals (0 and 5 volts)?

(*ii*) Is an interface necessary if we wish to switch on or off, under computer control, an electric motor which runs on 5 volts?

You have now completed the introductory material on microprocessors and should now be ready to consider how they may be programmed. If you are unsure of any of the points covered or you or if you want further information consult the literature indicated in the Study Guide.

1.3. METHODS OF PROGRAMMING MICROCOMPUTERS

As an illustration of the different methods of programming we shall examine the trivial problem of getting the computer to count down from 255 to 0 by steps of 1.

1.3.1. Machine Code Programs

The microprocessor executes machine code instructions and so let us start with programming directly in machine code, which is the lowest level of programming language. The actual coding depends on the microprocessor so that a machine code program for the Z80 microprocessor will not run on a microcomputer which is based on the 6502.

Fig. 1.3a shows an example of a machine code program for the 6502 microprocessor. The column on the left gives the addresses of the memory locations in which the machine code instructions have been placed. The column on the right gives the hexadecimal code which makes up the program instructions and associated data. There is no way that we can tell what the individual instructions do without reference to the literature provided by the manufacturer of the 6502 microprocessor. To understand how the program works you need to know that the 6502 microprocessor has, amongst other things, three 8-bit registers called A, X and Y, each of which can be used for the temporary storage and manipulation of a byte of information. These registers can be thought of as equivalent to the storage registers of a hand calculator for holding intermediate results during a sequence of calculations, except they can only hold 1 byte at a time.

The machine code program given in Fig. 1.3a works by first loading 255 into register A and then transferring this value to register X. (It is not possible to load X directly). The value in X is then decremented by 1 until zero is reached and then the program stops.

This illustrates that machine code programs are difficult for us to read, even if they are perfectly comprehensible to the computer. To find out what the program does we would have to look up the manufacturer's literature to find out what happens as the program is executed from the first instruction stored at the hexadecimal address 19B7.

Memory address	Machine code instructions
19B7	A9
19B8	FF
19B9	AA
19BA	CA
19BB	DO
19BC	FD
19BD	OO

Fig. 1.3a. *Storage in memory of the machine code program to count down from 255 to 0 by steps of 1. All numbers are in hexadecimal*

In principle we could use a hexadecimal keyboard to key in program as the following sequence of 9 bytes with the first one placed at memory location 19B7:

A9FFAACAD0FD00

This is a very cryptic sequence of code and illustrates the difficulty of working directly with machine code. It is very difficult to understand how the program works from just the listing and very easy to make mistakes when writing the program. Furthermore a change in one part of the program may unintentionally alter some other parts. This is because many machine code instructions have data bytes which refer to the memory addresses of other instructions and these data bytes may need to change if instructions are inserted or deleted.

No serious programming is in fact undertaken in machine code. A different approach is adopted in which the program is defined by a higher level programming language which is easier for us to use. These will be examined in the next section.

SAQ 1.3a

(*i*) What is the start address of the program in Fig. 1.3a?

(*ii*) By choosing T(true) or F(false) comment on the validity of the the following statements.

(1) A machine code program designed to run on one type of 8 bit microprocessor can be executed by any other 8 bit microprocessor.

(T / F)

(2) Machinc codc programs execute at the maximum possible speed for a given computer system.

(T / F)

(3) Machine code programs are difficult to edit.

(T / F)

(4) Machine code programs are difficult to read and understand.

(T / F)

1.3.2. Assembler Programs

This is just one step up from machine code programming. Each machine code instruction is replaced by a mnemonic code which is a reminder of the function of the code. Consider the 6502 machine code instruction which causes a copy of the contents of register A to be placed in register X. In hexadecimal this code is AA. It is represented in 6502 assembly language by the mnemonic code TAX, which stands for Transfer A to X. In addition to replacing each machine code instruction by an appropriate mnemonic, assembler language also allows labels to be used to identify points within a program. These are useful in specifying the route taken through the program depending on the nature of information being processed.

Consider the assembler program given in Fig. 1.3b which is equivalent to the machine code program discussed earlier for counting down from 255 to 0 by steps of 1. You should be able to see how the program works by reference to the explanatory comments on the right and relate the mnemonic code used to the explanation given. Notice how the labels are defined and used. A particular point in the program is labelled by a '.' followed by the characters which make up the label (eg .LOOP). In a machine code program actual addresses of memory locations which contain instructions have to be used in place of labels. For example if the start address of the program were &19B7 then the memory address of the instruction labelled as .LOOP would be &19BA. If in some part of the program we wanted to jump to the point labelled .LOOP we would simply write JMP LOOP, where JMP is the mnemonic code for the instruction code &4C which means jump. In machine code the same would be accomplished by the three-byte sequence &4C &BD &19. Here &4C is the jump instruction and &BD, &19 define the address which is the destination of the jump (notice this address is written in reverse so in fact it means &19BD). Apart from the advantages of using mnemonics in place of machine code instructions, assembler languages usually incorporate other facilities, such as the provision of labels, as programming aids.

Assembly code	Explanatory comment
LDA ?255	Load register A with the number immediately following the ? symbol (ie decimal 255).
TAX	Transfer the contents of register A top X.
.LOOP DEX	Decrement value of X by 1.
BNE LOOP	If the previous instruction gave a non-zero (Branch if Not Equal to zero) result, go to the instruction labelled LOOP.
BRK	Halt (Break) execution.

Fig. 1.3b. *A 6502 assembler program to count down from 255 to 0 by steps of 1*

Although we can understand the mnemonics in which assembler programs are written, the microprocessor cannot. The assembler program has to be converted to machine code before it can be executed. This is done by means of a special program known as an assembler. Assemblers are available for many types of microprocessor. As well as producing the required machine code stored in memory or on disc, assemblers usually give a listing of the original assembler program and the corresponding machine code. They also pick up and report programming errors which arise from incorrect use of assembler language. Fig. 1.3c shows the output obtained from the assembler program available on the BBC microcomputer.

Assembly level programming is best left to the professional programmer. For scientific work it has the limitation that the program has to be written in the mnemonic codes corresponding to the microprocessor's instruction set which does not include mathematical

Memory address	Machine code		Assembler code
19B7	A9 FF		LDA ?255
19B9	AA		TAX
19BA	CA	.LOOP	DEX
19BD	DO FD		BNE LOOP
19BD	OO		BRK

Fig. 1.3c. *Assembler output from the BBC microcomputer*

functions. For example 8 bit microprocessors have no multiply or divide instructions. Although the assembler language used for 16 bit microprocessors do allow multiply and divide, the do not have mathematical functions such as logarithms which are often needed by the analytical chemist. Although there are ways around this for experienced programmers, as a general rule scientific programming is best carried out in a higher level language such as those discussed briefly in the next section.

SAQ 1.3b

The 6502 assembler program below counts down from decimal 100 to 0, by steps of 1, repeatedly.

```
.START  LDA  ?100
        TAX
.LOOP   DEX
        BNE  LOOP
        JMP  START
```

(*i*) With reference to the above program, match the terms in the list on the left, below, with the correct explanatory comment in the list on the right. ⟶

SAQ 1.3b (cont.)

(1) .START	(*a*) Branch to the instruction labelled LOOP
(2) TAX	(*b*) Jump to the instruction labelled START
(3) DEX	(*c*) Transfer the contents of A to X
(4) JMP START	(*d*) Load register A
	(*e*) Increment the contents of A by 1
	(*f*) Decrement the contents of X by 1
	(*g*) A label used to identify a point in the program

(*ii*) By selecting T(true) or F(false), indicate which of the following statements are true or false.

(1) Assembler code, like machine code, can be directly executed by the microprocessor.

(T / F)

(2) An assembler program for one type of microprocessor can be assembled and run on another type of microprocessor.

(T / F)

(3) Once an assembler program has been converted to machine code it will run as fast as one which was written directly in machine code.

(T / F)

1.3.3. High Level Languages

To provide easier methods of programming computers so-called 'high level languages' have been developed. Common examples include BASIC, PASCAL, FORTRAN, ALGOL and COBOL. These languages all allow the program to be expressed in terms of English words or phrases and mathematical expressions as used in algebra.

Having written the program in a high level language it has to be treated in such a way as to make it comprehensible to the microprocessor. There are two approaches in common use. On the one hand we can convert the coding in the high level language into machine code by using a specially written program called a compiler. Each high level language needs its own compiler. Furthermore, since the machine code depends on the microprocessor, a separate version of the compiler is needed for each type of microprocessor. The output from the compiler is machine code which can be stored on disc for later use. The binary code produced as a result of compilation (the use of the compiler) would then have to be loaded into the computer main memory for execution.

An alternative approach is to use an interpreter. This is a machine code program which scans the program written in the high level language and interprets each line and takes the appropriate action. If a line requires material to be printed on the screen, then the interpreter brings this about. Because each line has to be interpreted before it can be executed, high level languages which use an interpreter are about a 100 times slower than the corresponding program run in machine code. BASIC is an example of an interpreter. For simple development work it has proved popular because the program is tested by the interpreter as it is programmed and invalid instructions are rejected straight away. With a compiler the only way errors can be found is by attempting to compile the program. Any incorrect instructions are usually detected at this stage by the compiler. Having successfully compiled the program and produced the binary version for execution further testing may reveal programming errors. Each correction requires that the program be recompiled to produce the revised binary program for execution.

For the purpose of comparison here are a few lines of program equivalent to the assembler and machine code programs given earlier, in the sense that the content of a main memory location labelled X is decremented from 255 to 0 by steps of 1.

(*a*) **FORTRAN** – This is a well-established, compiler-based language for scientific computation. Mathematical expressions are similar to standard algebra and a full range of mathematical functions is available. The language requires instructions to be placed starting in column 7, numeric labels are used to identify particular instructions in the program. Execution is sequential from the top of the program down, apart from any branching imposed by the programmer. Column 1 is used to identify comment lines (denoted by a C in this column) which increases readability of the program but are ignored by the FORTRAN compiler. The program lines given below in Fig. 1.3d are almost self-explanatory except for the IF statement which introduces conditional branching. It means IF X is Greater Than (GT) 0 then go to the line labelled 100 or else the next line, where the program halts.

```
Column  number
1       7

C       A simple example of a few FORTRAN instructions
        X=255
    100 X=X-1
        IF(X.GT.O) GOTO 100
        STOP
        END
```

Fig. 1.3d. *A few lines of FORTRAN to count down from 255 to 0 by steps of 1*

(*b*) **BASIC** – Most microcomputers offer the BASIC language in the form of an interpreter. It is an interactive language and errors in the spelling and use of the key words and phrases which make up the language are mostly picked up as the program is entered from the keyboard. Also if errors occur when the program is executing, the precise instruction where the fault occurred can usually easily be found. A few lines of BASIC are given in Fig. 1.3e to make the microcomputer count down from 255 to 0 by steps of 1 again. Notice that each line is numbered and execution proceeds from one line to the next in increasing line-number order unless branching (as in line 40) transfers execution to line 30 if X is still greater than 0. Otherwise execution passes to line 50 where the program halts. A detailed introduction to BASIC is given in Part 2 of this unit, but the operation of the program should be clear if one realises that line 30 means take the contents of the main memory location labelled X, subtract 1 and put the result back in the location labelled X. Thus each time the program executes line 30 the content of location X is decremented by 1. Line 10 is a 'Remark' (REM) statement which is used to increase the readability of the program but is ignored by the BASIC interpreter.

```
10 REM A Simple example of a BASIC program
20 LET X=255
30 LET X=X-1
40 IF X>O THEN GO TO 30
50 END
```

Fig. 1.3e. *A simple BASIC program*

SAQ 1.3c

> The above FORTRAN and BASIC programs both decrement the value of X from 255 down to 0 by steps of 1. Which program will execute the fastest, FORTRAN or BASIC?

SAQ 1.3c

1.3.4. Programming Applications

We can now look at some of applications of computers in the Analytical Laboratory. Most of the programs we are likely to come across will fall into one or more of the following categories.

- Calculations to obtain analytical results;
- Computer representation and manipulation of calibration plots;
- Data capture from analytical instruments;
- Control of analytical equipment;
- Storage, retrieval and display of analytical information.

In some cases programming is very straightforward but in others it can be complicated and beyond the scope of the beginner. The general approach to the programming of the first three of the above is discussed briefly below. The control of analytical equipment will be discussed in the case studies in Parts 4–6 of this Unit. The storage, retrieval and display of analytical information tends to be achieved using commercial software. This can be used as an aid to laboratory management by providing progress reports on samples, results sheets, and even costings. The software involved is beyond the scope of this introduction as it can be quite complicated, particularly if it runs on a central machine which is on-line to several instruments at the same time.

(*a*) *Calculations to Obtain Analytical Results*

As a general rule, if you can specify how a calculation can be performed then you can program it. Many analytical calculations are fairly simple and involve the multiplication of some measured quantity (titre, weight, absorbance, transmitttance etc) by one or more factors to obtain the analytical result.

eg %Cu = factor × (Weight of precipitate)/(weight of sample)

An expression of this type, in which the result can be written explicitly in terms of an arithmetic or mathematical expression involving known quantities, is always easy to program. Problems can sometimes arise if the expression on the right is poorly formulated so that errors are magnified. For example if we divide by the difference between two close numbers as in the following example.

$$Z = 122.8/(A - B)$$

If A = 25.00 and B = 24.00 then Z = 122.8. If A was an experimental measurement subject to experimental error of about 1%, then, without going into statistical detail, we can see that repeated measurement of A should give values in the range 24.75 to 25.25. This in turn would give rise to Z values in the range 163.7 to 98.4, or percentage errors of about +33% and −20% respectively.

This is an artificial example, but occasionally problems like this can arise in the solution of simultaneous equations in multicomponent analysis using spectrophotometric measurements. It is not a bad idea to examine expressions that you program for this type of numerical instability. Ask yourself what happens to the result of the expression if the measured quantity changes slightly and check that the changes in the result are reasonable.

Some care also has to be taken in programming mathematical functions which are available in high level languages. Take logarithms for example. Often we wish to compute 'log to the base 10' but sometimes the programming language provides natural logarithms only. The answer is to divide the natural log (represented here as LN) by LN(10) which is approximately 2.303.

$$LOG10(X) = LN(X)/2.303$$

A similar point applies to trigonometric functions, although these are less likely to be used in analytical applications. Here the functions such as sine and cosine normally require the angle to be expressed in radians rather than degrees. There are 2π radians in 360 degrees so the conversion is simply,

$$\text{Angle in radians} = (\text{angle in degrees}) \times 2\pi/360$$

Other problems that may occur in programming explicit mathematical expressions include attempts to calculate a number which is too large for the computer (typically about 10^{39}) or to try to calculate the value of a mathematical function using values for which the function is not valid. Examples include the square root, or logarithm, of a negative number.

SAQ 1.3d

Examine the following pair of simultaneous equations and decide if the calculation of the unknowns A and B, from the two measured quantities M1 and M2, would be a well formulated problem. Assume that M1 = 10.1 and M2 = 60.0 in a typical measurement.

$$M1 = 5 \times A + 7 \times B \quad (1.1)$$

$$M2 = 31 \times A + 41 \times B \quad (1.2)$$

(Hint: For M1 = 10.1 and M2 = 60.0 obtain an expression for A by multiplying Eq. (1.1) by 31/5 and subtracting Eq. (1.2) from the result. This will give an equation which can be solved for B. Having obtained B, substitute its value back in Eq. (1.1) to obtain an expression which can be solved for A. To test for instability, assume that on remeasurement, M1 became 10.2 and M2 = 60.4, and recalculate the results for A and B).

SAQ 1.3d

(b) Computer Representation and Manipulation of Calibration Plots

As analysts we are quite familiar with the use of calibration plots for a wide range of measurements. We draw the best straight line or curve usually quite accurately by eye. However, we need a different approach with the computer. The first problem in computerising a calibration plot is to find some way of representing it in the computer.

The simplest case is the linear plot, which can be represented by the general equation of a straight line as follows.

$$R = M.X + C$$

Here R could represent a reading such as absorbance, and X could be the concentration of solute. The slope of the plot is M, and C is the intercept on the R axis as shown in Fig. 1.3f(*i*). The value of M and C can be determined from just two measurements. In practice we would use perhaps five or six calibration points in defining the calibration plot as in Fig. 1.3f(*ii*). Here we see some experimental error so that some of the points do not lie on the best straight line.

In this case it would be wrong to select just two points to obtain the values of M and C. We would in fact use all the calibration points and obtain the equation of the best straight line. For this discussion it is sufficient to note that explicit expressions for M and C exist for the best straight line through a set of data points. Rather than write our own program to obtain the equation for the best straight line, it is usual to incorporate a ready-written subprogram, called a subroutine, into our own program. Once the constants M and C have been found the equation of the best straight line can be used to obtain the concentration of an unknown solution from a measured absorbance.

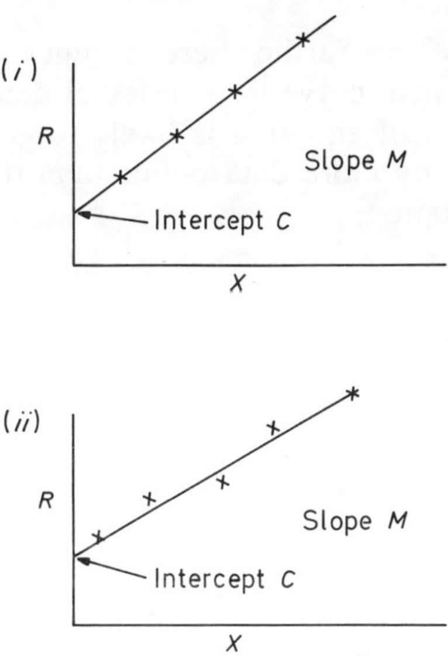

Fig. 1.3f. *Linear calibration plots (i) without experimental error, (ii) with experimental error*

If the calibration plot is curved, the approach is the same, namely to fit an equation to the data points and obtain a mathematical representation of the plot. For example the curve shown in Fig. 1.3g could be fitted to an equation of the type:

$$R = a + b.X + c.X^2$$

where a, b and c are constants and X is the concentration for a particular reading R. As with the linear plot, we must obtain the values of the constants a, b and c to give the best curve which can be drawn through the data. This equation is described as a polynomial in X of second degree, as the highest power of X is 2. The least squares method can be used to obtain the best values for a, b and c, as for the linear equation discussed above. We do not have to worry about writing our own programs for this as subprograms are readily available for a least squares fit to a polynomial function of arbitrary degree. In this way we can store in the computer's memory the equivalent of a calibration curve which relates the instrument reading to the concentration of the analyte.

We shall not pursue this further here, except to note that it is in general possible to fit a curve to a series of data points provided one knows what sort of equation is likely to give a good fit, and you should have many more data points than there are unknown constants in the equation.

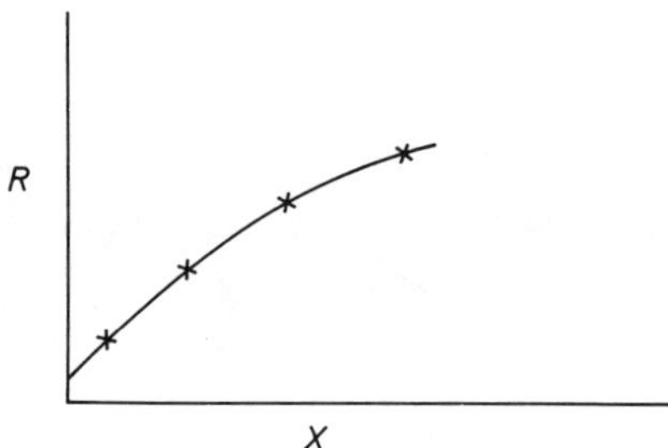

Fig. 1.3g. *A calibration plot showing curvature*

A somewhat simpler method of representing a calibration curve is to store it in the computer in tabular form, as shown in Fig. 1.3h. Here a list of readings R is paired with concentrations X. The entries in the table have to be determined experimentally by making up standard solutions and measuring them. Suppose we obtain a reading of 15 (arbitrary units). This is midway between the tabulated values of 10 and 20 with corresponding concentrations 0.8% and 1.4%. If we assume that there is minimal error in the readings, the concentration corresponding to R = 15 is 1.1% (midway between 0.8% and

1.4%). This is an example of linear interpolation (assuming minimal error in R). Essentially, we assume that the calibration points are sufficiently close together that the interval between them can be represented by a straight line. More sophisticated interpolation is possible using non-linear equations but these are beyond the scope of this introduction.

Reading	Concentration X(%)
0	0
10	0.8
20	1.4
30	2.0
40	2.5

Fig. 1.3h. *Tabular representation of a calibration plot*

SAQ 1.3e

Which one of the following equations would you use to fit the curve given below assuming the values of the constants M, C, A and B are non-zero in the equations which include them?

(*i*) $R = M.X + C$

(*ii*) $R = M.X$

(*iii*) $R = A.X^2 + B.X + C$

(*iv*) $R = A.X^2 + B.X$ ⟶

SAQ 1.3e (cont.)

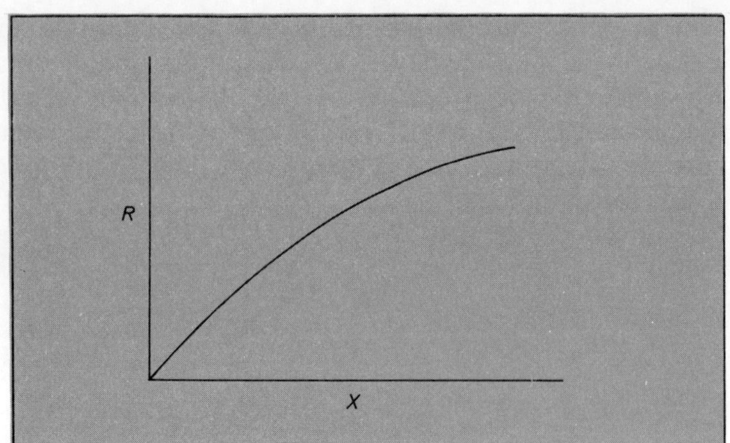

(*c*) *Data Capture from Analytical Instruments*

As a general rule any instrument which gives an electrical signal as output can be interfaced to a microcomputer providing the proper interface is available. Two problems can arise in acquiring such data.

Firstly we need to consider the rate at which the signal is to be sampled. For readings which change rapidly, a high sampling rate is required. The question then arises as to whether the computer can read data quickly enough, particularly if some computing is needed between successive data points. Many applications in analytical chemistry do not require fast data capture, but some do, for example the reading of mass spectral data. Problems can arise even with infra-red spectroscopy if fast scan rates are needed for spectra with sharp peaks. Each case has to be judged on its merits and for the moment it is sufficient to be aware of the potential problem.

Secondly, we have the problem of data storage. One must use common sense in deciding the amount of information that is to be retained. Not only is there a temptation to accumulate data for their own sake but there is a real possibility of running out of storage space if the sampling frequency and length of run are not considered together. To run out of space in this way causes an error, the unscheduled termination of the run and possibly the loss of all the data up to the point at which the error occurred. Careful program design is needed to avoid these difficulties.

SAQ 1.3f

The output from a colorimeter which is part of an auto-analyser system is monitored by a microcomputer. A sample is taken every 0.5 second, the run lasts for 20 minutes and each data value read takes 6 bytes of memory. There is 10K of memory available in the computer to store the data. Can the program be designed to accumulate all the data in memory and then transfer the whole lot as a batch to disc, or will each data item have to be transferred to disc as it is obtained?

1.3.5. Choice of Language

Several factors need to be considered in choosing a language to program a given task. A very practical consideration for many analytical chemists is the range of languages that they feel competent to use. Leaving this aside for the moment let us assume we can select a language by considering the needs of the problem alone.

Programs for control tasks with little or no mathematical processing are written in assembler language. This produces compact code and the program runs very rapidly. For a task of any size however care must be taken to produce a well-designed program which is well documented with explanatory comments included in the listing of the program. A badly structured and poorly documented program can be very difficult to understand. Furthermore, errors can then be difficult to find and greatly increase the time to produce a working program.

Many analytical chemists prefer to program in BASIC because it is simple, flexible and highly interactive. Also most BASIC interpreters allow the programmer to communicate with external devices from the program in a simple manner. This makes BASIC quite attractive for control operations which do not require rapid computations. The main problem with BASIC as an interpreted language is its slowness in execution. It means that rapid data capture from an instrument is virtually impossible unless a special interface is used in which data can be stored whilst BASIC gets round to reading them. Another problem with BASIC is that it does not protect the programmer as well as other high level languages from accidental errors. This is not the place to discuss specific programming examples, but as a general rule the bigger the program the more difficult it is to avoid making errors in BASIC. BASIC is probably best therefore for slow applications which do not require large and complicated programs (generalisations are difficult, but in this context a program of the order of 1000 lines of BASIC would be regarded as large).

High level languages such as FORTRAN are best used for scientific calculations. If you want to use it for control or data collection applications, ensure that the version of FORTRAN you are using does allow you to communicate with external devices. The advantage with a compiler based language such as FORTRAN is that once you have succeeded in getting the program to compile, the code produced is machine code and should therefore execute rapidly. It may not run as fast as the equivalent code produced from an assembler program because the generated code is not as efficient.

SAQ 1.3g

For each of the following tasks, which programming language would you use (MACHINE CODE, ASSEMBLER, BASIC, FORTRAN)?

(*i*) The solution of a set of 10 simultaneous equations to determine the concentration of 10 unknowns from 10 measurements of the absorbance of a mixture at different wavelengths. The computation has to be repeated every second to provide rapid continuous analysis of a reagent stream.

(*ii*) The monitoring of the number of times a microswitch closes in a control application involving the dispensing of liquids.

(*iii*) The measurement of pH every 5 seconds in an effluent stream from a chemical plant.

You should now be ready to embark on the detailed study of BASIC programming which starts in the next section. You can do a lot with BASIC, particularly if you develop good programming habits and produce programs which are both well-designed and documented. However BASIC has its limitations and as you progress to more demanding applications the time may come when you wish to use a more powerful language. Hopefully, the confidence you will gain from BASIC will encourage you to master more powerful and more rigorous high level languages such as FORTRAN or PASCAL.

Summary

Section 1 dealt with binary numbers, how they are related to both decimal and hexadecimal. It also introduced basic terms such as the weight of a binary digit, the byte as a group of 8 binary digits and also the nibble as a group of 4 bits. This section also introduced Binary Coded Decimal as a means of coding individual decimal digits, and also the specification of memory addresses in terms of 16 bit numbers or as 2 bytes.

Having established some of the basic terminology in section 1, and hopefully a competence in handling binary and hexadecimal numbers, Section 2 examined some of the fundamental features of the microcomputer. The microprocessor was established as the heart of all microcomputer systems, but consideration was also given to the wide variation in associated hardware which which is needed to produce a working microcomputer system.

Section 3 considered the different ways that microprocessors could be programmed. Machine code programming was discussed along with the disadvantages at working at this the lowest level of programming. In order of increasing ease of use, assembler programming was considered next, followed by two examples of high level languages (FORTRAN and BASIC). This section then went on to consider the applications of programming in the Analytical Laboratory and terminated with a discussion of the choice of language in relation to the needs of the programming task.

Objectives

On completing this Part you should now be able to:

- Relate binary numbers to their decimal and hexadecimal equivalents, and interconvert decimal and binary-coded decimal numbers. (SAQs 1.1a, 1.1b, 1.1c, 1.1d)

- Express 16 bit memory addresses in binary, hexadecimal or decimal. (SAQ 1.1e)

- Recognise the importance of the microprocessor in a microcomputer system, describe its function, and distinguish between 8 and 16 bit systems. (SAQs 1.2a, 1.2b)

- Describe ways in which long term storage of programs and data can be achieved. (SAQs 1.2c, 1.2d)

- Relate the needs of a task to the available facilities eg computer, keyboard, display medium and available memory. (SAQs 1.2e, 1.2f)

- Describe why an electronic interface is necessary between a computer and an external device. (SAQ 1.2g)

- Describe the relative merits of machine code, assembler language and high-level languages as methods of programming microcomputers. (SAQs 1.3a, 1.3b)

- Distinguish between an interpreter and a compiler for a high level language. (SAQ 1.3c)

- Indicate in general terms the type of application, relevant to the analytical chemistry laboratory, which may be programmed, and have an elementary awareness of the sort of problems which can arise. (SAQ 1.3d, 1.3e, 1.3f)

- Relate the choice of a programming language to the programming task. (SAQ 1.3g)

2. Introduction to Programming

Overview

In this introduction to programming we help you to develop some skill in the computer language BASIC so that you can use a simple computer to perform useful tasks and, often equally important, so that you can speak to computer specialists with some confidence.

The course is quite self-contained; you need not have studied Part 1 of the Unit and very little mathematics and chemistry are required. We assume of course that you have a computer, preferably a microcomputer, readily available and we start the course with a 'hands-on' section to get you accustomed to the feel of your own machine. We then study relatively simple programs in BASIC taking examples that are of some analytical interest where this is possible. By the end of Section 4 you should be quite competent in the elements of the BASIC language. Much of the instruction is through examples which you are expected to follow *on your computer.*

After Section 4 you might find it convenient to adopt your own order of study. You could look at slightly more advanced programming techniques in Section 5 or you might go to Section 6 and learn how to plot points and draw lines on the screen. If you want to collect and store data then Section 8 gives a brief introduction to filing systems. If your primary interest lies in using computers linked to instruments, you might pass directly to Section 9 where we look more closely at how the computer 'memory' works and introduce ideas and operations which will be of use in Part 3 of the Unit. If you do change the order of study please note that it would be unwise to omit any section completely or to proceed too far without studying program structure in Section 7.

Different microcomputers have different characteristics and consequently it is not possible to write instructional material which is applicable to every machine. The text of this Part has therefore been written as generally and as simply as possible to concentrate on those parts of BASIC that are indeed basic and that are applicable to virtually all available machines. Since it is difficult to guess how much you know already we make few assumptions about knowledge and experience. If certain passages seem rather elementary you will apreciate that someone else may welcomc an opportunity for a little revision.

2.1. USING A COMPUTER

2.1.1. Introduction

Computers are instructed or programmed by means of a sequence of numbers known as *machine code*. The computer is able to understand and execute instructions given in this way but the preparation of a machine code program is not an easy task. Because of this a number of more readable *languages* have been developed; these range from *assembly* languages which consist of a number of mnemonics for machine code instructions to *high level* languages such as FORTRAN, PASCAL and BASIC. Different approaches to programming a microprocessor or computer were outlined in Part 1 of this Unit and Part 2 is devoted to a more comprehensive account of the language known as BASIC.

We assume that you have available a microcomputer with a visual display unit (VDU) and either a magnetic disc or cassette system for storing programs. Because of differences in language some of the examples may not be in exactly the right form for your microcomputer but you should find little difficulty in adapting them to suit your particular machine. From time to time material is given for the sake of completeness rather than because it is absolutely essential. At these points some indication will be given so that you can pass to a section of more immediate importance. You should, however, study the omitted material at a later reading.

One of the books on programming listed in the Unit bibliography or any other book which you find readable and suitable to your circumstances should be adequate to support the course. Apart from a book on programming, the manual or guide for the computer must be available. The manual will be required when you have to find out something which is peculiar to your own situation, eg how to switch on the machine or when to use a semicolon (;) rather than a colon (:). We prompt you to find this information at suitable points throughout the course, indicated by the symbol Π. You must yourself check answers to these exercises since the answers will depend on the machine you are using. You should write down the answers in a systematic manner so that you can refer to them at any time.

The symbol Π is also used when certain examples are being discussed to indicate where it is important for you to follow the examples on your machine and to note any points of difference between our procedures and yours.

One of the first things you must do is read enough of the manual to permit you to:

— connect up and switch on your machine;

— enter characters from the keyboard;

— delete a character or characters typed;

— clear the screen.

Very soon you will also have to find out how to:

- start a new program;
- list a program (ie make it appear on the screen);
- stop a program which is being run;
- save a program on tape or disc.

While a program is running it often happens that it has to be stopped and this is usually done by pressing a key marked ESCAPE or BREAK. You should find out how this is done on your machine as soon as possible. At the same time find out whether or not a program is lost when you ESCAPE or BREAK. Of course switching off is always a last resort if there is trouble but the program is usually lost when the power is disconnected.

It may be that you have to find out how to make your machine operate using BASIC. This will be no problem with most microcomputers but some may require a special instruction to be typed in or some special program to be loaded.

Since this is a course in programming many questions ask you to write programs. *Every time* you write a program you should type it into your computer and run it. If your will is strong enough you might even do this before looking at the answer but do not worry if you steal a look after writing and before typing. The answers to the questions often contain useful information. You should also type and run the example programs and program segments in the text.

Our prime objective is to guide you in your efforts to achieve some skill in programming applications of interest to an analyst. You should look upon this part of the Unit as a practical course. *Use your computer as much as possible* and when trying something new keep it simple. Remember that complicated computer programs are usually just a collection of simple procedures.

Throughout the course DOING MUST DOMINATE READING. If you read these notes for more than about ten minutes without either writing or running a program you have stopped following the course.

TYPE AND RUN EVERY EXAMPLE

WRITE, TYPE AND RUN EVERY QUESTION

2.1.2. Initial Practice

If you are not familiar with a computer it is a good idea to use it first as a calculator, provided of course that you machine can operate in this way. When the machine is used in this way it is usually said to be operating in the *immediate mode or direct mode.*

The following example is a good start. Type it into your machine by typing each line in turn and then pressing the RETURN key.

```
LET X = 12

LET Y = 10

LET Z = X*Y

PRINT X, Y, Z
```

These four lines are instructions which tell the machine to do the following:

Assign the value 12 to the *variable* called X.

Assign the value 10 to the *variable* called Y.

Multiply the value of X by the value of Y and assign the result to the *variable* called Z.

Show on the screen the values of the *variables* X, Y and Z.

You can learn three important points from these four lines:

The equals sign (=) has a somewhat different meaning from usual. Here it means *become equal to*. When this sign is used the variable on its left is *assigned* a value.

The asterisk or star (*) means *multiplied by*.

The word PRINT in capital letters means *show on the screen.*

Each of the four lines contains an *instruction* or a *statement*.

The *syntax* or structure of a statement is very important. For example the words LET and PRINT must usually be in capital letters though some machines may allow them to be typed in lower case.

Most computers allow the word LET to be omitted so that the first line above could be simply X = 12.

The word PRINT is used above as a statement of the BASIC language. This word is unfortunate, having been introduced before VDU screens became common and the only way of seeing output from a computer was to have it printed on paper. For real printing on paper the word LPRINT or WRITE is usually used.

While you must be quite rigorous as to spelling, punctuation and structure when typing statements you should feel free to use your imagination in any way that will help you form a picture of what is going on. For example, you may find it useful to think of a variable as a box or location in memory which is labelled with the name of the variable. Then the statement LET X = 12 can be taken as meaning *place the number 12 into the box labelled X*. Similarly, LET Z = X*Y means *multiply X by Y and place the result in box Z*.

Many microcomputers have keys that reduce the amount of typing required; thus a single key press might result in the word PRINT appearing on the screen, while another key might produce the word INPUT. This kind of facility depends on the particular machine and you must therefore discover any special aids for yourself. The same may be said about any editing facilities. There is usually some means of correcting wrong words or lines typed on the screen but the systems employed are so machine dependent that it is not possible to give general instructions.

If you have little experience of using a computer it is a good idea to do a few calculations like the one above, perhaps finding the sum (+) or the difference (−) of two numbers, or dividing one by the other (/). This will familiarise you with the keyboard and with any special keys.

2.1.3. Scientific Notation

You probably make use of a simple hand calculator and so are already familiar with what is usually called *scientific notation*. For example, the number 6.02E23 is the same as 6.02 multiplied by 10 to the power 23.

While we usually use this system when dealing with very large or very small numbers most computers use it quite often and so we must become quite confident with the notation. In case you feel that you need some revision here is a question. Try it mentally at first!

SAQ 2.1a

In the tables below, the entry in column X, row A is 2.0 and that in column Y, row B is 4.0. The product of these is 8.0. That is, AX times BY gives 8.0. What other combinations produce the same result, viz 8.0? ⟶

SAQ 2.1a (cont.)

	X	Y
A	2.0	0.4E-1
B	2.0E1	4.0
C	200E-2	4.0E3
D	0.2E3	40.0E-1
E	0.002E0	4.0E-1

2.1.4. Printing Words

While we shall be mainly concerned with numbers and computation it is also necessary to make the computer store and print characters. This usually requires the use of quotation marks which will be single or double depending on your computer.

Type in the following line and press RETURN:

```
PRINT "JOHN SMITH"
```

If an error is signalled try again using single quotation marks. The characters contained inside the quotation marks form a *string*.

Just as we used X and Y as variables with numeric values, so we can have variables with 'values' which are strings of characters. Type the following lines into your machine, pressing the RETURN key at the end of each line. (We assume double quotes though you may need single):

```
LET A$ = "JOHN"

LET B$ = "SMITH"

LET C$ = A$ + B$

PRINT C$
```

The variables here all end with the dollar sign ($). This indicates that they are *string* variables. That is, they represent strings of characters rather than numbers.

2.1.5. Variable Types and Names

You can probably read this discussion fairly lightly at first since its utility will depend on the microcomputer you use and on your applications. Do not omit it completely, however, as you will need some of the concepts later.

The two main types of variable are the numeric and the string. However, many computers permit a distinction to be made between *real variables* and *integer variables*. Both of these are numeric but the latter can only have values which are whole numbers. When this distinction is allowed in BASIC it is usual to indicate an integer variable by means of the percent sign (%) as the last character of the variable name, eg:

Names for Real variables: X A

Names for Integer variables: X% A%

Names for String variables: X$ A$

Valid assignments of values to two of the examples given above would be:

LET A = 5.62

LET A% = 65

An attempt to assign a value 32.9 to the variable A% would be unsuccessful. The fractional part of the number would be discarded and A% would be assigned the value 32.

There are certain advantages in using integer variables. These include:

When integers are used in calculations there are no rounding-off errors because the result of the calculation must also be an integer.

With integer variables you can rely on the last figure given whereas with real variables the number of significant figures is usually less than the number actually displayed. You will have to refer to your manual for details.

Many programs run faster if integer variables are used.

There is some saving in space because integers are usually stored in a smaller number of memory locations. For example with the BBC machine an integer uses four locations or bytes whereas a real variable uses five.

Against these advantages it must be remembered that integer variables cannot be used when fractional values occur. Also, the range of integer variables is normally much smaller than the range of real variables. (For the BBC machine the highest possible real and integer values are respectively 1.7×10^{38} and 2 147 483 647).

SAQ 2.1b

What type of variable might be employed for the following:

(*i*) the date of a month;

(*ii*) the name of a day;

(*iii*) the weight of a sample.

In the example programs above simple capital letters like X and Y were used as numeric variables. The use of single letters is always safe but most computers allow a wide choice of variable name. Some examples of possible names are:

Real variables:	A3	SUMM	pay	my_age
Integer variables:	YEARS%	QUANTA%		
String variables:	Names$	TITLE$		

Since practice depends on the particular computer only two general rules can be given. These are:

(*a*) The name of a variable must always start with an alphabetical character, never with a number or a punctuation mark.

(*b*) The name must not include a space. (And probably certain other characters).

While it is often useful to be able to use words as variable names care must always be taken not to include some reserved word in the name. For example, the word PLOT usually has a special meaning associated with graphics and therefore must not be used as part of a variable name. This excludes names like XPLOT or OLDPLOT for variables though the corresponding words in lower-case letters may be acceptable. It may be that your machine only objects if a reserved word forms the first part of the variable name (eg it allows XPLOT but not PLOTX); this is another point you can only discover for yourself.

One pitfall is that some computers allow variable names to be of any length but only pay attention to the first two or the first four characters. When this is the case it is advisable to use names of no more than two or four characters.

Π Find out how variables may be named for your computer.

Does your computer permit a distinction between real and integer variables?

What are the lowest and highest numbers that can be stored (*a*) as real variables and (*b*) as integer variables?

SAQ 2.1c

Some of the following statements are unacceptable and some will lead to errors. Comment on each statement. ⟶

SAQ 2.1c (cont.)

(*i*) LET X% = 12

(*ii*) LET X + Y% = 12

(*iii*) LET residue = 0.456

(*iv*) LET PPT = 1.347g

(*v*) LET B% = 3.456

(*vi*) LET 3rd% = 4.67

(*vii*) LET sample% = 3.92

(*viii*) LET Name$ = "Tom Brown"

(*ix*) LET PRINTER$ = "Caxton"

(*x*) LET FIVE = 6

2.2. SIMPLE PROGRAMMING

2.2.1. Computer Programs

A *program* is a series of instructions which the computer obeys one after the other. In a BASIC program each line of instruction has a number. When the program runs it takes the line numbers in order and carries out the instruction on the line. The program always goes from one line number to the next higher number unless there is an instruction directing the program to go to a certain line.

The line numbers do not have to go up by one each time. It is usual to number the lines 10, 20, 30 ... so that additional lines can be inserted if desired. This is an important point. You should always assume that more lines will be necessary after you have tested a program. Never use line numbers 1, 2, 3 ... if it can be avoided.

Most computers have a facility for renumbering lines automatically. It is usually also possible to type a line at any point of a program, the computer itself placing the line into its proper position within the program.

Many computers allow several instructions to be placed on one line, the instructions being separated from each other by a colon (eg LET X = 5: LET Y = 7). However, since not all computers permit more than one instruction per line we shall always place only one on each line. This is always a safe procedure. Various short cuts suitable for your own computer are probably available and you will learn these as you gain confidence in simple tasks.

The big advantage of a computer program is that it can be used time after time with different values of the variables.

The following is a simple program:

```
 9 REM ** A 3-LINE PROGRAM
10 LET X = 12
20 PRINT X
30 END
```

Line 9 is simply a reminder. Any line that starts with REM is ignored by the computer. The two stars help to make the reminder stand out when the program listing is printed.

At line 10 a variable X is assigned the value 12.

At line 20 the value of the variable X is printed on the computer screen.

At line 30 the program stops.

Π Type this program into your computer and run it.

Even though it may seem extremely simple it will get you familiar with the keyboard and with entering program lines. After the program has been typed in you should LIST it; that is, make the complete program appear on the screen. You can then make any necessary alterations before running the program. The usual method of running a program is to clear the screen and then enter the word RUN.

If a program is stored on a magnetic tape or disc it must first be loaded into the computer before being run. Your instruction manual will describe how a program is loaded from or saved to tape or disc. The keywords LOAD and RUN are normally used but the word CHAIN may mean *load and run*. The name of the program will probably have to be within quotation marks. The keyword SAVE is usually used when transferring a program to tape or disc so that it can be recalled at a later time.

Π Find out how to LIST, SAVE, LOAD and RUN a program.

Experiment with a simple program like the one above before it is necessary to save an important program.

2.2.2. Output to Screen

The value of a variable is displayed on the screen when the program meets a statement like:

```
20 PRINT X
```

A single statement can output more than one value. Alter the program above by adding line 12 and changing line 20:

```
12 LET Y = 25
20 PRINT X, Y
```

In line 20 there is a *list* of two variables, X and Y, which are to be printed or displayed on the screen. The variables are separated by a comma. Most computer obey this instruction by printing the two numbers in different *columns* or *zones* or *fields*, eg:

```
12          25
```

Other separators besides the comma may be used with different results. As practice differs considerably you should find the punctuation used by your own machine, paying particular attention to the use of the comma, the semicolon and the colon.

Π Find out how the numbers 12 and 20 are output by your computer.

What instructions are needed to print the numbers:

(*a*) with 1 space between,

(*b*) in columns or fields,

(*c*) with no spaces (to give 1220)?

Make your computer actually print these numbers in the forms suggested and note carefully the answers to the questions. The three types of format mentioned, *viz* columns or fields, one space, no spaces, are the formats most commonly used. With the column format the number of columns across the screen depends on the particular computer but it may be possible to alter this number.

It is a good idea to experiment with various statements using simple numbers. You could try this program:

```
10 LET X = 12
20 LET Y = 10
30 PRINT X, Y
50 END
```

and run the program with various statements at lines 30 and 40, eg:

```
30 PRINT X, Y          30 PRINT X ;Y

30 PRINT;X;Y           30 PRINT;X,Y

30 PRINT X;            30 PRINT X,
40 PRINT Y             40 PRINT Y
```

and indeed any other combination of PRINT statements that occurs to you.

Later on we shall have to consider how to tabulate data and how to round off numbers. If a column can display no more than 7 characters a number like 121.7346 will encroach on the next column. Unfortunately several computers print out numbers with a large number of figures and as analysts who understand about significant figures we must make any necessary adjustments. We shall deal with this later.

2.2.3. Arithmetical Operations

The computer was developed to do calculations and this is still one of its principal functions. Many program instructions or statements are concerned with arithmetical operations.

Consider this program:

```
 9 REM ** MULTIPLY
10 LET X = 0.5
20 LET Y = 25
30 LET Z = X*Y
40 PRINT X, Y, Z
50 END
```

At line 30 the numbers represented by X and Y are multiplied together and the result assigned to the variable Z. At line 40 the values of X, Y and Z are printed in three different fields.

In algebraic expressions the symbols '+' and '−' have their usual meanings but the following must be noted:

* means *multiply*

/ means *divide*

^ means *raise to the power of*

Be careful to use the correct sign (/) for divide. The reverse sign (\) called the *backslash* must not be used.

Brackets are often necessary to ensure that operations are performed in the correct order. eg Suppose X = 4 and Y = 8. Then

```
Z = (3*X + Y)/2      gives Z = 10

Z = 3*X + Y/2        gives Z = 16

Z = 3*X^2            gives Z = 48

Z = (2*X)^2          gives Z = 64
```

Any expression within brackets is evaluated first so that the brackets can be removed. Otherwise the order of operations is:

Raise to power

Multiply and divide

Add and subtract.

Sometimes the sign '**' can be used in place of the upward arrow for raising to a power, eg:

A*X**2

A*X**(A + B)

This use of '**' is not general. You will have to find out if your machine allows it.

It is often necessary to use more than one pair of brackets as in

Z = ((A + B)*(C + D))/(X + Y)

If an expression looks complicated it is always a good idea to count the total number of left and the total number of right brackets to ensure that a pair has been used.

SAQ 2.2a

Problems (*i*) to (*iii*) below should be done by hand (or mentally) and the results checked by computer.

(*i*)

Z1 = 100/(X + Y)   Z2 = 100/X + Y

If X = 5 and Y = 15, what are the values of Z1 and Z2?

(*ii*)

Z3 = 100/X/Y   Z4 = 100/X + Y
Z5 = 100/(X + Y)

If X = 5 and Y = 2, what are the values of Z3, Z4 and Z5? ⟶

SAQ 2.2a (cont.)

(iii)

Z6 = 100/X*Y   Z7 = 100/(X*Y)

If X = 2 and Y = 10, what are the values of Z6 and Z7?

(iv)

Write computer-type expressions corresponding to:

$Z8 = Ax^2 + Bx + C$

$Z9 = Ax^{(a + b)} + Bx^2 + 1/C$

2.2.4. Input from Keyboard

Values of variables are often assigned by being typed in at the keyboard. A program will stop at a statement like:

```
10 INPUT X
```

and wait for a number to be entered. The user does this by typing the number and then pressing the RETURN key. While it is waiting for this to happen the computer usually gives a prompt, that is, it displays a character like ! or > or ? or it displays a flashing line or rectangle.

Π Try this on your machine:

```
10 INPUT X
20 PRINT 2*X
30 END
```

Run this program and when a *prompt* character appears type in a value for X. The number only must be typed, no words like "input" or other characters like "X=".

Note particularly how the computer shows that it is waiting for you to type in the value for the variable X. The prompt that it gives here is likely to be different from the prompt given when it expects a line of program to be entered.

When the computer expects input it is important that the correct kind of information is given. For example, if a number is expected an alphabetical character will not be acceptable. It is a good idea to make some deliberate errors early in your use of a computer to find out how it reacts. If you type in a non-numerical character when input is expected at line 10 above some computers will simple ask you to re-enter, others will stop the program completely, and others will assign a value of zero to the variable. This last result is probably the most dangerous as you are not warned of your error.

It is usually advisable to print a line to remind the user what input is expected and what the output means. We can do this in the program above by including two more lines as *prompts*:

```
8 PRINT "ENTER A NUMBER"

18 PRINT "DOUBLE THE NUMBER IS ";
```

Not only do these lines make the program more comprehensible to the user, they help the programmer in the development of the program.

The program can be modified to accept two numbers and print out the product. The simplest modification would be to include another line 12 INPUT Y and alter line 20 to PRINT X*Y. Of course the prompts would have to be changed. It is usually possible, however, to enter or input more than one value in a single statement as in the following program:

```
10 PRINT "ENTER X AND Y"
20 INPUT X, Y
30 PRINT "THE SUM OF X AND Y IS ";
40 PRINT X + Y
50 END
```

X and Y form a *variable list* at line 20. The two values entered are separated by a comma (eg you would type 12,10).

At line 40 the value of the expression X + Y is computed before being printed. The two PRINT statements could be combined in a print list:

```
40 PRINT "THE SUM OF X AND Y IS ";X + Y
```

You should modify the program above to make it operate on the numbers entered in different ways, eg make it print the product or the difference of the numbers or the square of one plus the other.

Let us take an example involving a little chemistry.

Suppose we want a program to calculate the molarity (ie mol dm^{-3} concentration) of a sulphuric acid solution after titrating a sample with standard sodium hydroxide.

The *first* thing that must be done is to sort out the chemistry. From the chemical equation for the reaction we arrive at the conclusion that the unknown molarity is given by the expression:

$$\frac{\text{molarity of base} \times \text{volume of base}}{2 \times \text{volume of acid}}$$

(We leave you to justify this result by your own method).

It cannot be emphasised too strongly that the chemistry must be done before the programming. We can now use symbols. For clarity we shall represent variables by two characters rather than one as we have been doing: volume 'VA' of acid requires volume, 'VB' of standard base. If the molarity of the base is 'MB' then the unknown molarity, 'MA' is given by:

MA = MB*VB/2/VA

If your computer permits it you might prefer to use names like MBASE, MACID, VBASE, VACID for the variables. We are now in a position to write a program which does three things, *viz*:

accepts the necessary data;

calculates MA from the equation;

prints out the result.

A suitable program is:

```
 99 REM ** SULPHURIC ACID - SODIUM HYDROXIDE
100 PRINT "ENTER MOLARITY & VOLUME OF
BASE";
110 INPUT MB, VB
120 PRINT "ENTER VOLUME OF ACID";
130 INPUT VA
140 LET MA = MB*VB/2/VA
150 PRINT "MOLARITY OF ACID IS "; MA
160 END
```

Run this program and perhaps modify it to deal with any acid–base titration by introducing a variable to represent the ratio base/acid in the chemical equation; the ratio is 2 in the example taken since this is the mole ratio $NaOH/H_2SO_4$ in the neutralisation reaction.

SAQ 2.2b

In the standardisation of acid solutions a known weight of anhydrous sodium carbonate is titrated with an acid. One mole of monobasic acid neutralises 53.00 g of Na_2CO_3.

Write and run a program to calculate the molarity of a hydrochloric acid solution from the weight of sodium carbonate (W g) and the volume of hydrochloric acid (V cm^3) required for neutralisation.

Reminder:

molarity = number of moles per litre of solution.

The inclusion of prompts when input is required is so useful that many computers allow a prompt as part of the INPUT statement. An example is:

```
10 INPUT "ENTER VALUE OF X",X
```

The words in quotation marks form a *prompt-string*. When the statement is reached these words are printed on the screen and the computer waits for the value of X. In this example a comma is placed between the prompt-string and the variable name but practice differs considerably.

Π Does your computer support prompt-strings?

If it does support prompt-strings find out the punctuation. In particular note the answers to the following:

Are single or double quotation marks used?

What punctuation mark, if any, comes between the prompt-string and the variable list?

If a question mark appears, does it always appear or can it be suppressed?

It is a good idea to repeat your answer to SAQ 2.2b, but this time employ prompt-strings if your computer supports them.

When the input required is a single character it may be possible to make use of the string function GET$. This is available in the BASIC of most microcomputers.

GET$ produces a string variable of a single character which is assigned merely by pressing a key. While we shall make little use of the function a brief mention may be of interest.

You will probably have used GET$ or an equivalent function when running a commercial package or game which invites you to select from a 'menu' by pressing a key; as soon as the correct key is pressed the appropriate program runs.

The use of GET$ is machine dependent. With the BBC machine the statement:

```
30 A$ = GET$
```

has the effect of assigning the *character* of the next key pressed to the string variable A$. If the key 'K' is pressed then A$ becomes 'K'. If key '3' is pressed A$ becomes *character* 3. The 'character' assigned may be a space.

When GET$ is used it is not necessary to press RETURN but only one character can be accepted and the character does not appear on the screen. The following simple program will show how GET$ works:

```
10 A$ = GET$
20 PRINT "YOU PRESSED KEY "; A$
30 GOTO 10
```

This will allow you to try various keys though you will have to press ESCAPE or BREAK to stop the program.

2.2.5. Mathematical Functions

All computer languages have facilities for working out certain standard mathematical *functions*. Your computer manual will list those available in your particular version of BASIC.

The most frequently used functions are:

V = LOG(X)	V becomes the logarithm of X. (See 2.2.7 for discussion on common and natural logs.)
V = EXP(X)	V becomes the exponential of X.
V = SQR(X)	V becomes the *square root* of X.
V = SIN(X)	V becomes the sine of angle X. COS(X) and TAN(X) similarly. (NB Angles must usually be in *radians*.)

V = ASN(X)	V is the angle whose sine is X. ACS(X) and ATN(X) similarly.
V = RAD(X)	V becomes the value in *radians* of X degrees. (Possibly not always available)
V = ABS(X)	V becomes the *absolute value* of X. (ie V is the same as X but always *positive*.)
V = INT(X)	V becomes the integer less than X.

Be careful when using INT when X is negative as some computers respond to it by returning the integer nearer to zero.

Π Type PRINT INT(5.4),INT(-5.4) into your machine to find how it operates.

The functions, above give a selection of the more important ones usually available. You computer language may not have all of them but will probably have a number of others. There is often a function for giving a random number and one for the number of characters in a string. Many versions of BASIC also permit the programmer to define *user* functions within a program. This is a useful facility but is unlikely to be used in elementary programming.

You will observe that the *argument* of the function is always enclosed by brackets. In the examples above X represents the argument. A program segment might be:

```
100 INPUT "ENTER A NUMBER ",X
110 LET Y = SQR(X)
120 PRINT "SQUARE ROOT OF ";X;" IS ";Y
```

The argument of a function may itself be a function. A simple example occurs when an angle is known in degrees but a trigonometric function requires radians. If your computer language has the RAD function then the cosine of an angle D degrees is given by:

```
30 V = COS(RAD(D))
```

Here the argument of the COS function is RAD(D). Normally angles have to be in radians when using the SIN, COS and TAN functions and angles are returned in radians by the inverse trigonometric functions. You should check this for yourself.

The argument of a function may be an expression which must be evaluated before the function operates. Here is a simple example:

We require a program to calculate the length of the long side (C) of a right-angled triangle from the lengths (A and B) of the other two sides.

Simple geometry gives:

$$C = \sqrt{(A^2 + B^2)}$$

and this leads to the program:

```
 99 REM ** PYTHAGORAS
100 PRINT "LENGTHS OF SHORT SIDES? ";
110 INPUT A,B
120 LET C = SQR(A^2+B^2)
130 PRINT "LONG SIDE = ";C
140 END
```

The expression in brackets at line 120 is evaluated and then the square root of the result is returned in variable C. Try this program with a few numbers. (A triangle with sides in the ratio 3:4:5 is right-angled).

The program can be extended to return the two other angles since the sines of the angles must be A/C and B/C:

```
132 X = DEG(ASN(A/C))
134 Y = DEG(ASN(B/C))
136 PRINT "ANGLES ARE ";X;" AND ";Y;
138 PRINT " DEGREES"
```

Here DEG is a function which converts radians to degrees. We have not used the word LET at lines 132 and 134. This is permissible because the use of the word is usually not essential when assigning values to variables.

A more chemical example is the calculation of the hydrogen ion concentration of a weak acid solution.

If the dissociation constant of a weak acid is K_a and the solution concentration is C mol dm^{-3} then the approximate H^+ concentration is given by:

$$[H^+] = \sqrt{(K_aC)}$$

Written as a BASIC statement this becomes:

```
110 H = SQR(K*C)
```

The significance of the variables should be obvious. You should use your computer to check that a 0.001 mol dm^{-3} solution of an acid of $K_a = 10^{-5}$ has a hydrogen ion concentration of 10^{-4} mol dm^{-3}.

2.2.6. Rounding and Truncation

One of the less attractive features of many computers and electronic calculators is a tendency to display too many figures. When we know that the result of a calculation is 4.52 it is disconcerting to be presented with 4.519998 or, worse, .4519998E01.

The reason for this kind of behaviour is to be found in the way the computer stores numbers. We need not go into this in any detail but we must know how to get the number of significant figures we want. Some languages have a built-in facility for formatting output but this is not general and so we shall now address ourselves to this problem. Even if your machine can produce output as you want it you will still find this discussion instructive.

The key to the control of numerical output is the function INT. This function discards the fractional part of a decimal number. The number 54.67 is changed to 54. The syntax is

30 Y = INT(X)

The value of Y is equal to the integer which is less than X. That is, X is *rounded down* to give Y.

As mentioned above, there may be a problem with negative numbers and so it is safer to assume that this discussion applies only to positive numbers.

It is most important to remember that the number is rounded *down* or *truncated* to the next lower integer and not to the nearest integer. Thus, following the use of INT:

2.001 becomes 2

2.999 becomes 2

Clearly this can cause problems unless something is done.

To get the *nearest* integer we simply add 0.5 to the number before we use INT:

INT(2.001+0.5) = INT(2.501) = 2

INT(2.999+0.5) = INT(3.499) = 3

It will be found that this rule works generally, viz., to obtain the nearest whole number add 0.5 and then truncate. The procedure gives the next higher whole number if the fraction is 0.5 or greater and the next lower if the fraction is less than 0.5.

A very important use of the INT function is in the adjustment of the number of decimal places in a number before printing. Suppose we only want two figures after the point. For example, we might require:

12.764 to become 12.76 (round down)

12.757 to become 12.76 (round up)

Let us take A = 12.764. The procedure involves 4 steps:

multiply by 100	(A = 1276.4)
add 0.5	(A = 1276.9)
use INT to truncate	(A = 1276)
divide by 100	(A = 12.76)

If only one figure is required after the point you would multiply and divide by 10 instead of by 100.

SAQ 2.2c

Write a program which accepts numbers from the keyboard and prints out the reciprocal of the number correct to 3 decimal places.

2.2.7. Using Logarithms

Of the available functions the one most frequently employed in analytical chemistry is probably LOG or LN. It is particularly important to note whether LOG means the *common* (base 10) logarithm or the *natural* (base e) logarithm. Practice differs: if your computer has both functions LOG probably means *common* log while LN refers to the *natural* log. You will have to find this out for yourself either by reading the manual or by trial. If only one of the log functions is available the following conversions will be found useful:

$$\log_{10}(X) = 0.4343*\log_e(X)$$

$$\log_e(X) = 2.303*\log_{10}(X)$$

A simple application of the LOG function is to the conversion of transmittance to absorbance:

```
 9 REM ** T to A
10 PRINT "ENTER TRANSMITTANCE (FRACTION)"
20 INPUT T
29 REM ** ASSUME LOG TO BASE 10
30 LET A = LOG(1/T)
40 PRINT "ABSORBANCE = "; A
50 END
```

The value of 1/T is calculated from the known T value. This result becomes the argument of the LOG function. If T = 0.5 then 1/T = 2 and the log (base 10) of 2 is 0.301. The value of A is therefore 0.301.

The recovery of a number from its logarithm involves exponentiation. If X is the common log of a number then the number is 10^X. A computer statement would be:

```
110 Y = 10^X
```

A pH of 6.5 means a hydrogen ion concentration of $10^{-6.5}$ mol dm^{-3} and so the statement:

```
110 H = 10^(-P)
```

gives variable H the value of the H^+ concentration of a solution having a pH equal to variable P. The transmittance of a solution can be calculated from its absorbance in a similar manner.

While the sign '^' (or perhaps '**') must be used to recover the 'antilog' from a common or base 10 logarithm, the function EXP is usually used to recover a number from its natural logarithm. If X is the natural log of a number and E the base of natural logs (E = 2.7183 ...) then the number is given by either of the two expressions:

```
Y = EXP(X)        Y = E^X
```

Though these two should be equivalent, the second is never used.

SAQ 2.2d

Write programs to calculate:

(*i*) absorbance from percent transmittance;

(*ii*) percent transmittance from absorbance;

(*iii*) pH from hydrogen ion concentration;

(*iv*) hydrogen ion concentration from pH.

Reminder:

absorbance = $\log_{10}(100/\%\text{transmittance})$

pH = $-\log_{10}[H^+]$

SAQ 2.2d

2.2.8. String Functions

Several functions use a string as argument. Many computer languages include a function which produces a new string which is part of the old (argument) string. Thus, if one string is "CHEMISTRY" a function may produce a new string, "CHEM", by selecting the first four characters of the original string. You can find what your own computer does if you are interested in these things; we shall only mention three functions that involve strings.

We sometimes want to regard a sequence of numerical characters as a string rather than a number or *vice versa*. The conversion from one variable type to the other is accomplished by means of the functions STR$ and VAL.

To illustrate, suppose we have the number 22.4 and that this is the value currently assigned to variable V. We can assign a value to a string variable, A$, by means of either of the statements:

A$ = "22.4" or A$ = STR$(V)

It must be remembered that A$ represents the *characters* of 22.4 and not the actual number.

Provided the characters of a string represent a number it is possible to convert the string into a numeric variable. The function VAL does this:

X = VAL(A$)

If A$ = '22.4' then the value of numeric variable X would become 22.4.

SAQ 2.2e

We wish to write a program which asks a child his/her name and age and then prints the message

"(NAME) WILL BE (Y) NEXT YEAR"

with the correct name and number.

(*i*) Write the program on the assumption that the child will always enter the age as a number.

(*ii*) Write the program to allow the age to be entered as a number followed by a word like 'years'. For this case assume that the VAL function can be used to return a number which appears before any non-numeric character.

SAQ 2.2e

The string function GET$ has already been mentioned in connection with the input of data at the keyboard. This function produces a string variable of a single character which is assigned merely by pressing a key.

The GET$ function can be useful if you are having trouble with a program and you cannot establish where the difficulty is located. At certain points of the program you arrange that variable values are printed and place the statement Z$ = GET$. When the program reaches this point it waits for a key to be pressed before proceeding. Z$ should not have any particular significance. In this way you introduce a pause which allows you to check variable values as you step through a program. This is part of the process known as "debugging".

2.3. PROGRAM CONTROL (1)

2.3.1. Simple Loop

It is often necessary to repeat a section of a program. Consider a program to calculate the concentration of a substance from the absorbance of a solution.

We know by experiment that a certain solution containing an iron complex has an absorbance of 0.6 for a 3 mg dm^{-3} concentration of iron when measured using a 1 cm cell. Beer's Law holds and so absorbance (A) is proportional to concentration (C) and to path length (b):

$$A = E*C*b$$

Putting $A = 0.6$, $C = 3$ mg dm^{-3} and $b = 1$ cm we find the proportionality constant $E = 0.2$ dm^3 mg^{-1} cm^{-1}. (Note that the concentration is expressed in *milligrams* per dm^3.)

We can now use this E value to determine the percentage iron in different samples.

Suppose that a weight W g of sample dissolved in 1 dm^3 of solvent produces a solution which has an iron concentration of C mg dm^{-3}. The percentage of iron in the sample is:

$$\%\ \text{iron} = \frac{10^{-3}\ C}{W}\ 100$$

or:

$$\%\ \text{iron} = \frac{0.1*C}{W}$$

Replacing C by the ratio A/(E*b) from the relationship between concentration and absorbance we get:

$$\%\ \text{iron} = \frac{0.1*A}{W*E*b} = K*A/W$$

where K = 0.1/E/b is a constant for the particular system. The value of the constant must be found by experiment.

At this point the problem passes from chemistry to computing. We represent percent iron by a variable P and then write a skeleton program:

INPUT values for K, W and A

Calculate

PRINT value of P

From this a simple program is produced:

```
 9 REM**PERCENT IRON
10 PRINT "ENTER CONSTANT K";
20 INPUT K
30 PRINT "ENTER WEIGHT, ABSORBANCE";
40 INPUT W, A
50 LET P = K*A/W
60 PRINT "PERCENT IRON = ";P
70 END
```

Π Type this into your computer and run it to see that it gives correct results.

This program suffers from a disadvantage that it must be run every time a calculation is required and the value of the constant, K, must be entered each time. If several calculations are to be done with different values for W and A each time the simplest way of repeating the program is to include a line like:

```
65 GOTO 30
```

After printing the value of P the program returns to line 30 where it receives the next pair of W and A values. It continues round the loop again and again. Note that the value of K is entered before the loop starts because its value remains the same for every pass through the loop. Put the modified program into your computer and see that it runs as described.

While the program keeps repeating lines 30 to 60 there is no way of stopping it other than by using the ESCAPE or BREAK keys or by switching off. One way of avoiding this is to include a *conditional* statement like one of the following:

```
35 IF W>90 GOTO 70
35 IF W>90 THEN 70
35 IF W>90 THEN GOTO 70
```

It is then possible to stop the program under control by entering 99,99 when input is requested at line 30. When the conditional statement is reached the computer applies the appropriate test (in this case it tests to see if W is greater than 90) and then acts as instructed.

Of course the actual number is selected to suit the situation. For example, one could use IF W<O ..., the loop being stopped by entering a negative value for W.

It is sometimes necessary to avoid conditions which test for equality (eg IF W = 2) because it may be difficult to ensure that two variables have *exactly* the same value. The reason for this lies in the manner in which a computer stores numbers.

When a variable has a value 10 the computer may hold this in its memory as a number 9.99998 or 10.00002. If integer variables can be employed it is possible to ensure that the value is exactly 10 but meantime suppose that the value held is not an exact integer. When a test is applied to find if the value of some other variable is *equal* to this value the computer may take the result quite literally and register a failure even if the difference is only 0.00002. It is as well to remember this any time a test is included in a program and make sure that the test allows a little latitude. For example, IF W > 2.2 is a safer test that IF W = 2 when W is being assigned values which are whole numbers.

Many computer scientists and programmers do not like the statement GOTO as it appears in line 65 above. They do not mind too much if it is accompanied by a conditional statement as in line 35 but otherwise they recommend that GOTO should not be used. As far as we are concerned we shall not worry too much as all our programs are relatively short and are unlikely to contain many such statements.

2.3.2. FOR NEXT Loops

The FOR-NEXT loop is a common method of repeating a section of program. When appropriate, it is a much more elegant method than using the IF ... GOTO structure as in the last program.

In its simplest form a FOR-NEXT loop causes a variable to increase by a certain amount each time the loop is executed:

```
 9 REM ** FOR-NEXT
10 FOR I=1 TO 10 STEP 1
20     PRINT I
30     NEXT I
40 END
```

Π This program will print the numbers 1, 2, 3 ... 10 (one per line) and then stop. Run it on your computer.

In the program listing the lines of the loop following FOR ... are indented to make the program more readable.

The variable I is the *index* or *control variable*.

The *end value* or *limit* of the *index* is 10.

The *step* or *increment* is 1 because the index increases by 1 each time the loop is traversed.

When you have it working correctly, alter the program to make it print out both I and the square of I in column format. Changing line 20 to 20 PRINT I , I^2 should achieve this.

The last line of the loop is line 30 NEXT I. Many microcomputers allow the variable name to be omitted so that the line could be simply 30 NEXT. Since not every computer allows this to be done however, and since the omission of the name can often lead to confusion, we shall make it our practice to include the variable name.

The step size may be a whole number or a fractional number, and it may be positive or negative. With most computers it is not necessary to state the step size if it is +1 but any other size of step must be stated.

There are sometimes restrictions on the name and type of the control variable. One popular microcomputer requires that the control variable be represented by a single letter like I or X or P. Another insists that the control be a *real* variable and not an integer despite the fact that this is the reverse of the requirement of some high level languages. You will soon discover any restrictions imposed by your machine.

Sometimes it is convenient to replace the start value or the end value or the increment by an expression which yields the proper value when evaluated. Though this is not recommended for the beginner an example is:

```
110 FOR I=1 TO 3*P STEP 2
```

A loop starting with this statement would require to have a value assigned to variable P before the statement is reached. The computer calculates the value of 3*P before the loop is entered and the control variable, I, is compared with this value between passes round the loop.

While it is usually possible to use expressions instead of the index or the limits or the step size, you should keep things as simple as possible until you have considerable experience in programming. Use either proper constants or variables with assigned values.

Finally, remember that a test is applied before the start of each cycle round a loop and remember how a computer stores its numbers. On occasion it may be necessary to write a statement like:

```
120 FOR I=2 TO 20.1 STEP 2
```

Here the end value is made slightly higher than 20 in case the value of I is held as, say, 20.00002 when the test before the last cycle is made. The problem can be avoided by using integer variables since there would then be no doubt about exact values:

```
10 A%=2
20 B%=20
30 S%=2
40 FOR I%=A% TO B% STEP S%
...
...
```

Of course your machine must allow the control variable to be an integer for this to be practicable.

This is a good point at which to pause and assess progress. Try SAQ 2.3a and study the answer carefully. Then review the ground we have covered until you feel quite at home with all the parts of programming we have looked at. When you can answer SAQ 2.3a without help you are ready to continue.

SAQ 2.3a

Write programs to display on the screen:

(*i*) the squares of the numbers 1 to 10 one per line;

(*ii*) the squares of the numbers 1 to 10 in a row with a space between each number shown;

(*iii*) the *even* numbers from 0 to 10 and their squares in two columns (take 0 as an even number.)

(*iv*) as (*iii*) but starting with 10 and finishing with 0.

SAQ 2.3a

2.3.3. Loops and Calculations

The FOR-NEXT loop is particularly useful when it is necessary to operate on a series of numbers, as for example in doing a simple addition. As always you should remember that the sign '=' is best interpreted as meaning *become equal to*.

Here is a simple adding program:

```
 9 REM ** ADD 10 NUMBERS
10 PRINT "PROGRAM TO ADD 10 NUMBERS"
20 PRINT
30 LET S=0
40 FOR I=1 TO 10 STEP 1
50    INPUT "ENTER NUMBER ",X
60    LET S=S+X
70    PRINT S
80    NEXT I
90 END
```

A REM statement is simply a reminder. The program ignores this.

The variable X may take a different value each time the loop is traversed depending on the value input at line 50.

The variable S represents the sum of all the X values. S starts with the value zero at line 30.

At line 50 a value for X is entered and at line 60 this value is added to S to give a new value for S. The form of line 60 often causes problems because we have been conditioned to expect the sign '=' to link two equal values. In the present context it is best to translate the effect of line 60 as: *Add the value found in box X to the value found in box S and place the result in box S.*

At line 70 the value of S is printed. Lines 50, 60 and 70 are repeated ten times so that S finally becomes the sum of ten numbers.

Π Run the program, entering your own numbers when prompted. Keep the numbers simple so that you can check the result.

The program as written can only deal with 10 numbers because it goes round the loop 10 times. To make it more general the number of items to be added could be entered before the loop commences. The following changes will achieve this:

```
10 PRINT "HOW MANY NUMBERS?"
15 INPUT N
40 FOR I=1 TO N STEP 1
```

Π Alter your program in this way and run it again.

Another method of exerting some control is to use a *conditional statement* so that when a certain value of X is entered the program stops. Look back to Section 2.3.1 (program line 35) and then answer the next question.

SAQ 2.3b

(*i*) Alter the adding program of Section 2.3.3 so that the program stops and prints the sum when a negative number is entered. The negative number should not be included in the summation.

(*ii*) Alter the program so that the number of items to be added is entered at the keyboard before the loop commences. The program must add this number of items and finish by printing not only the value of the sum, S, but also the value of the control variable, I.

Is the value of I the same as the number of items added?

If I is different from the number of items added why is this?

Sometimes it is necessary to check that an operation within a loop does not lead to disaster. The following is an example worth studying.

You will recall that the factorial of an integer N is the continued product:

$$N! = N(N-1)(N-2)(N-3) \ldots (3)(2)(1)$$

The program below will calculate a factorial for any value of N greater than zero. It has been written in such a way that it keeps asking for an integer and printing the factorial until you type in 0 (zero). This saves having to type RUN for every factorial you want.

```
 99 REM ** FACTORIAL PROGRAM
100 PRINT "FACTORIAL CALCULATION"
110 PRINT
120 PRINT "ENTER INTEGER (0 TO STOP)";
130 INPUT N
140 IF N<1 GOTO 210
150 F=1
160 FOR I=1 TO N
170     F=F*I
180     NEXT I
190 PRINT "FACTORIAL ";N;"="F
200 GOTO 110
210 END
```

The step size has been omitted at line 160 since the increment is +1.

Π Run the program on your computer and use it to evaluate factorials. Enter larger and larger numbers each time input is requested until the computer gives up because it is being asked to compute a number that is outside its range. (This will give you a very rough idea of the range of your machine). You can then alter the program so that if such a number is entered it simply repeats a request for input.

You will notice that the program as written continues to request input until it receives a value of zero. It then moves to line 210 and stops. An alternative stopping method would be omit line 210 and re-write line 140:

```
140 IF N<1 THEN END
```

To make the program refuse a number that is too large we first find the number which makes the program "crash" by entering larger and larger numbers. If this number is M we can make the program refuse a number larger than M−1 by inserting the lines:

```
145 IF N>(M-1) GOTO 220

220 PRINT M;" IS TOO LARGE"

230 GOTO 100
```

Π The calculation part of this program could be done in reverse. Instead of starting with F = 1 and then multiplying by 1, 2, 3, ... N we could start with F = N and multiply by N−1, N−2, ... 1. The principal change would be to line 160. Why not rearrange the program to do this and see if you get the same answers?

Here is an SAQ which you can regard as a key assessment exercise. That is, you can judge your progress so far by the ease with which you can write and run the program.

SAQ 2.3c

Write a program which:

(*i*) accepts a maximum of 20 numbers;

(*ii*) computes the sum of the numbers;

(*iii*) computes the sum of the squares of the numbers; ⟶

SAQ 2.3c (cont.)

(*iv*) stops when a negative number is entered;

(*v*) calculates the total number of items;

(*vi*) prints out the sum, the sum of squares, the total number of items and the mean of the numbers entered.

Run the program using the following numbers:

89 82 67 72 86 80 86 91 80 80

2.3.4. Limiting Loop Size

In some of the examples above the programs jump out of loops. In general this procedure is frowned upon though it does no great harm provided the program stops almost immediately.

While jumping out of a loop may be simply bad practice 'jumping in' is a major error. A GOTO statement must *never* direct a program to go into a loop from outside the loop.

When the size of a FOR-NEXT loop cannot be determined before the loop is entered it may be possible to use a different type of loop. Some computers support loops that are governed by the word UNTIL. One possibility is:

```
REPEAT

...

(lines to be repeated)

...

UNTIL N<1

...
```

As with FOR-NEXT loops some condition is checked each time the loop is traversed but the REPEAT-UNTIL loop has the merit of adjusting the size of the loop automatically to the size required. Since loops of this type are not so universally used as are FOR-NEXT loops, we shall not employ them in this course except for illustrative purposes. It is not difficult to see when they can be employed to advantage and if your computer language supports this kind of structure the manual will give the proper syntax. Some languages also support loops which are similar to the REPEAT-UNTIL but in which the test UNTIL is replaced by the test WHILE. Again, these can be very useful but we shall not employ them in the course.

When only the conventional FOR-NEXT loop can be used it is good practice to avoid jumping out in case problems result through an accumulation of incomplete loops. This is best done by setting the proper end value of the index before the loop is entered. It may be that this is not possible however, and then another strategy must be used. We give below one method of making sure that all passes round a loop are completed and, for comparison, an example of a REPEAT-UNTIL loop. After studying these methods you might care to modify your answer to SAQ 2.3c; proper operation of the program will provide assessment.

Let us suppose that we are going to add some numbers and are pretty confident that we shall not need to add more than 100 items. Also, no number will be greater than 998.

(*a*) All passes are made.

In the following program a decision is made at line 190 in every pass round the loop:

```
 99 REM ** COMPLETING LOOPS
100 PRINT "ADD UP TO 100 NUMBERS"
110 PRINT " LESS THAN 999"
120 PRINT
130 PRINT "TO STOP ENTER 999"
140 PRINT
150 S=0
160 N=0
170 INPUT "ENTER FIRST NUMBER ",X
180 FOR I=1 TO 100
189     REM ** CHECK FOR TERMINATION
190     IF X>998 THEN 240
200     S=S+X
210     N=N+1
220     PRINT "NEXT NUMBER ";
230     INPUT "(999 TO STOP) ",X
240     NEXT I
250 PRINT
260 PRINT "SUM OF ";N;" ITEMS = ";S
270 END
```

Variable S represents the sum, variable N the number of items added. Input is requested at line 170 initially, and then at line 230 inside the loop. Provided the number input (X) is 998 or less all operations in the loop are completed. If, however, the number entered is greater than 998 the conditional statement at line 190 ensures that lines 200–230 are missed in the remainder of the 100 loop cycles.

This kind of program can be wasteful of computer time if the number of steps set is much greater than the number actually required. It could be improved by using integer variables if this is possible, particularly at line 190 where a test is made in every pass round the loop.

(*b*) Loop size not specified.

The REPEAT-UNTIL or REPEAT-WHILE structure is better than (*a*) if your computer language supports one of these. Here is the same program written for a language which allows REPEAT-UNTIL:

```
 99 REM ** UNTIL LOOP
100 PRINT "ADD UP TO 100 NUMBERS"
110 PRINT "LESS THAN 999."
120 PRINT
130 PRINT "TO STOP ENTER 999"
140 PRINT
150 S=0
160 N=0
170 INPUT "ENTER FIRST NUMBER ",X
180 REPEAT
190     S=S+X
200     N=N+1
210     PRINT "NEXT NUMBER ";
220     INPUT "(999 TO STOP) ",X
240     UNTIL X>998
250 PRINT
260 PRINT "SUM OF ";N;" ITEMS = ";S
270 END
```

In the REPEAT program a test is applied automatically at the end of each cycle (line 240). This is clearly a much tidier program. It does not employ a GOTO and its construction requires much less thought than the previous version.

You should note the feature which is common to both these methods of control:

Input is taken before entering the loop and then again at the end of each loop cycle.

A third method of controlling loop size which is sometimes used involves changing the index or control variable inside a loop. This method cannot be recommended generally as some languages do not allow any of the loop parameters to be changed after they have been specified. If it is permissible line 190 (and 189) of program (*a*) would be omitted and a new line 235 inserted:

```
235 IF X>998 then I =101
```

After X is assigned a value greater than 998 this line changes the control variable so that when it is tested against the limit or and value all passes appear to have been completed and the program passes to line 250 and out of the loop.

We must repeat, however, that changing the control variable cannot be recommended as a general method for controlling a loop.

2.4. DATA CONTROL

2.4.1. Read, Data, Restore

The DATA statement is a convenient way of supplying data to a program. It is particularly useful when a program requires a lot of data or uses the same set of data many times.

The idea is that instead of the program prompting you to input a number or string at the keyboard the program reads the item itself from a list that you have prepared. One of the great advantages of this method of supplying data is that you can check the data at your leisure before starting the program. In this way, if you make a mistake when typing the data, you have an opportunity of correcting the mistake before the program runs. Also, the data supplied stays with the program and can be checked for accuracy if there are any doubts about what the program actually read. Here is a simple example:

Π We want to add several numbers and find the mean. Type and run this program. Do not destroy it as we shall develop it later.

```
 99 REM ** READ DATA
100 S=0
110 READ N
120 FOR I=1 TO N
130     READ X
140     S=S+X
150     NEXT I
160 M=S/N
170 PRINT "MEAN VALUE=";M
300 END
399 REM ** DATA LIST
400 DATA 9
410 DATA 10,11,12,13,14
420 DATA 15,16,17,18
```

The data items are on lines 400 to 410. The comma is used to separate the items. It does not matter how many lines are used for the data so long as each line starts with the word DATA.

DATA statements may be located at any point in the program; it is usual to place them at the end of the program.

In the above program the first item of data (line 400) gives the number of items to be added. This number (9) is read into variable N at line 110 by a READ statement.

The READ statement at line 130 causes each succeeding number in turn to be read into the program as though it had been typed in at the keyboard.

Instead of reading the value of N one could use a large loop (eg FOR I=1 TO 1000) and a conditional statement after line 130 to test for, say, a negative item being read. The last item of the list must then be negative and care would have to be taken not to count this item as one of the numbers to be added.

When the computer reads from a DATA statement a 'pointer' moves from one item to the next. The pointer starts at the first item when the program begins. If the same data list is to be used more than once it is necessary to restore the pointer to point to the first item again. This is done by the simple statement RESTORE.

RESTORE by itself returns the pointer to the very first DATA item. However, many versions of BASIC permit the pointer to be moved to the first item of a particular line by a statement like:

```
174 RESTORE 410
```

This has the effect of making the next READ statement take in the first item on line 410 of the program and continue to read the following items as before. RESTORE without a following line number would move the pointer to the very first item on line 400 of the program.

Π This facility to re-use items of data can be quite useful. To check how it works let us add the following lines to the program:

```
172 PRINT
174 RESTORE 410
176 S=0
178 FOR I=1 TO N
180     READ X
182     D=X-M
184     PRINT X;" DEVIATES FROM MEAN BY ";D
186     S=S+D
190     NEXT I
200 PRINT "SUM OF DEVIATIONS= ";S
```

The effect of these additional lines is to read again the data items (but not N) and subtract the mean value from each one. The deviation of each item from the mean is calculated and printed. At the same time variable S is used to sum the deviations. As a final check the value of S is printed.

Following this example it only needs a few more steps to calculate the *standard deviation* of the data items. Before tackling the next question recall that the standard deviation of a number of measurements is defined as the square root of the *variance* and that the variance is obtained by dividing the sum of the squares of the deviations by one less than the number of items:

variance = V = (sum of D^2)/(N − 1)

standard deviation = square root of V

SAQ 2.4a

Write a program which includes DATA statements and which:

(*i*) reads the number of items from the data list;

(*ii*) calculates and prints out the mean of several numbers;

(*iii*) restores the data pointer;

(*iv*) calculates and prints out the standard deviation of the numbers.

Test the program using the following numbers which were obtained by 11 analysts for the concentration (g dm^{-3}) of copper in a solution:

49.89 49.82 49.67 49.72 49.86 49.80
49.86 49.96 49.80 49.80 49.83

SAQ 2.4a

2.4.2. Tabulation of Output

Sometimes the columns or fields or zones which are obtained by using the comma to separate items of the print list are not the right length for the material we have to print. It is then necessary to use a TAB statement. This takes various forms depending on the computer, eg TAB(5) or TAB5, to place the first character of the item in position 5 from the left (starting at zero).

After TAB(5) or its equivalent one usually places the punctuation mark that means 'leave no space' but practice differs and it may be that no mark is required. In this example the semicolon is used though it may not be necessary:

```
110 PRINT "SAMPLE "; N$;TAB(20);"WEIGHT=  ";W
```

This statement would cause the following to be printed:

```
SAMPLE .............WEIGHT= ...
```

with the string N$ in the first space and the value of variable W in the second.

Since the syntax is critical and as there are many variations you must find details for your own machine.

The TAB statement can also be used to start printing data at any character position on the screen. A statement like

```
110 PRINT TAB(10,15);X
```

would have the effect of placing the value of X on the screen starting at character position 10 horizontally and 15 vertically. Text character positions normally start at top left. That is, the first character at the top left of the screen is normally in position 0,0. It may be worth checking this for your machine in case its first screen position is 1,1.

String variables may be used like any other in tabulation as, for example, in the headings of a table:

```
110 R$="READING"
120 P$="CONC IN PPM"
130 PRINT TAB(15);R$;TAB(25);P$
```

A last word on tabulation concerns the use of the word *column*. A 40 column screen means a screen which displays 40 characters horizontally. The *columns* are usually numbered from 0 to 39. Similarly, an 80 column display means that 80 characters are shown horizontally. This kind of display is often used by word processors.

SAQ 2.4b

In a book on quantitative analysis there are tables of data on common acids. Typical entries are:

	% w/w	kg/litre	mol/litre
Hydrochloric acid	35	1.18	11.3
Sulphuric acid	96	1.84	8.0

Write a program which employs TAB statements to produce a table like this from data in DATA statements.

SAQ 2.4b

2.4.3. One-dimensional Array

It frequently happens that you want to take a series of readings and store them for later computation. For example, you might take a series of pH or absorbance readings over a period of time and then investigate the pattern of behaviour by plotting against time.

For the moment, let us not worry about how the data are collected but simply consider how we might store and manipulate the data. This is where an *array* is useful.

You will be familiar with labelling a series as:

X_1 , X_2 , etc.

An array does the same thing but the subscripts are in parentheses, eg

X(1), X(2), X(3), ...

Each of these elements represents a variable. We can write a program statement:

```
30 LET X(9)=A
```

The value of the variable X(9) would then be the same as variable A. Again, a variable can be used to assign a value to an array element as in:

```
40 LET X(I)=B
```

The value of element X(I) becomes equal to the value of variable B. Of course the computer would require to know the values of subscript I and variable B when the program reaches this statement.

An array is very convenient when a set of numbers is to be stored or is to be operated on in a repetitive manner. The same segment of a program can be used repeatedly on different numbers simply by changing the subscript within a FOR-NEXT loop.

But let us start at the beginning. We must first decide the maximum number of elements the array can have. The computer will then reserve the correct amount of memory space. Since it will not recognise more then the number of elements we decide upon we must make sure that the size of the array is big enough before we start to use it. Suppose we know that we shall need no more than 100 elements. We then *dimension* the array by the statement

```
10 DIM X(100)
```

This makes the computer set aside memory space (boxes!) for, usually, 101 variables called X(0), X(1), X(2), etc. We say *usually* because some computers will set aside only 100, the variable X(0) not being allowed. This is a point that you should check from your computer manual.

If the numbers to be stored are integers and your computer can make a distinction between real and integer arrays it is often a good idea to use an integer array. This save time and space. The normal convention for naming integer variables is followed, ie the name of the array has a % sign as the last character, eg DIM X%(100).

Your computer probably supports *string arrays*. That is, a series of strings can be held as variables with names like A$(1), A$(2), etc. However, we shall not make use of this kind of array.

With some languages the dimension statement must be made at the start of the program while with others it may be made at any point before the array is used. You should check this point but in any case it is always good practice to dimension arrays early in a program to ensure that enough memory space is made available. The same array must not be dimensioned more than once in a program.

Π Suppose we want to collect a set of numbers. Let us start with X(1) rather than X(0). Here is a segment of program:

```
 99 REM ** SIMPLE ARRAY
100 DIM X(100)
110 PRINT "HOW MANY VALUES?";
120 PRINT " (UP TO 100) ";
130 INPUT N
140 FOR I=1 TO N
150     PRINT "ENTER NUMBER ";I
160     INPUT X(I)
170     NEXT I
180 END
```

Run this program and check it using a few simple numbers. The loop should collect N values and place them in locations labelled X(1), X(2), ... X(N). To see if this happens add the next segment (line 180 being replaced):

```
179 REM ** CHECK ENTRIES
180 FOR I=1 TO N
190     PRINT X(I)
200     NEXT I
210 END
```

The numbers entered should now be printed out in the same order as they were entered. Make sure this happens and then alter line 180 to

```
180 FOR I=N TO 1 STEP -1
```

Run the program again. The numbers should now be printed in the reverse order.

It is necessary to include the step size in the altered line because it is not +1.

A FOR-NEXT loop and an array are often used together as in the program developed above. You should become quite familiar with the technique as it will prove to be extremely useful.

SAQ 2.4c

Write a program to accept up to 10 real numbers into an array. After testing the program with simple numbers (eg 2,4,6, ...) alter the latter part of the program to make it print out:

(*i*) the numbers and their cubes in column format;

(*ii*) the numbers, and their reciprocals correct to 3 decimal places

2.4.4. Arrays and Data

It is often useful to use DATA statements when assigning data items to the elements of an array. When an array, a FOR-NEXT loop, and DATA statements are used together in a program you have a very powerful combination. Items from the data list are read into the array and then computations are done as required. The next question illustrates a typical application.

SAQ 2.4d

In a gas chromatography experiment the solvent peak was recorded at a retention time of 0.85 minutes and successive peaks were recorded at retention times of 3.96, 4.33, 5.20, 6.72, 7.58, 9.28 and 10.75 minutes. Write a program which will:

(*i*) read the experimental data from DATA statements, placing the peak times in an array;

(*ii*) compute retention times relative to peak number 4 (6.72 minutes); that is, for each peak compute the ratio

$$\frac{\text{retention time} - 0.85}{\text{peak 4 time} - 0.85}$$

(*iii*) display the retention times and the relative retention times in a table.

The array is particularly useful when several calculations require the same data at different points in a program. To follow up this point you might like to look again at SAQ 2.4a which required the use of the RESTORE statement in the calculation of a standard deviation. The program could be written to read the data into an array and then calculate the sum of squared deviations without restoring the data pointer.

2.5. PROGRAM CONTROL (2)

2.5.1. IF ... THEN Control

The material of Section 2.5 is not essential to the study of elementary programming. You should read the section over lightly to see what it is about and return for more detailed study later.

In several programs we have employed statements that start with the word IF. As this word implies these are *conditional* statements.

Every conditional statement involves at least one *test* by one of the *relational operators* ‘=’ or ‘>’ or ‘<’. Following the test a certain course of action is taken by the program. This action is usually easily seen:

```
IF X<0 THEN GOTO 250

IF X>0 THEN END

IF X<22 THEN X = 22

IF X% = Y% THEN PRINT "EQUALITY"
```

We have already used statements like the first two. The third statement is a means of ensuring that the value of X does not fall below a certain value (in this case 22). This kind of limitation is very useful when plotting points on the screen to guard against attempting to plot a point outside the correct range.

Let us discuss the last example in more detail. The test can be stated in words '*Is the value of variable X% the same as that of variable Y%?*'. If the answer to the question is 'yes' the test is successful and the program prints the word "EQUALITY". If the answer to the question is 'no' the test fails. What is the program to do in this case? In most cases the program simply moves to the next *numbered* line.

It is very important to note that control passes to the *next numbered* line.

The BASIC of most microcomputers allows more than one statement to be placed on a numbered line, the statements being separated from each other by colons. When this is the case the IF statement acts as a 'gate' to the rest of the line. For example, consider a program line which contains two statements:

```
110 IF A%>B% THEN PRINT A%;" IS BIGGER THAN
    ";B% : M% = A%
```

Provided A% is greater than B% the message is printed *and also* M% is made equal to A%. (Perhaps the intention is to find the maximum of a series of numbers.)

It is important to note that the second instruction, M% = A%, will not be executed unless A% is greater than B%.

Here is an example which illustrates a common trap:

Suppose a program includes the two lines:

```
100 IF X>0 THEN Y = SQR(X)

110 A% = B%
```

In the first of these lines care has been taken to avoid trying to find the square root of a negative number.

If it is possible to write more than one statement on each line there is sometimes a temptation to make the program neater with a line like:

```
100 IF X>0 THEN Y = SQR(X) : A% = B%
```

Unfortunately, the second statement can only be reached if X is greater than zero because of the 'gate' effect of the IF statement. As this example demonstrates it is essential to consider the consequences of placing more than one statement on a line. This is particularly important when a conditional statement is involved.

To illustrate the use of IF statements let us write a program to find and print out the maximum and minimum of a sequence of numbers entered at the keyboard.

```
 99 REM ** HIGH & LOW
100 PRINT "TO FIND HIGHEST & LOWEST OF N
               NUMBERS"
110 PRINT
120 PRINT "HOW MANY NUMBERS ";
130 INPUT N
140 PRINT
150 PRINT "ENTER FIRST NUMBER ";
160 INPUT X
168 REM ** H = HIGHEST L = LOWEST
169 REM ** MAKE FIRST NO. BOTH H AND L
170 H = X
180 L = X
190 FOR I = 2 TO N
200     PRINT
210     PRINT "NEXT NUMBER ";
220     INPUT X
229     REM ** TEST AND CHANGE
230     IF X>H THEN H = X
240     IF X<L THEN L = X
250     NEXT I
260 PRINT
269 REM ** NOW REPORT
270 PRINT "HIGHEST = ";H
280 PRINT "LOWEST = ";L
290 END
```

The first number (N) input at line 130 gives the total number of items to be entered. The first of the numbers is taken to be both the highest (H) and the lowest (L) initially (lines 170 and 180). As each subsequent number is input these are altered as necessary (lines 210 to 240).

2.5.2. Combining Operators

The relational operators can often be combined as in the statement

```
110 IF X >= A THEN ...
```

The use of combinations helps to simplify programming. But be careful. Check your manual before you use a combination like this since your machine may be fussy about whether '=>' or '>=' is used to test for 'equal to or greater than'.

Again be careful. The computer takes things literally. It may consider a number like 1.99999 to be less than 2 even though you would like it to be taken as equal to 2. Problems of this nature are more likely when using real variables rather than integer variables and you always have to be on your guard.

Sometimes the NOT operator is used to reverse the effect of a test. An example is:

```
100 INPUT X
110 IF NOT (X = 6) THEN PRINT;X;" IS NOT SIX"
```

An alternative to this example uses the combination <> which means *not equal to*:

```
100 INPUT X
110 IF X<>6 THEN PRINT;X;" IS NOT SIX"
```

A common use of the conditional statement is to ensure that the computer is not asked to perform an impossible calculation. Common errors that must be guarded against include:

trying to find the square root of a negative number;

trying to divide by zero;

trying to take the logarithm of zero or a negative number.

SAQ 2.5a

Write a program which prints out the reciprocal and the common logarithm of any number entered at the keyboard and which warns without crashing when the task set is impossible. All output should be correct to two places of decimals.

Apart from using combinations like >= or <=, two or more conditions may be tested in one statement:

```
110 IF X>0 AND X<10 THEN ...
```

This will ensure that the action following THEN is taken only if X has a value which is greater than zero and less than 10. The AND must be in upper-case.

Another example uses OR:

```
110 IF X<5 OR X>10 THEN ...
```

The action which follows THEN will take place if X is less than 5 (but not 5) or if X is greater than 10 (but not 10). If we wanted to include the value 5 and 10 we must use the <= and >= signs.

While AND and OR can be very useful some care must be taken when using them. Frequently the NOT operator comes in handy as an aid to our thinking. The meaning of the following line is quite clear:

```
110 IF NOT (X = 0 OR X = 10) THEN ...
```

It is normally permissible to use more than one AND or OR:

```
110 IF A>B AND X>Y AND C<>D THEN ...
```

We shall meet AND and OR again when we study memory locations.

One application of conditional statements is to test a string before attempting to use the VAL function. Every keyboard character has a code number, generally referred to as its ASCII code number (from American Standard Code for Information Interchange). Thus, the ASCII code for the letter 'A' is 65 and that for 'a' is 97. Within a program the code for the first character of a string X$ can be obtained by means of a statement like:

```
102 C = ASC(X$)
```

The reverse operation of obtaining the character corresponding to a given code number uses a statement like:

```
104 X$ = CHR$(C)
```

You should check the syntax required by your machine for both of these statements.

SAQ 2.5b

Write a program which accepts words or numbers greater than −1000 from the keyboard and which:

(*i*) prints the message 'POSITIVE' if the first character of the input string is one of the 'numeric' characters 0 to 9;

(*ii*) prints 'NEGATIVE' if the first character is the minus sign (−);

(*iii*) prints 'FRACTION' and the character of ASCII code 7 if the first character is the period (.);

(*iv*) stops accepting input when the character 'Q' is entered;

(*v*) reports the total number of entries made, the sum of all the numerical entries and the highest negative numerical entry.

SAQ 2.5b

2.5.3. IF ... THEN ... ELSE

It was stated previously that the program proceeds to the next line number when a test fails. This default action can be prevented by the use of ELSE. Consider the program line:

```
120 IF X<0 THEN END ELSE GOTO 30
```

The program stops if X is negative, otherwise it jumps to line 30.

This kind of structure can be very useful and leads to some neat programming. If we did not use ELSE in the line above two lines would be required to produce the same effect:

```
120  IF X<0 THEN END
130  GOTO 30
```

It is often possible to place several statements after the THEN but before the ELSE. This is one way of overcoming the 'gate' effect of IF mentioned earlier. Also, more that one IF ... THEN may be 'nested' as in:

```
110 IF X<0 THEN IF X>Q THEN Q = X
```

Before attempting to nest conditional statements you should consult the computer manual to see what is allowed. It is also a good idea to run a short experimental program to check that nested statements have the effect you want.

You will appreciate that the use of these structures can lead to some very complicated tests. Properly used, they can impose all sorts of control on a program. They may, however, lead to programs which are difficult to understand and you are therefore well advised to avoid such complications if at all possible. If you find yourself writing a statement involving several tests and branches it is a good idea to pause and consider if they are all really necessary.

SAQ 2.5c

Provided your computer allows the structure answer SAQ 2.5a again but using IF ... THEN ... ELSE.

2.5.4. Nested Loops

If a program contains several loops it is very important to ensure that the loops are self-contained. If all steps of one loop have not been completed when the next loop starts the computer thinks that the new loop is inside or *nested* in the previous one. Provided the second loop is completed before the first this causes no problem. There is, however, a limit to the number of nested loops permitted and if this number is exceeded an error will be signalled. The number of loops which may be nested depends on the particular version of BASIC supported by the computer.

Π As a simple illustration of nesting here is a program which produces three multiplication tables:

```
 9 REM ** NESTING
10 FOR I = 1 TO 24
20    FOR J = 1 TO 3
30       PRINT I*J;
40       NEXT J
50    PRINT
60    NEXT I
70 END
```

Try this program. It should produce something like:

```
1   2   3
2   4   6
3   6   9
.   .   .
.   .   .
```

Each of the three tables should extend from 1 to 24. You may have to modify the punctuation at line 30 to make the output suit your computer. Integer variables would be better than real for a program like this.

SAQ 2.5d

The potential of an ion-selective electrode may depend on the activities of two ions:

E = E° + 58*LOG(A1 + K*A2) (mV at 292K)

A1 is the activity which the electrode is designed to measure, A2 the activity of an interfering ion. K is the selectivity ratio or selectivity constant.

Write a program to create a table showing, for any value of K, how the second term on the right depends on A1 and A2. Suitable ranges might be:

A1: 0.001 to 0.005 mol dm^{-3} in 5 steps
A2: 0 to 0.01 mol dm^{-3} in 5 steps

Output should be given to the nearest millivolt.

....

An example of output for K = 10 and A2 = 0.002 mol dm^{-3} is given below:

```
INTERFERING ION AT 2 MMOL

ION 1 AT 1 MMOL     TERM = -97
ION 1 AT 2 MMOL     TERM = -96
ION 1 AT 3 MMOL     TERM = -95
ION 1 AT 4 MMOL     TERM = -94
ION 1 AT 5 MMOL     TERM = -93
```

SAQ 2.5d

2.5.5. Double Array

A *double array* or an array of two dimensions is often convenient because it saves some programming. However, as such an array is hardly ever essential and as the associated programming requires careful thought it would not be unreasonable to skip this section until you feel ready for it. The subject is dealt with here simply for completeness.

Suppose that we have a series of ten measurements to make each day for five days. We could dimension an array A(5,10) and place the data in five DATA statements, one for each day. The following program segment could then be used to read the data into a program:

```
199 REM ** DAILY TEN
200 DIM A(5,10)
299 REM ** SELECT DAY
210 FOR D = 1 TO 5
219     REM ** TEN READINGS
220     FOR I = 1 TO 10
230         READ A(D, I)
240         NEXT I
249     REM ** TEN TAKEN
250     NEXT D
259 REM ** FIVE DAYS READ
499 REM ** KEEP DATA AT END
500 DATA M1,M2,M3,... M10
510 DATA Tu1,Tu2,Tu3, ... Tu10
520 DATA W1,W2,W3, ... W10
530 DATA Th1,Th2,Th3, ... Th10
540 DATA F1,F3,F3, ... F10
```

Here M1, M2, etc. represent the measurements made on Monday, Tu1, Tu2, etc. those made on Tuesday and so on.

Once all the data have been read in they may be used in calculations. The alternative to using the two-dimensional array would be to use five different one-dimensional arrays, M(10), T(10), etc. and this would lead to a larger program.

To check if the program segment works all right it might be followed by a segment to print out the data entered in another format. This may seem trivial but it can be quite useful as a check on the program:

```
259 REM ** CHECK READ LOOPS
260 FOR D = 1 TO 5
270     PRINT "DAY ";D
280     FOR I = 1 TO 10
290         PRINT A(D, I),
300         NEXT I
310     PRINT
320     NEXT D
330 END
```

This should print out the data in column or zone format:

```
DAY 1
M1     M2    M3    M4
M5     M6    M7    M8
M9     M10
DAY 2
Tu1    Tu2   Tu3   Tu4
... etc.
```

Note the PRINT statement at line 310. This ensures that the next day starts on a new line.

The most important rule in using arrays of more than one dimension is to make sure that the loops do not intersect. This has already been emphasised in connection with nested loops. In the program above NEXT I must come *before* NEXT D. That is, the I loop must be properly *nested* within the D loop. There will be chaos if proper nesting is not ensured. Those computers which allow the statement NEXT but do not require the index variable to be specified would appear to have certain advantages in this respect. This facility must be used with great caution however because errors due to incorrect nesting may occur without any indication being given. The computer cannot rectify poor thinking.

SAQ 2.5e

To assess a new analytical method four samples of a product were taken and each sample was divided into two parts. Four different analysts determined the purity using the old method for one part and the new method for the other. The results were listed as they were reported by the analysts, always in the order:

method, sample, % purity. ⟶

SAQ 2.5e (cont.)

The results so listed were:

2	2	98.8	1	4	98.4
1	3	98.1	2	4	98.1
1	2	98.1	2	3	98.9
1	1	98.6	2	1	98.6

Write a program to read these results into a two-dimensional array and present them in something like this table:

SAMPLE	1	2	3	4
METHOD 1	98.6	98.1	98.1	98.4
METHOD 2	98.6	98.8	98.9	98.1

2.6. GRAPHICAL OUTPUT

2.6.1. Graphs Using Text Characters

If you do not have a facility for plotting points on the screen it is still possible to produce a rough graph by means of text characters. The technique employed is also useful for histograms and for obtaining some kind of pictorial output by means of a printer.

The following is an example:

```
**********
************
**************
***************
*************
***********
*********
*************
****************
**************
************
```

This is a very rough graph but it gives an idea of a spectrum or a histogram though it would appear more conventional if turned through ninety degrees. A suitable program for this graph is:

```
 99 REM ** STAR GRAPH
100 READ N
110 FOR I=1 TO N
120     READ A
130     L=40*A
140     FOR J=1 TO L
150         PRINT "*";
160         NEXT J
170     PRINT
180     NEXT I
190 END
200 DATA 11,0.25,0.3,0.35,0.38,0.33
210 DATA 0.28,0.22,0.33,0.4,0.35,0.3
```

Notice that the punctuation mark after the string at line 150 is the one which ensures continuation of the printed line. On the other hand the PRINT at line 170 makes the line terminate so that the next printing starts on a new line. The scale chosen allows 40 character positions for an A value of 1 (line 130).

The next question asks you to improve this program.

SAQ 2.6a

Write a program to display a graph like the example shown below. The figure produced should have a wavelength scale from 400 to 700 printed at 20 unit intervals on or at the base line and each line should represent 5 wavelength units. You may assume that DATA statements contain the correct number of data items.

```
400 :     *
    :          *
    :             *
    :          *
420 :       *
    :
    :
```

2.6.2. The Graphics Screen

Most microcomputers are able to place marks on the screen at points which can be specified by a program. Unfortunately, every computer has its own statement which means *place a mark at position given by X and Y* where X and Y are variables representing coordinates.

When it is possible to specify the colour of a mark as well as its position the number of possible plotting statements becomes quite large.

Following our usual practice we shall keep things as simple as possible and look at the fundamentals of graphics. We shall not bother with colour at all since every microcomputer will have *default* colours for text or graphics and for background. Usually the default situation is white or green on a black background.

You may have noticed that we have avoided the use of the word *point* so far. This is because we must be careful in our use of this word as will become clear shortly.

A graphics screen can be considered as a grid of *addressable points* which can be addressed in the same way as we identify a point on graph paper. For example 450,200 specifies a point which is 450 units in the X (horizontal) direction and 200 units in the Y (vertical) direction. But the question arises: *where do we measure from*? While many microcomputers have the origin (point 0,0) at the bottom left of the screen some start at top left and measure vertically in a downward direction.

It is relatively easy to find the answer to the last question (where is the origin?). It is less easy to answer the next question: *is every point really different*? The answer to this lies in the *pixel*.

A *pixel* is a *picture cell* or *picture element*. It may be the same as an addressable point or it may contain several such points. Look at this diagram:

```
. . . . . . . .
. . . . . . . .      one pixel =
. . . . . . . .      8 × 4 points
. . . . . . . .
```

If this block of points represents a pixel there are $8 \times 4 = 32$ points in the pixel. If *any one* of the 32 points making up this pixel is plotted the whole pixel is marked or lit up. The smallest 'point' which can be marked would therefore encompass 32 addressable points. This particular example applies to the BBC machine in modes 2 and 5; this microcomputer can operate in several modes: mode 0 has a pixel size of 2×4 points, mode 1 of 4×4.

From this example we can see that the *resolution* of a graphics screen depends on the number of pixels on the screen rather than on the number of addressable points.

When plotting points on a screen you must decide on a scale in much the same way as when plotting on graph paper. Just as you must know the size of a graph paper so you must know how many points there are in each direction of the screen. It is wise to find this out early and build checks into programs to ensure that points are not plotted off the screen. Some microcomputers will give an error and stop the program if this is attempted while others will simply ignore futile attempts. On balance, an error message is to be preferred because it does let you know that there is a problem.

Here is an exercise to help you collect together the various factors which must be known before graphics programming can begin.

Π For your computer find answers to the following questions:

How many points are there in each direction and where is the origin?

What is the maximum X value and the maximum Y value for addressing points? (These numbers will probably be one less than the number of points in the X and Y directions because the first point is usually 0,0.)

How many points are there per pixel in each direction?

What is the syntax for plotting a point? Check by actually plotting several points.

2.6.3. Plotting Points

The only way to learn how to produce a graph on a computer screen is to do it. We have to worry about drawing and labelling the axes of course but the programming to draw and label axes and to mark out a grid on the screen is usually done after we are sure that we have our mathematics right and our plotting procedures correctly programmed.

It has been said that every physical chemist who starts to study computing writes a program to plot acid–base titration curves. If this is the case it shows how sensible the physical chemists are because there is no better way of getting to know your microcomputer and revising certain aspects of chemistry.

To illustrate how a plotting procedure is set up let us go through a typical argument in preparation for computing and plotting a titration curve.

We shall assume that a pixel consists of a block of 4 × 4 addressable points and that plotting coordinates refer to points rather than to pixels.

Along the X axis we want, say, 50 cm^3 in total (we hope our end point comes earlier). If there are 320 pixels in the X direction this means 6 pixels or 24 points per cm^3:

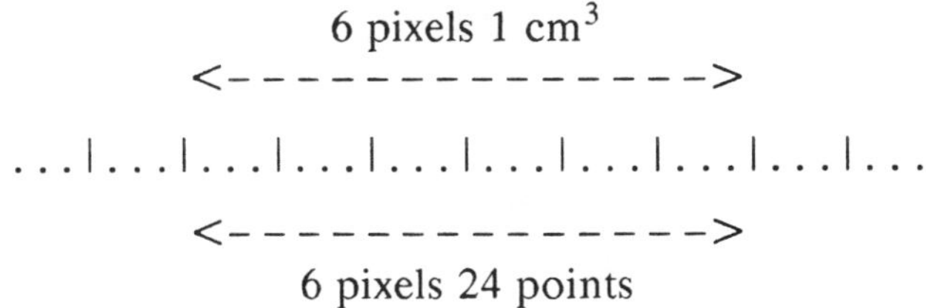

It follows that our scale in the X direction should be based on

24 points (X) per cm^3

We must however recognise that 4 points (X) will mark the same pixel so that the resolution expressed in terms of the titration is 4/24 or 1/6 cm^3. Nothing can be gained by trying to work with increments less than this.

In the Y direction we plot from, say, pH 2 to pH 12. If we have 200 pixels vertically we need 20 pixels per pH unit. Taking 4 points per pixel we have:

80 points(Y) per pH unit

Again we recognise that the resolution in terms of pH is 4/80 or 1/20 of a pH unit: between one plotting and another the pH change will only be noticed if it is greater than 0.05.

An exercise of this kind is always necessary when planning a plotting program. The conclusions can be checked by using test data before doing any serious calculations. We shall return to the discussion of a titration curve in 2.6.5.

SAQ 2.6b

> Write a program to read plotting coordinates from DATA statements and plot points on the screen. Confirm the program by plotting several points.

SAQ 2.6b

2.6.4. Drawing Lines

Most microcomputers make it relatively easy to draw lines on the screen. Usually you have to move a *graphics cursor* to one point and then draw the line to another point. In very general terms two statements are necessary:

PLOT A,B

DRAW X,Y

Together these statements might mean either of two things:

(*i*) Go to point A,B and draw a straight line from this point to point X,Y.

If this is the meaning the line is drawn using *absolute* coordinates. That is, X and Y are actual coordinates of a point on the screen.

(*ii*) Go to point A,B and draw a straight line to a point which lies X units in the horizontal direction and Y units in the vertical direction from point A,B.

If this is the meaning then X and Y are *relative* coordinates since they are used as if the last point (A,B) is the screen origin.

An important point to note is that with relative plotting it is possible to use negative values for X and Y.

With some microcomputers the statement DRAW is replaced by the statement PLOT along with some parameter which determines whether the line is drawn *to* a new point (absolute) or is drawn relative to the first point. (eg In BBC BASIC the statement PLOT 5,X,Y means draw a line from the last position of the graphics cursor *to* point X,Y *absolute*, while PLOT 1,X,Y means draw a line to a point which is X and Y *relative* to the last position of the graphics cursor.)

Some microcomputers have facilities for drawing parts of a circle, an extra parameter being included in the DRAW statement to define the circle radius.

For most of our purposes it is sufficient to draw straight lines from one point to another. Though it is seldom necessary to do this when producing a graph on the screen it is usual to draw lines when defining axes and scales.

Π Find out how to draw straight lines on the screen.

Draw X and Y axes suitable for the plotting procedure discussed in 2.6.3. You should try to indicate the scale by short lines at 10 cm^3 intervals and at each pH unit. Label the axes if possible.

When plotting graphs there is often a temptation to join up the points as we are accustomed to do when using graph paper. Often, however, the graph looks better with the points alone. The reason for this lies in the way lines are drawn on the computer screen. Any line must be made up of pixels which are lit up and since a pixel is effectively a small rectangle any curve is a series of short straight lines. This becomes very obvious when a line is only slightly curved or when it is very curved.

2.6.5. A Titration Program

In 2.6.3 we talked about titration curves. At this stage you should be in a position to write a complete program to produce the curve for a weak base–strong acid titration. If you feel sufficiently confident to do this now you may move immediately to the next SAQ but if you would like some guidance just continue to read.

We must be absolutely clear on our chemistry before starting to program. The first question we must therefore ask is: 'How can the pH be calculated at the various points of the titration?' There are in fact four different calculations for the case of the weak acid – strong base titration. In the following analysis we consider a weak monobasic acid at molar concentration C (mol dm^{-3}) being titrated with sodium hydroxide of the same concentration.

(*i*) Before titration starts the hydrogen ion concentration depends on the concentration and the dissociation constant, K_a (mol dm^{-3}), of the weak acid. The approximate equation normally used is:

$$[H^+] = \sqrt{(K_aC)}$$

It is convenient to take negative logarithms:

$$pH = pK_a/2 - \log(C)/2$$

(*ii*) After the start of titration but before the end point is reached the pH depends on the extent of neutralisation. Provided sufficient alkali has been added it is usual to use a form of the Henderson equation:

$$pH = pK_a + \log T/(E - T)$$

In this form the equation gives the pH when a volume T of titrant has been added, E being the volume necessary to reach the end point. In the present case, E is equal to the volume of acid taken initially since the concentrations of acid and base are equal.

This equation is often called the *buffer* equation because it is applicable in the *buffer region*; that is, it holds quite well in the region of pH from about $pK_a - 1$ to $pK_a + 1$. It must be remembered, however, that the equation is an approximation which must not be used too near the start or end point of a titration.

(*iii*) At the end point the hydrogen ion concentration depends on K_a, on K_w (the ionic product of water), and on the concentration of salt according to:

$$[H^+] = \sqrt{\frac{K_a K_w}{C/2}}$$

(C is divided by 2 because the volume has been doubled at the end point.)

Taking K_w as 10^{-14} $mol^2\ dm^{-6}$ this leads to:

$$pH = 7 + pK_a/2 + \log(C/2)/2$$

(*iv*) After the end point the pH depends on the excess of strong base:

$$[H^+] = \frac{C(T - E)}{100 + T}$$

$$pH = -\log \frac{C(T - E)}{100 + T}$$

The chemistry required for the next SAQ should now be reasonably clear. The programming will provide some useful revision of much of our earlier work.

SAQ 2.6c

Write and run a program to display on the screen the neutralisation curve for 100 cm^3 of a weak monobasic acid being titrated with sodium hydroxide of the same molar concentration. Calculate the pH at increments of 2 cm^3 titrant from 10 to 90 cm^3, and then in increments of 0.5 cm^3 to 120 cm^3. Make the usual approximations.

SAQ 2.6c

2.7. PROGRAM STRUCTURE

2.7.1. Subroutines and Procedures

We started off in this course with extremely simple programs which could be read and understood almost at once though we introduced REM statements at an early stage to help us remember what was happening at the various stages of a program.

As you will probably appreciate by this time most programs consist of a number of sections linked together. A program also contains a number of tests to determine which part of the program is to be done next.

As programs become more and more complex so the programmer finds it more and more desirable to construct programs from parts which have been used before and which are known to have the correct syntax. Ideally, each little section of program should be self-contained and capable of being called upon to do its job when required. Some of the more advanced high-level languages have been developed in this way and some versions of BASIC have been designed to allow programs to be written in sections which are reasonably self-contained. One of the chief criticisms of many versions of BASIC, however, is that the language tends to be relatively unstructured or loose.

We took a step along the road to structured programming when we prepared a program for plotting points in Section 2.6.3. If this program is built into other programs it could become a *procedure* or a *subroutine* for plotting points.

Procedures and *subroutines* are sections of a program that can be called upon by the main program to perform repetitive tasks. The use of a procedure or subroutine makes it unnecessary to repeat a section of program every time the task is performed. Most versions of BASIC support subroutines and some support both subroutines and procedures.

In BASIC a subroutine is usually called by its line number while a procedure is called by its name. Typical statements are:

```
50 GOSUB 1000
50 PROCplot
```

The first of these statements directs the program to the subroutine which starts at line 1000. The second tells the program to carry out the procedure named 'plot'.

A small program like that developed in 2.6.3 could form a subroutine for the plotting of points and be called upon at various stages of a large program to perform this task. If we do this however we shall probably have to re-organise the line numbers to suit each program and perhaps also make sure that the variable names do not conflict with names in the main program. For this reason a procedure is better than a subroutine since the former is called by its name without reference to line number. With some versions of BASIC it is possible to use variables which are *local* to the procedure; that is, the value which the variable takes inside the procedure does not affect the value of a variable of the same name in the main program.

Let us look at the question of structure in a general way.

Suppose that a program is to receive data and perform several calculation and plotting sequences. Typically you might want to plot two concentrations against time, the concentrations being obtained from measurements of absorbance. The essentials operations are:

Read two absorbances, A and B, and a time, T, from a DATA statement.

Calculate the concentration of substance P and plot concentration of P against time.

Calculate the concentration of substance Q and plot concentration of Q against time.

Repeat the steps a given number of times.

The outline of a suitable program is:

```
 99 REM ** SKELETON PLOT
100 READ N
110 FOR I=1 TO N
120     READ A,B,T
130     REM ** CALCULATE CONCN. OF P
140     REM ** AND WORK OUT PLOTTING
150     REM ** COORDINATES X AND Y
160     C=A
170     GOSUB 400
180     REM ** CALCULATE CONCN. OF Q
190     REM ** AND WORK OUT PLOTTING
200     REM ** COORDINATES X AND Y
210     C=B
220     GOSUB 400
230     NEXT I
240 END
399 REM **** PLOTTING S/R STARTS AT 400
400 REM ** CALCULATE X FROM T
410 REM ** CALCULATE Y FROM C
420 REM ** PLOT X,Y YOUR WAY
    ...
    ...
490 RETURN
499 REM ** DATA
500 DATA ...
```

This hardly counts as a program but should show why a subroutine is useful. At line 160 and again at line 210 the value of an absorbance is transferred to a variable C and then the program goes to the subroutine to calculate and plot a point. After plotting, the program returns to the next line number after 170 or 220. A subroutine always ends with the statement RETURN.

If a procedure is used instead of a subroutine the main program structure would be similar but the calling statement (lines 170 and 220) would be different and the statements marking the beginning and end of the procedure would be different. Thus, with the BBC microcomputer we might use a procedure and re-write the following lines:

```
170 PROCplot
    ...
220 PROCplot
    ...
399 REM **** PLOTTING PROCEDURE
400 DEF PROCplot
    ...
490 ENDPROC
```

Note that the line number of the start of the procedure is not stated when the procedure is called by PROCplot.

It is sometimes possible to pass *parameters* to a procedure; that is, the variables to be used within a procedure may be specified when the procedure is called. Thus, instead of the two lines 160 and 170 it may be possible to write

```
160 PROCplot(A,T)
```

and to define the procedure by

```
400 DEF PROCplot(C,T)
```

The values of variables A and T would then be assigned to two variables in the procedure and the calculation carried out. Lines 210 and 220 could be replaced in a similar manner.

Π Find out whether your computer supports procedures or subroutines or both.

Write a short program which includes a procedure or subroutine for plotting points.

When using a procedure or subroutine is it necessary to use the same names for variables as in the main program? If it is not necessary to do this how are the variable values transferred?

2.7.2. Flow Diagrams

You will have noticed by now that it is becoming more and more difficult for us to give you specific tasks which can be checked properly because we do not know the details of your computer. At the same time a program outline like the one above should be comprehensible to any computer programmer. Even better, a well-constructed *flow diagram* can represent a program structure no matter what language is being used.

A *flow diagram* or *flowchart* is a graphical representation of a program which gives a pictorial display of program structure without too much concern for proper syntax, variable names, etc. It is therefore an extremely good means of communication between programmers who may use different languages. This, of course, means people like us.

A full flow diagram is not necessary at the early stages of program planning. When a program is first contemplated it is important to understand the science and mathematics involved. Then it is necessary to think in terms of a very simple block diagram:

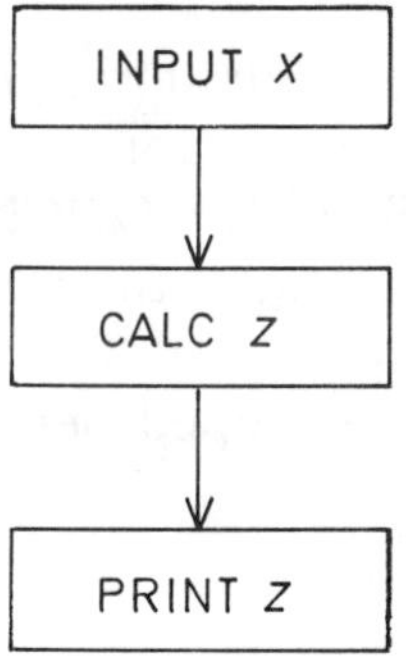

Obviously the program is going to calculate Z from a value of X entered at the keyboard. But have you all the information for the calculation and can you write the calculation program? Questions like these must be answered and the proper preparations made before real programming starts.

Of course this program is trivial. But suppose we were going to repeat the calculation and output several times. We would incorporate a loop as we have done so often:

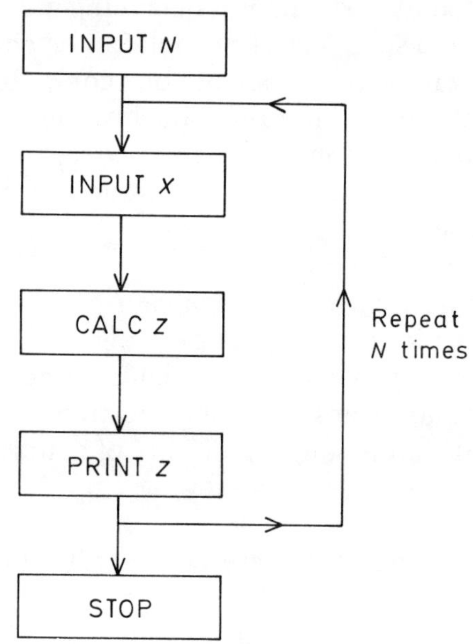

This kind of diagram helps the planning process without being too formal. You could include blocks for procedures or subroutines, eg 'TO PROCplot', 'TO S/R FOR CONCN. OF A', and generally build up a picture of what you want the program to do. In this way the *logical structure* of the program can be determined and algorithms for the calculations developed before proper programming is commenced.

Once the program is complete and you want to tell someone else how it works or simply keep a record of its structure for future reference a full flow diagram should be prepared. This is similar to the block diagrams above but is more formalised.

Most programs include tests, branches and loops. As discussed in Sections 2.3 and 2.5 a *test* is made every time a loop is traversed or a program branches. The next example shows how a test and the action which follows are incorporated into the flow diagram:

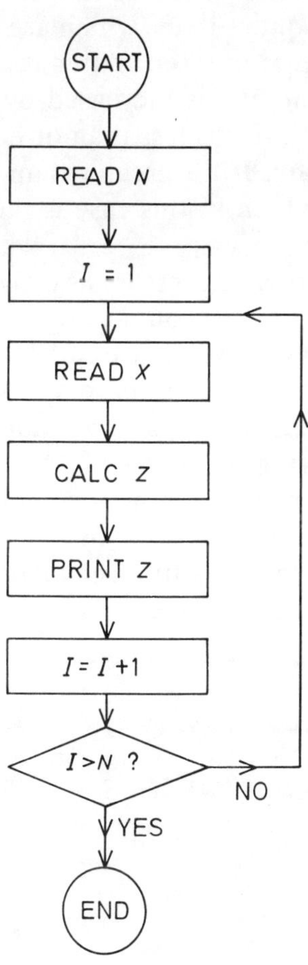

The conventions are simple:

> The start and end of the program are indicated by circles or ellipses.
>
> A rectangle contains some operation or series of operations.
>
> A diamond shape indicates a test and decision. The result of the test decides subsequent action.

In the example shown above data are read from DATA statements by means of a READ statement. The first item which is read gives the number, N, of subsequent items which are to be read. A counter, I, is set to 1 and a loop is entered. After reading X, calculating Z and printing Z, the value of I is increased by 1. Then I is tested to see if it is greater than N. If it is less than or equal to N the program goes round the loop again. If I is greater than N the program passes on to the next section which in this case is the end of the program.

Flow diagrams become more necessary as programs increase in complexity because the diagram sets out a program in general terms and allows you to concentrate on structure rather than on details of syntax. While it is not necessary to prepare a full diagram initially it is often useful to draw a diagram for a particularly complex segment of program to help in solving problems of programming. When many tests, decisions and re-routings must be included in a program a flow diagram is invaluable. Of course the translation of the diagram into a proper program requires full knowledge of the syntax of the particular language being employed.

SAQ 2.7a

Construct a flow diagram for a program which repeatedly accepts a value of an electromotive force from the keyboard and prints out a concentration. The program must stop when a negative value is entered. (Calculation details are not expected.)

SAQ 2.7a

2.7.3. Complex Programs

Flow diagrams really come into their own when a program incorporates a number of tests, branches, loops and procedures or subroutines. Each procedure or subroutine can be regarded as a separate little program which is called upon by the main program as required. The diagram for the main program may include rectangles for the calling of a procedure or a subroutine, eg

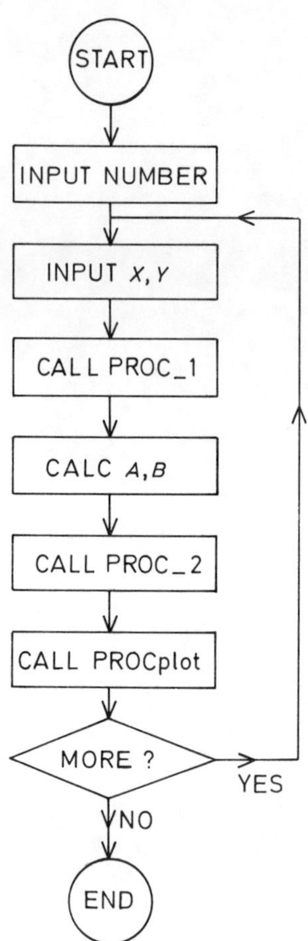

This can probably be regarded as the skeleton of a skeleton. The procedures may themselves be represented by flow diagrams.

SAQ 2.7b

Construct a flow diagram for a program which successively reads three data items from data statements, performs two different calculations and plots a point after each calculation. The 'end of data' is indicated by negative values of three data items. Assume that a subroutine is available for plotting points.

SAQ 2.7b

2.8. FILE HANDLING

2.8.1. Sequential Data Files

As its name suggests a *file* is essentially a store for information. A computer program is usually held in a file with a particular name. From the file the program can be loaded into the computer and then run. The filing system is either a magnetic disc or cassette.

Using files is relatively easy if you have a disc system. If you use a cassette system you probably know that the tape must be set at the start of a file before any operations are performed and that, if a file is to be saved on tape, there must be enough blank tape available.

Data to be used by a program may also be held in a file and read into the program in much the same way as is done by the READ statement. Similarly, instead of the program displaying information on a screen by a PRINT statement it can be made to send the information as data to be stored in a file on the disc or on tape. Files which are used in either of these two ways are called *data files* when it is necessary to distinguish them from files which contain programs.

Since this course is introductory we shall deal chiefly with *sequential* files. That is, when we read a file or write to a file we start at the beginning and work through to the end. Other types of file will be mentioned briefly in Section 2.8.4.

We can picture a file as a series of DATA statements with one data item per line, a data item being a number or a string of characters. Unlike data statements however, the data pointer of a sequential file cannot be 'restored'. The pointer always moves from one data item to the next.

2.8.2. Writing To Files

We want to send data to or write data into a file. A program is going to place the data items in the file.

The first thing we need to decide is the name of the file. This is usually a normal name without spaces and with some restrictions on punctuation marks. Let us use the name "tester". We inform the program of this name by a statement which assigns a number or a variable to the file. The same statement may indicate that the file is for printing or writing *to* rather than for reading *from*. Every computer language has its own syntax but an example might be

```
10 X=OPENOUT "tester"
```

This is BBC syntax. The statement tells the program that "tester" is a file which will accept data coming *out* of the micro through *channel* or *unit* X. In this example the computer decides the value of X and you do not need to know it (just do not use variable X for anything else). With another computer it may be necessary that *you* decide the value of X.

As well as allocating a channel to the file this statement *opens* the file. This means that the file is made ready to accept data.

Each file has a *pointer* which moves along or down the file as each data item is written. When the file is first opened the pointer is set to the top or start of the file. Again, there is a close analogy with DATA statements.

Once a file has been opened to receive data it may be written to by a statement like line 40 below:

```
30 INPUT A,B
40 PRINT#X,A,B
```

The values of variables A and B are entered at the keyboard (line 30) before being written to the file. The use of the symbol # to indicate an operation on a file is quite general but some microcomputers may use WRITE# instead of PRINT#. The effect of the statement is to place the values of variables A and B into successive positions in file "tester".

As each data item is written to the file the pointer moves along (or down if you prefer so to imagine it). Therefore, if the next output statement is

```
70 PRINT#X,"THAT WAS A AND B"
```

then this string is the next data item to be written to the file. There is normally a limit of 255 characters on the size of a string.

When the last data item has been written the file must be *closed*. A typical statement is

```
120 CLOSE#X
```

If a file is not closed properly it cannot be re-opened when required and is in danger of being corrupted.

∏ For your computer find out the syntax for:

(*i*) opening a file for writing;

(*ii*) writing to a file;

(*iii*) closing a file.

2.8.3. Reading Files

Before data can be read from a file it must be opened. Again, there are many variations but usually one statement opens the file, assigns a unit or channel number to it, and sets the file pointer to the first data item. A typical statement is:

```
10 Y=OPENIN "tester"
```

After the file is opened for input or reading, data are taken into the computer by statements like

```
40 INPUT#Y, A,B,R$
```

Here two numerical values are read in from the file and assigned to variables A and B. The third item is a string which is assigned to string variable R$.

It will be obvious that the kind of variable being read must be known by the program. It *may* be that a number can be read as a string (possible with some versions of BASIC) but a sequence of non-numeric characters can never be taken as a number.

When reading data from a file it is usually best to know how many items are to be read but if this is not known it may be possible to use the *end of file* signal which is a feature of many versions of BASIC. This usually takes the form of a function EOF#. The following program, written in BBC syntax, illustrates how EOF# is used:

```
 99 REM ** FILING
100 Y=OPENIN "TESTER"
120 REPEAT
130     INPUT#Y, A,B,R$
140     PRINT A,B,R$
150     UNTIL EOF#Y
160 CLOSE #Y
170 END
```

This program makes use of a REPEAT-UNTIL loop and exits from the loop when the EOF#Y function is 'true'. That is, when the end of the file on channel or unit Y has been reached. A FOR-NEXT loop could also be used provided the number of passes is large enough and a conditional statement is included to allow the program to jump out or terminate the loop.

As with writing to a file, the file should be closed as soon as operations on it are finished.

If a file is re-opened after it has been closed the pointer is set to the first data item.

Π Find out how to:

(*i*) open a file for reading;

(*ii*) read data from a file;

(*iii*) close a file.

Because of the differences in syntax from one computer to another it is not possible to set a detailed self-assessment question on files. You must therefore check your answers by actually preparing a file and then reading from it. The example lines of this and the previous section may be used with any necessary modifications to suit your microcomputer. The manual for your machine will probably give examples to follow.

2.8.4. More Advanced Files

Filing systems can be quite complex and very powerful. We have concentrated on the simplest sequential files and it is strongly recommended that you become conversant with this type before using any other. However, some brief notes on other aspects of filing systems may be useful. To find more details of what is possible for your own particular computer you will have to refer to the manual.

It is sometimes possible to re-set the file pointer to the first data item of a sequential file without first closing the file and in some cases it may be possible to re-set a pointer for numeric data items and a separate pointer for string items.

Another type of file is known as a *random access* or *direct access* file. With this type it is possible to set the file pointer to any position in the file and then to read or write at the chosen position. Obviously, one must really know the structure of a file to use it in this way. Random access files can be very useful but are not easy to handle.

Some versions of BASIC allow a file to be opened for both reading and writing and sometimes a file can be opened for writing in such a way that data are *appended*; that is, the new data are added to the end of the old file.

When the append option is not available and files can be opened only for reading *or* writing but not for both. The following technique is often used to add new data to the end of an old file:

1. Open the old file for reading.

2. Open a new file for writing, giving it a new name (eg " temp").

3. Read old data from the old file and write it to the new until 'end of file' is reached. Then close the old file.

4. Continue writing to the new file but now with new data. Close the new file when all data have been written

5. Delete the old file and rename the new file, giving it the old name.

SAQ 2.8a

> Three instruments send temperature readings in the range 0 to 100 degrees to a computer which later transfers the readings to an *append* file " temp" on a disc. ⟶

SAQ 2.8a (cont.)

The first entry on the file is the time the file is started. Once started, each instrument reads temperature every three minutes and the three readings are spaced at one minute intervals. Immediately before each series of three readings the name of the operator on duty is written to the file. The following shows typical entries:

1630
JOE
60
55
56
MIKE
63
54
55
TOM
...
...

Construct a flow diagram for a program which uses the filed data to produce a graphical display of the temperature variations over a period of one hour. You may assume that points can be plotted in any of three colours or that three characters are available for use as plotted points.

2.9. BITS, BYTES AND MEMORY

2.9.1. Bits and Bytes

In this section we shall look at how the microcomputer stores numerical data and at a few examples of how the memory is used. Before doing this however it might be useful to have some revision of the binary number system discussed in Part 1. We shall work with examples.

Take the decimal number 13.

This number is represented by a sequence of *binary digits or bits*:

$$13 \text{ (decimal)} = 1101 \text{ (binary)}$$

The sequence of bits 1101 means that we add together:

one 8 one 4 no 2 one 1

$$= 1 \times 2^3 + 1 \times 2^2 + 0 \times 2^1 + 1 \times 2^0$$

Adding 1 to decimal 13 gives decimal 14 or binary 1110.

If we are restricted to four binary digits or bits the highest possible number is 15 or 1111 or $2^4 - 1$. However, most microcomputers work with a byte of 8 bits so that the maximum number in a byte is 255 or $2^8 - 1$:

$$255 \text{ (decimal)} = 11111111 \text{ (binary)} = 2^8 - 1$$

It is often convenient to think of eight boxes numbered from 0 to 7 starting from the right:

7	6	5	4	3	2	1	0
0	0	0	0	1	1	0	1

The pattern shown represents decimal 13. Bits 0, 2 and 3 are said to be set. The others are *clear*.

Note how we refer to the bits: bit 0 is on the extreme right, then comes bit 1, then bit 2 and so on. Bit 7 is on the extreme left. Bit 0 is also called the *least significant bit* (LSB) and bit 7 the *most significant bit* (MSB) of the byte.

To obtain a number greater than 255 we must use more than one byte. The number 258 uses two bytes. This number is represented by:

	one 256	+	one 2
=	1×2^8	+	1×2^1

Decimal 258 therefore has 1 in the *high* byte and 2 in the *low* byte:

high byte		low byte
00000001		00000010
$(256) \times 1$	+	$(1) \times 2$

It is useful to remember the relationship between a number and the contents of the two bytes:

number = 256 * number in high byte + number in low byte

2.9.2. Data Storage

The microcomputer is an excellent store for numbers. When we define an integer variable in a BASIC program the system allocates several bytes to the variable. If, say, we dimension an integer array of 100 elements and the system allocates 4 bytes per element then the amount of memory set aside is 400 bytes (404 if the zero element is counted).

If we know that the integer values we are going to store will never be greater than 255 it is really only necessary to allow one byte per element and therefore we need only use 100 bytes instead of 400. It follows that a lot of memory can be saved if we are able to place values into memory locations and read them back into ordinary variables as required. Further, it is frequently possible to store numbers in memory locations that the BASIC system does not normally use.

The microcomputer memory is often used to hold *look-up tables*. These are tables of data held in known memory locations. When an application calls for the same kind of calculation very frequently the use of a look-up table can reduce execution time. Examples are tables of squares or cubes or a much used multiplication table. (The BBC machine has a 640 times multiplication table in its read-only memory!) In spectroscopy a table is sometimes used for the conversion of transmittance to absorbance. While the use of tables saves computing time, any increase in speed is only obtained at the cost of the memory space needed to store the tables.

SAQ 2.9a

(*i*) Convert these binary numbers to decimal:

00011000 10101010 01010101

(*ii*) Represent the following numbers in binary notation:

13 35 68 129 212

299 4,000 60,000

(*iii*) What are the maximum numbers that can be stored in (*a*) 2 bytes and (*b*) 3 bytes?

SAQ 2.9a

2.9.3. Memory and External Links

The interfacing of an analytical instrument to a microcomputer involves circuits to convey signals from the instrument to the microcomputer and, if the microcomputer is to exert some control, from the microcomputer to the instrument. Details of circuitry are outside the scope of this Unit though the principles of the methods used to send and receive signals are considered in Part 3. We must, however, look at how a program can handle information supplied by an external device and send information to a device.

Data are normally received by a microcomputer in the form of bytes which the interface places in certain memory locations of the micro. A program which reads the contents of such a location is in effect reading information supplied by the device. Similarly, data which a program places in certain other locations pass to the device. From the point of view of the programmer therefore interfacing involves reading from and writing to memory locations.

Before studying how to read from and write to memory we shall look briefly at a few examples to illustrate the objectives of interface programming.

(*a*) Input and Output Ports

When a microcomputer communicates with instruments it usually does so by detecting or setting the voltage levels on the pins of conductors in *ports*. A port is addressable. That is, data can be sent to it or read from it as if it were an ordinary memory location. If a port has eight pins or lines a pattern representing a binary number can be placed on the pins by setting the voltage levels of the pins: a high voltage (about 5 volt) normally corresponds to a bit with a 1, a low voltage (about zero) to a bit with a 0. The voltage pattern on the pins of a port therefore communicates a number. If the external device sets the pattern to be read by the microcomputer the port acts as an input. If the microcomputer sets the pattern to be communicated to the device the port acts as an output.

(*b*) Configuring Pins

Sometimes individual pins of a port can act to receive signals from an external device or to send signals to a device. The signals are in the form of voltage pulses or voltage changes. The mode of operation of a particular pin (ie whether the pin receives or sends) is usually determined by the status of a particular bit of a particular byte in memory.

Suppose that the number in a certain byte is 13 (decimal). This is held as the binary number 1101 and the pattern in the byte is 00001101. Bits 0, 2 and 3 are said to be *set* or are *on* or are *high.* If this byte controls a port the pattern formed by the bits determines which pins of the port receive signals and which send out signals: pins 0, 2 and 3 will be outputs, all the other inputs.

(*c*) Bits as Flags

In many applications the state of a particular bit acts as a signal or *flag* to a device outside the computer or to a procedure within the computer. For example, when a microcomputer is sending data to a printer the state of a certain bit of a certain byte lets the printer know that the microcomputer has data to send. When the printer is ready to receive the data it advises the microcomputer by setting another bit.

A very important application of this kind of signal is in the servicing of *interrupts*. It is possible to arrange that when a signal is received at a certain pin of an input port the microcomputer executes some pre-determined program. In a typical case, a temperature-measuring device might be part of a circuit which causes the voltage on a port line to go low when the device detects a temperature which is above a certain value. The microcomputer responds by stopping its current program, *servicing* the interrupt (eg it sounds or prints a warning), and then continuing where it left off.

(*d*) Analogue–Digital Conversion

The microcomputer can act as a very efficient data collector when it is used in conjunction with an instrument or sensor which can supply information in digital form.

To use a microcomputer in this way the data must be supplied in the form of binary numbers which enter the microcomputer through a port as outlined above. If the instrument which is the source of the data does not give a digital output it is necessary to use some device to convert the signal from the instrument into a digital form. A good example is a pH meter or spectrophotometer which gives output to a recorder. The output might vary from 0 to 1 volt full scale. To read this information into a microcomputer rather than send it to a recorder an *analogue–digital converter* (ADC) is used. This is a device which accepts an analogue signal in the form of a continuously varying voltage, and converts the signal into a binary number which the microcomputer can read. Many modern analytical instruments are manufactured with converters already built in so that they can present measurements in digital form.

The *resolution* possible with an ADC depends on the number of bits used to store the binary numbers and on the maximum voltage which the device can accept. For example, many ADCs are able to accept potentials from zero to 5 volts. If it is an 8-bit device then a potential of 5 volts must correspond to the number 255 and the resolution is 5/255 volt. Again, if the ADC converts 5 volts into a 10-bit number the resolution would be 5/1023 volt.

No matter how it operates the net result of the action of an ADC is that a number is read into the computer. This number may be stored in an array element but, as discussed above, it is often more efficient to place it in one or two bytes of memory. Then a program can read the data from memory for subsequent computations.

The reverse kind of operation is achieved by a *digital–analogue converter* (DAC). This is used to send data out of a computer in analogue form. In spectroscopy an ADC might accept analogue signals from a spectrophotometer, perform some calculations, and then send analogue output through a DAC to a recorder to display derivatives of the original spectrum.

SAQ 2.9b

An instrument is interfaced to a microcomputer by means of a 5-volt, 12-bit ADC. What resolution is nominally possible and how many bytes are required to store each reading?

2.9.4. The Memory Map

The memory of a microcomputer consists of a large number of memory locations. Each location is capable of holding a number between 0 and 255 but only a small part of the memory is freely available to the user. The two major types of memory are *Read Only Memory* (ROM) and *Random Access Memory* (RAM). These are considered in more detail in Part 1. As far as the programmer is concerned the only memory locations that can be altered are those designated as RAM since the machine has been designed to make it impossible to write data into ROM.

The capacity of a computer is usually expressed in *kilobytes*, one kilobyte (or 1K) being 2^{10} or 1024 bytes. A microcomputer with 32K of RAM has 32 times 1024 memory locations or bytes to which the user may write data. However, while it is possible to change its contents not all of RAM is available for the storage of programs and variables. When a program is running the *operating system* itself makes use of some RAM. The computer manual will give some detail about the allocation of memory to various tasks such as plotting on the screen, storing variables, operating filing systems, and perhaps making sound.

It is common practice to use hexadecimal notation when discussing memory allocation. We shall indicate a hexadecimal number by means of the prefix '&'. Some authors use the symbols '$' and 'H'.

As a brief reminder the hexadecimal (hex) system has number base 16 where the binary system has base 2 and the decimal system base 10. The hex numbers corresponding to decimal 10 to 15 are represented by the letters A to F:

0 1 2 3 4 5 6 7 8 9 A B C D E F

The number &2C8 is equivalent to decimal 712:

$$
\begin{aligned}
\&2C8 &= 2 \times 16^2 + 12 \times 16^1 + 8 \times 16^0 \\
&= 2 \times 256 + 12 \times 16 + 8 \times 1 \\
&= 512 + 192 + 8
\end{aligned}
$$

A common method of showing how memory is allocated is by means of a *memory map*. This is a diagram in which a block of memory is represented by an area. Here is a general example:

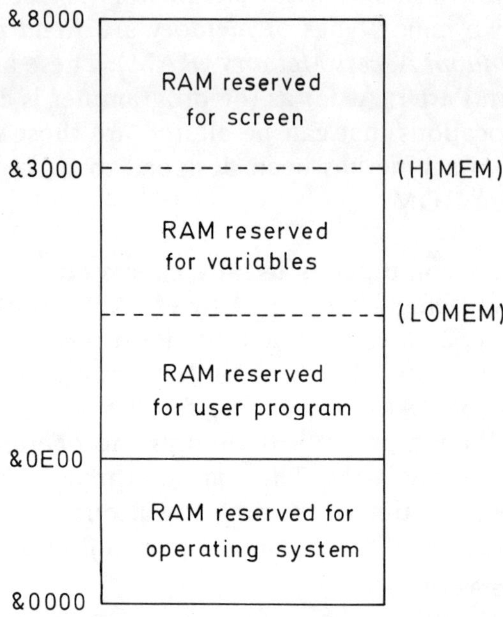

This diagram illustrates how 32K of RAM might be assigned. The numbers on the left represent memory addresses (in hexadecimal). Addresses of ROM locations would follow on from the highest RAM address (eg from &8000 to &FFFF for a microcomputer having a total memory of 64K).

In the example, a BASIC program prepared by the user would be stored in memory starting at address &OEOO. Data in an address between &3OOO and &7FFF would affect the screen display.

The lower boundary of the screen area is shown as a dotted line because this boundary may depend on the particular mode selected (assuming that the microcomputer can operate in more than one mode).

The boundary between the user program area and the variable area is movable. It is usual for variables to be stored in addresses between the last address used by the program itself and the start of screen memory. The lowest address for a variable is normally referred to as LOMEM and the highest address as HIMEM. In the example map these addresses are represented by the two dotted lines.

It often happens that the user wishes to reserve a part of memory for data received from an ADC or for data required by a machine code program. One method of doing this is to use an area immediately below screen memory and change the setting of HIMEM so that variables used by the program do not intrude into this area.

In the next section we shall find out how to read data from memory and place data into memory. No harm can be done by reading a memory location but if the data in a location is altered you must be sure that that location is not being used by the operating system or for the storage of a program. Unfortunately it is not possible to give any general rules to decide 'safe' locations. The best thing to do is to consult the microcomputer manual and make intelligent deductions. For example, the BBC model B assigns locations between &0C00 and &0CFF to user-defined character definitions. If you do not intend to define special characters it is safe to change any of these locations.

Π Prepare a rough memory map for your microcomputer and make a note of some memory locations to which data may be safely written.

SAQ 2.9c

> You have a program to accept input from a 10-bit ADC and you need to make 100 readings. You want to store the data immediately below the screen memory which starts at &3000. Where would you place HIMEM?

2.9.5. PEEK and POKE

In principle the programming necessary to read data from a memory location is easy. Instead of statements like:

30 INPUT X (from keyboard)

40 READ X (from DATA statement)

we simply use a statement like:

50 LET X = (value in location ...)

That is, a numerical value in a memory location or locations is transferred to a location (box) labelled with the name of a variable. The only difficulty which arises in doing this is that the method of transfer varies from one computer to another.

A very common method of making the transfer uses a PEEK statement:

30 LET X = PEEK mmmm or 30 LET X = PEEK (mmmm)

Here mmmm represents the number or address of a location in memory. As in normal assignment statements the word LET can usually be omitted.

Colloquially one might say '*peek into the memory byte(s) at address mmmm, read the number there and put that number in the box labelled X*'. Paraphrasing the operating like this helps you to remember an important point: reading a memory location does not change the contents.

Not all microcomputers use PEEK but we hope you will bear with us if we continue to use it. The popular BBC microcomputer employs a question mark to mean something like *the contents of memory location* .... For example, with this microcomputer the statements above would be written as:

40 LET X = ?mmmm

A mainframe computer may not permit easy reading of memory locations but then such a machine is unlikely to be used for interfacing laboratory instruments. Assuming that you are using a microcomputer do the following exercise:

Π Find out how to read data from a memory location or byte of your computer. Note the following points in particular:

(*i*) Must the address be in decimal or hexadecimal or is either allowed?

(*ii*) If hexadecimal is allowed, how is this indicated?

(*iii*) Must the address be in parentheses?

Test your conclusions by reading from a location into a variable and then printing the value of the variable.

POKE does the opposite of PEEK. The effect of POKE is to place a number in a memory location. A typical statement might be:

50 POKE mmmm,X

The effect of this would be to place the value of X into the memory location mmmm. With 8-bit bytes the maximum value of X is 255. X may be an actual number or a variable with a value.

As with PEEK the exact syntax depends on the computer. Again taking the BBC microcomputer as an example, the question mark is used to mean something like *let the number in memory location mmmm be . . .:*

60 ?mmmm=X

Π Find out how to place a number into a memory location.

Refer to your notes made when constructing a memory map (see 2.9.4) to find a suitable address for data.

Place a number (less than 256) into a location and read it back as in the last exercise.

SAQ 2.9d

(*i*) Write a program to place the numbers 8 and 10 in two successive memory locations.

(*ii*) Alter the program to make it read the locations after the numbers have been placed in them.

2.9.6. DIV and MOD

When dealing with the contents of memory two operators are very useful. The operators DIV and MOD give the integer dividend and the remainder of a division. Take an example:

14 divided by 3 gives 4 with remainder 2

Using the operators

14 DIV 3 returns 4
14 MOD 3 returns 2

(*returns* means *gives the result*)

You will see that DIV gives the integer after division while MOD gives the remainder. Let us study a short program:

```
 9 REM ** DIVIDE BY 2
10 INPUT N
20 D = N DIV 2
30 R = N MOD 2
40 PRINT R
50 N = D
60 IF N > 0 GOTO 20
70 END
```

When a number is entered at line 10 a sequence of numbers is printed. Can you see the significance of the sequence without running the program?

The answer to the question is that the numbers printed represent the number entered in *binary* notation but reading from left to right.

Suppose we enter the number decimal 13 (ie N = 13 at line 10). The value of D will be 6 and the value of R will be 1. R is printed out. So, at line 40:

D = 6 R = 1 1 is printed

At line 50 the variable N takes the value of D (ie 6) and the process is repeated so that when line 40 is reached again:

D = 3 R = 0 0 is printed

Again N takes the value of D and since this value (now 3) is greater than 0 the process is repeated to give:

D = 1 R = 1 1 is printed

On passing through the sequence again we find that D is zero:

D = 0 R = 1 1 is printed

When N takes the value of D (now zero) and the test is made at line 60 the sequence stops and the program ends.

If the output is read starting with the *last* figure printed we get 1101 which is the binary representation of decimal 13.

To make the binary number worked out by this program appear in the correct order the values of the remainder could be stored and printed out in proper order just before the program stops. This needs an array.

SAQ 2.9e

(*i*) Write and run a program which will accept any number from the keyboard, divide it by 3, and print the result like this example:

14/3 = 4 + REMAINDER 2

(*ii*) Write and run a program to convert a decimal number less than 256 into a binary number and print out the result *in correct order*.

SAQ 2.9e

The operators DIV and MOD are particularly useful when data are to be placed in the memory locations of a microcomputer.

You will recall that each memory location can contain an integer between 0 and 255. For a number greater than 255 and less than 65536 it is necessary to use two locations, referred to as the high byte and the low byte. The number in the low byte must be added to 256 times the number in the high byte. That is,

N = 256*(contents of high byte) + (contents of low byte)

Suppose we use variable H for the contents of the high byte and L for the contents of the low byte. Then

N = 256*H + L

To place a number in memory we have to separate it into H and L. This is done as follows:

H = N DIV 256
L = N MOD 256

IF you are interested in such things you will recognise that we are effectively using a number base of 256. The low byte contributes L times 256 raised to the power 0. The high byte contributes H times 256 raised to the power 1.

SAQ 2.9f

(*i*) Write a program to place into two successive memory locations the low byte and the high byte of an integer entered at the keyboard. The low byte should be placed in the lower of the locations. Check that the program works by placing data and then reading the locations.

(*ii*) Outline how you would store the data from 100 readings from a 10-bit ADC in memory locations immediately below screen memory.

2.9.7. The AND operator

When we discussed conditional statements in Section 2.5 we saw that it was possible to make two tests in one statement as in, for example,

100 IF A = B AND X = Y THEN ...

The action following THEN is taken only if both conditions are satisfied.

Another, very different, use of the operator AND is to find out which bits of a byte are *set*. It often happens that the information contained in a byte lies in whether or not certain bits are set rather than in the numerical value that the byte represents. We therefore frequently need to focus attention on particular bits and ignore others. In other words we *mask* the byte so that we only examine the bits we want.

A typical statement takes the form

100 X = (contents of byte) AND 8

Following this statement X has the value 8 if bit 3 of the byte is set but otherwise X is zero. To see how this works we must write the numbers concerned in binary notation. Let us suppose that the byte being examined contains decimal 13. In binary notation this is

00001101

We note that bit 0, bit 2 and bit 3 are all set. The other number, decimal 8, in binary is

00001000

The only bit set in this case is bit 3.

The effect of the operator AND is to compare the two binary numbers *bit by bit*. If *both* of the bits compared are set then the corresponding bit of the result is set. Here it is in detail:

Write the two binary numbers one above the other:

```
00001101     13 decimal
00001000      8 decimal
--------
00001000      8 decimal
--------
```

Moving along the bits we write a 1 if *both bits* are 1, otherwise we write 0. As seen above the result is the binary representation of decimal 8.

Taking another example, suppose we AND decimal 13 with decimal 254. The result is decimal 12. This is shown below:

```
00001101    13 decimal
11111110   254 decimal
--------
00001100    12 decimal
--------
```

In many applications of microprocessors it is useful to regard a byte as made up of two half bytes or *nibbles*. To examine the low nibble (bits 0–3) we AND with decimal 15. To examine the high nibble (bits 4–7) we AND with decimal 240:

```
10011011            10011011
00001111   (15)     11110000   (240)
--------            --------
00001011            10010000
--------            --------
```

In the first case the bit pattern of the low nibble only is transferred to the result, in the second case the result holds the high nibble only.

SAQ 2.9g

Find the results returned by the following:

Z1 = 24 AND 194

Z2 = 3 AND 15

Z3 = 54 AND 240

SAQ 2.9g

2.9.8. The OR operator

The operator OR is similar to AND in that it can be used in conditional statements and in operations on bytes.

Whereas the AND operation masks out bits that are not of interest OR has the effect of *merging* the patterns in two bytes. The operation is used when we want to ensure that certain bits are set without disturbing other bits.

In the OR operation the numbers are written in binary and the bits compared. The result bit is set to 1 if *either or both* bits being compared are set. Take the statement

```
50 X = 13 OR 136
```

We write the two numbers in binary one above the other. Then we write a 1 in bits 0, 2, 3 and 7 to obtain the result:

```
00001101     13 decimal
10001000    136 decimal
--------
10001101    141 decimal
--------
```

You will notice that the result holds the combined pattern of *both* numbers; the two patterns have *merged*.

The strictly correct name for the OR operator is *inclusive* OR because a result bit is set if *either or both* test bits are set. Another operator called the *exclusive* OR, represented by EOR or XOR, is sometimes used when comparing bit patterns. With this operator a result bit is set if *either* test bit is set but not if *both* are set.

SAQ 2.9h

Find the results returned by the following:

Z4 = 24 OR 194

Z5 = 3 OR 15

Z6 = 48 OR 31

2.9.9. Setting and Changing Bits

It is often necessary to set or clear particular bits in a byte and to examine a byte to find out which bits are set. While assembly language is often used for operations of this kind they can also be performed through BASIC.

To configure a byte initially we simply write down the correct pattern of bits and then place the number obtained in the appropriate memory location. Thus to set bits 2, 4, and 6 in location mmmm while leaving the others clear we place binary 01010100 (decimal 84) in mmmm.

The OR operation is used to set selected bits without altering others. The following sequence of steps will set bits 0 and 3 in mmmm:

Read contents of mmmm into variable M

LET X = M OR 9

Place X in mmmm

This results in binary 00001001 merging with the byte initially in mmmm.

AND is used to test a particular bit of a byte. To find the status of bit 4 we AND the byte with binary 00010000 or decimal 16. A suitable program segment might be:

```
50 LET M = PEEK mmmm
60 LET Y = M AND 16
```

Y will have the value 16 if bit 4 of number M is set. Y is zero if bit 4 is clear.

AND is also employed to clear selected bits. For example, to clear bits 3 and 4 we use the binary having these bits clear and all others set:

Read contents of mmmm into variable M

LET X = M AND 231 (decimal 231 ≡ binary 11100111)

Place X in mmmm

SAQ 2.9i

Outline procedures to perform the following operations on memory location mmmm:

(*i*) set bits 0, 3, 5 and 7 to 1 and all other bits to 0;

(*ii*) clear bits 0 and 5;

(*iii*) set bits 4 and 6 without altering other bits;

(*iv*) test bits 0 and 2.

Summary

This Part started off by using a computer as a powerful calculating machine. We then wrote simple programs to make the computer do some mathematical manipulations and use string variables. Applications included calculation of pH, hydrogen ion concentration, transmittance and absorbance and how to control the number of figures output by the computer. At this point the important computer words were INPUT, LET and PRINT.

The elements of program control were introduced in section 3 with the FOR..NEXT loop and IF statements and this was followed by a section on the control of data through the use of data statements and one-dimensional numeric arrays (DATA, READ, DIM). At this stage of the course it became apparent that the computer could make a major contribution to the data handling requirements of the analyst; examples studied included the calculation of a standard deviation and chromatographic relative retention times. More powerful controlling tests (IF ... THEN ... ELSE, AND, OR, NOT) and double arrays were studied in section 5. Program control by means of REPEAT ... UNTIL and REPEAT ... WHILE techniques were mentioned though we did not make great use of these.

In Section 6 we looked in a rather general manner at how we might use the graphics facilities which are available on most microcomputers (MOVE, DRAW, PLOT), illustrating the techniques by devising a program for the construction of an acid-base neutralisation curve. This was followed by a section on program structure, subroutines (GOSUB, RETURN), procedures and flow diagrams. Section 8 was concerned with simple filing systems, again the treatment being quite general.

In the last section we revised the binary number system and introduced the hexadecimal system. We examined the structure of a typical microcomputer memory and looked briefly at how the microcomputer communicates with external devices. Finally we learned how to read from and write to memory (PEEK, POKE, DIV, MOD) and how to examine the contents of memory locations (AND, OR, EOR).

Objectives

Now you have completed Part 2 of this Unit you should be able to use a computer which supports a version of the BASIC computer language and be able to:

- perform calculations and use different types of variable (SAQ 2.1a–2.2a);
- write programs which accept input from the computer keyboard, perform simple calculations and display results on the screen (SAQ 2.2b–2.2e);
- incorporate tests, branches and loops into programs (SAQ 2.3a–2.3c);
- use data statements and simple numeric arrays in programs (SAQ 2.4a–2.4d);
- use more complex control tests, arrays of two dimensions and nested loops (SAQ 2.5a–2.5e);
- write programs to draw straight lines and to plot points on the screen (SAQ 2.6a–2.6c);
- understand the use of subroutines, procedures and flow diagrams (SAQ 2.7a, 2.7b);
- use a simple filing system (SAQ 2.8a);
- understand the use of the binary number system by a microcomputer (SAQ 2.9a–2.9c);
- read from and write to the memory of a microcomputer (SAQ 2.9d–2.9f);
- examine and modify the contents of memory locations (SAQ 2.9g–2.9i).

3. Microcomputer Interfacing

Overview

It is essential that you understand Parts 1 and 2 of this unit before studying Part 3. The first section introduces the concept of a computer port and how a particular port is selected during input and output operations involving external devices or instruments. To master the material of this section you will need to be familiar with binary, hexadecimal, BASIC programming and the use of logical operators such as AND, OR, NOT and EOR. Many specialised texts cover fundamental aspects of computer interfacing and those who would like a more in-depth treatment are referred to the following books.

E. Morgan, *Laboratory Computing*, Sigma-Technical Press, distributed by J.Wiley, Chichester 1984.

B. A. Artwick, *Microcomputer Interfacing*, Prentice-Hall, 1980.

D. J. Malcolme-Lawes, *Microcomputers and Laboratory Instrumentation*, Plenum Press, 1984.

R. A. Sparkes, *The BBC Microcomputer in Science Teaching*, Hutchinson, 1984.

Section 2 builds on the work of Section 1 by introducing programmable interfaces for digital input and output. The discussion is restricted to two commonly used chips and for details of other programmable interfaces the reader is asked to consult specialised texts such as that by Artwick (above) or the following book.

S. Libes and M. Garetz, *Interfacing to S-100/IEEE 696 Microcomputers*.

Section 3 concentrates on analogue input and output. Again the discussion of analogue to digital conversion is restricted to one operational type. A number of different types are available and for further details a specialised text must be consulted (all of the above mentioned books cover analogue to digital conversion). Section 4 covers the ways in which data can be passed between computers and between computers and instruments. Parallel data transfer is considered first and the principles explained. Space does not permit an introductory treatment to delve too far into this area. However both the programmable interfaces discussed in Section 2 can be used efficiently for parallel data transfer. The best source of information as to how this can be achieved, once you understand the basic principles, is the manufacturers' literature which describe the characteristics of each chip. The section closes with an overview of the IEEE-488 instrumentation interface. A detailed discussion of this system is given in the book,

E. Fisher and C. W. Jenson, *PET and the IEEE 488 Bus*, Osborne/McGraw-Hill,1980.

Another instrumentation system (S-100) is also briefly mentioned, but those wanting more detail should refer to the text by S. Libes and M. Garetz (detailed above).

3.1. COMPUTER INTERFACES FOR DIGITAL INPUT AND OUTPUT

3.1.1. The Concept of an Interface and Port Selection

As you may recall from Part 1, the microcomputer represents binary information in terms of one of two voltage levels, 0 and 5 volts approximately. Although some chemical instruments may produce output signals which are compatible with these levels, many do not. Some provide continuously varying signals which are not restricted to these voltages. For example we may wish to use a chart recorder output as a source of signal for the computer. Typically this may vary continuously through the range 0 to 100 mV during a measurement. Clearly some conditioning of the signal must be undertaken, so that each reading can be expressed as a binary number which can be read by the microprocessor.

Whatever the nature of the input signal, one of the tasks of the interface is to make it compatible with the voltage levels used by the computer. Similar considerations hold for microcomputer output. Many external devices require more than 5 volts to drive them and often a substantial current is also needed. Output interfaces may need to step up both the voltage and the current available to drive a device.

Irrespective of the external signal, communication between the interface itself and the microprocessor takes the form of single byte transfers. In the remainder of this section we will consider how a particular interface is selected by the computer to allow such transfer to be achieved.

Another important function of an interface is to ensure that data transfers are only permitted, or 'enabled', when the microprocessor either reads from or writes to the interface. Interfaces are distinguished either by assigning them an address, rather like a memory address, or a port number.

Port numbers are restricted to the range 0 to 255, whereas addresses can be in the range 0 to 65,535 although they should not coincide with the addresses of any memory locations used by the microcomputer.

External devices which are assigned addresses are said to be 'memory-mapped' because the address takes up part of the addressable range used by the microprocessor. In selecting a particular memory mapped device, the 16 address lines emanating from the microprocessor hold the address as a series of digital signals (0 or 5 volts). These 16 lines are known as the 'address bus' and connect the microprocessor to all memory chips and the interfaces for memory-mapped devices. The interface electronics for a particular device has to be designed so that interaction between the microcomputer and the external equipment connected through the interface is only possible when the address assigned to the device is on the address bus. This function of an interface is known as 'address decoding'.

A similar situation applies when devices are assigned port numbers. As these range only from 0 to 255 the port number can be fully specified by the lowest 8 lines of the address bus. This time the interface must be enabled when, amongst other things, the appropriate port number is on the address bus.

When an interface is selected by the appropriate address or port number, several control signals from the microprocessor must also become active. These specify the direction of information flow (from the computer to the interface or vice versa) and the timing of data transfer. These signals are carried from the microprocessor to each interface by the so-called 'control bus'.

Once an interface as been activated by the presence of the correct address (or port) and control signals, data are transferred to or from the microprocessor via the 8 signal lines known as the 'data bus'. As with the other busses, the data bus is connected to all interfaces with external devices and is also used for the data transfers between the microprocessor and locations in the main memory.

A schematic representations of the bus connections between the microprocessor and interfaces for external equipment is shown in Fig. 3.1a.

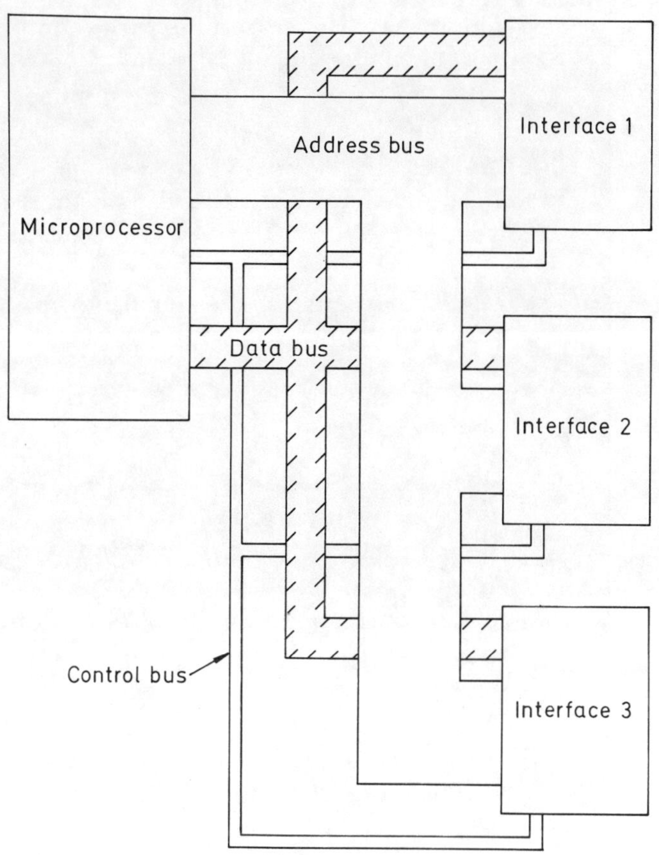

Fig. 3.1a. *A schematic representation of the bus connections between a microprocessor and interfaces to external equipment*

SAQ 3.1a

A microcomputer is connected to a pH meter which has BCD output with computer-compatible digital signals of 0 and 5 volts. The digits for the tenths and units are connected to an interface which is memory mapped with an address of &FE00. Answer each of the following questions about the interface and the signals involved.

(*i*) Match the name of the signals given in list 1 with the appropriate bus in list 2.

List 1	List 2
(A) Signals corresponding to the pH value from the pH meter.	(X) Address bus
(B) Signals corresponding to &FE00 from the microcomputer.	(Y) Data bus
(C) A microcomputer signal which indicates that data are to flow from the pH meter to the computer.	(Z) Control bus
(D) A microcomputer signal which indicates when the microprocessor is ready to receive data.	

(*ii*) By circling Y (for yes) or N (for no) indicate which of the following are the functions of the interface in this example. ⟶

SAQ 3.1a (cont.)

(1) To let data flow on to the data bus when the microprocessor has placed the correct address on the address bus and then is ready to accept data.

(Y / N)

(2) To allow data from the pH meter to reach the data bus whenever the address &FE00 appears on the address bus.

(Y / N)

(3) To protect the microcomputer system from the voltage levels output by the meter.

(Y / N)

3.1.2. Data Input from a Port

As an example, we shall consider the problem of reading the pH from a pH meter with BCD output. The interface connections are given in Fig. 3.1b which shows that the tenths and units of pH can be read as a single byte from a memory-mapped interface with the address &FE00. The units are contained in the most significant 4 bits and the tenths in the lowest 4 bits. To cope with pH values greater than 9.9, a 'tens' bit has been included in the byte read from a second memory-mapped port with address &FE01. If bit d0 is 1, then the pH is greater than 9.9, and if d0 is 0 then the pH is less than 10.

After the pH meter takes a reading, it is quite common for a small amount of time to elapse before the pH value is presented as valid binary data to be read by the computer. This will be discussed in more detail in Section 3.3.2 but it is sufficient to note here that the computer may have to wait until the datum presented at the interface is valid. But how can the computer know? The answer is that the pH meter must provide a signal, or 'flag' in computer jargon, to indicate that the datum is valid. Referring to Fig. 3.1b we see that this information has been made available through bit d1 of the input port at &FE01.

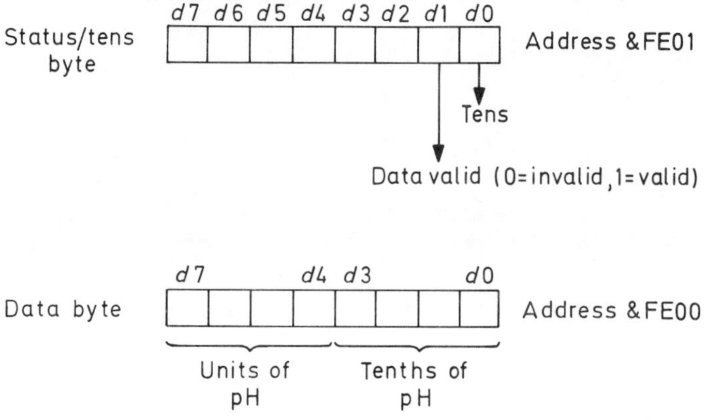

Fig. 3.1b. *Possible interface connections to read two data digits of pH (units and tenths) and a status byte which indicates the validity of the data. The tens digit is indicated by bit D0 of the status byte*

The computer can monitor this bit (d1) and when it has a value '1', the valid data can be read from the port at &FE00 and bit d0 of &FE01.

A program to monitor the pH continually could involve the following sequence, assuming the interface details in Fig. 3.1b:

(*a*) Read the datum byte from address &FE01.

(*b*) By using the logical AND instruction, check the value of bit d1.

(*c*) If d1 = 0 then the datum is INVALID so return to step (*a*) to check again.

(*d*) The datum is VALID at this stage, so we can check the value of bit d0. This bit gives us the number of tens in the measured pH (limited to 0 or 1). Store the number of tens in the variable TENS.

(*e*) Read the byte of data from &FE00.

(*f*) Deduce the tenths of pH by ANDing the data byte with 15 (binary 00001111) and store the result in the variable TENTHS.

(*g*) Deduce the number of pH units by dividing the datum byte (read in (*e*) above) by 16. The units value required is equal to the integer part of the result of this division. (Recall the DIV operator from Part 2).

(*h*) Combine the tens, units and tenths to give the measured pH using the formula:

$$pH = TENS * 10 + UNITS + TENTHS/10$$

SAQ 3.1b

All of the following questions refer to the interface given in Fig. 3.1b which allows the computer to read pH values in BCD.

(*i*) Bit d1 of the port with address &FE01 indicates that the BCD datum is valid (d1 = 1 valid, d1 = 0 invalid). ⟶

SAQ 3.1b (cont.)

Suppose the byte X, read from &FE01 has a value 32 and we wish to check that the datum is valid.

(1) What number would you use to mask off all but bit d1 of the byte X ?

(2) What is the result of ANDing the answer from (1) with the byte X = 32?

(3) Is the datum valid?

(*ii*) Bit d0 of the byte X, read from the port with address &FE01, indicates the number of 'tens' in the measured pH. If X = 32 ,is the pH above or below ten?

(*iii*) The byte Y, read from the port of address &FE00, contains the units and tenths of pH. If the value obtained for Y is 89 what are the values of:

(1) the tenths;

(2) the units of pH?

What is the total pH, assuming no tens are involved?

SAQ 3.1b

SAQ 3.1c

Recalling the flow-charting method discussed in Part 2, devise a flow chart for a program to read and print pH values repeatedly from a pH meter with BCD output from the interface connections as indicated in Fig. 3.1b. Assume that the pH meter is in 'free-run' mode so that a continuous stream of data is produced. Note that you must check the data valid signal before reading the pH.

SAQ 3.1c

You should now be able to understand the reading of data into the computer from virtually any interface chip. In general even the most complicated interfaces will not involve any more then checking a given status byte, reading in one or more bytes of data and then doing a little simple processing to either decode the data or recombine it to produce meaningful numbers. In the next section we shall examine the reverse problem of output to a port.

3.1.3. Output to a Port and the Control of External Devices

There are two applications of microcomputer output to a port which are often used in an analytical laboratory. One is to transmit data to another computer or instrument, and the other is to exercise control over external devices such as valves, pumps, motors associated with analytical chemical operations including dilution, liquid transport, sample selection and filter, prism or grating changes. The discussion of parallel data transmission will be deferred until Section 3.4.1, and attention here will be restricted to the control of external devices.

It is possible to switch an external device on or off using a microcomputer by connecting the device or a suitable relay to an appropriate output port (or memory-mapped output interface). This is shown in Fig. 3.1c for two devices.

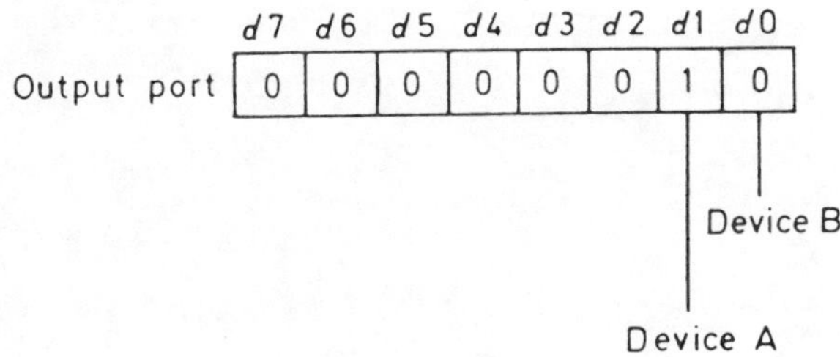

Fig. 3.1c. *Connections for two devices, A and B, to an output port*

To switch on device A we need to write a program which will cause the microprocessor to send the binary equivalent of 2 along the data bus to the interface. The 5v signal (corresponding to the digital value of '1' at d1 of the data bus) would normally be unable to switch on device A because it only appears at the data bus for a few microseconds and the current drive is usually far too small anyway. In addition to device selection, the interface in this case has the job of catching (or 'latching' as it is called) the microcomputer output and holding it until a subsequent output signal is sent. Finally the interface must increase the drive current available by appropriate electronic circuitry to meet the needs of the application.

SAQ 3.1d

It is proposed to operate some equipment under microcomputer control by sending output to an appropriate interface. For each of the following items indicate the main functions of the interface.

(*i*) A peristaltic pump operating at 240 volts A.C.

(*ii*) A gas valve operated by a 12 volt D.C. supply.

(*iii*) An indicator lamp operating on 5 volts.

(*iv*) Another computer.

SAQ 3.1d

Let us now consider a simple programming example involving sequence control of a number of devices. Automated discrete colorimetric analyses are becoming increasingly common in well-equipped analytical chemical laboratories. Here a particular colorimetric analysis can be undertaken on a sample by adding reagents in the required sequence to a small amount of sample drawn from a sample cuvette under computer control. Spectrophotometric measurements can then be made on the solution and the results fed back to the computer for processing.

We shall consider only one part of the procedure, namely the addition of reagents to the sample solution. A commonly used method for this is to store the reagents in pressurised bottles and control their addition by switching on and off in-line liquid valves (one line for each reagent). The apparatus represented schematically in Fig. 3.1d will serve to illustrate the principle of automated sequence control.

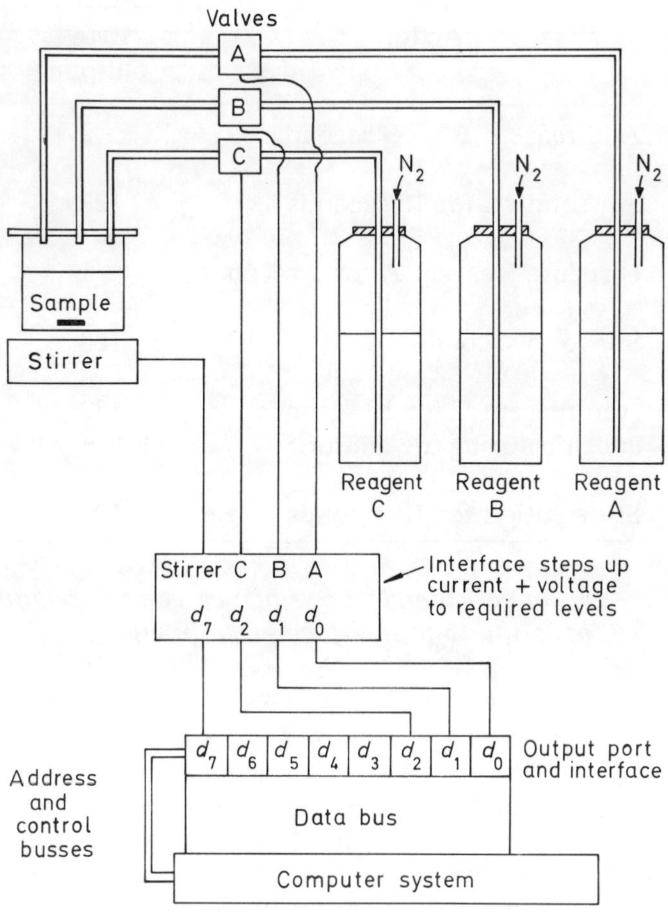

Fig. 3.1d. *Simplified apparatus for adding three reagents to a sample solution with computer-controlled stirring (digital 1 ≡ ON, 0 ≡ OFF)*

Here we have three reagents, A, B and C with in-line control valves and a computer-controlled stirrer. In this simple example we need only a single output port, to which reagent valves A, B and C are connected at bits d0, d1 and d2 respectively and the stirrer is connected to bit 7. The delivery of a given volume of liquid is achieved by opening the appropriate liquid-line control valve for the correct time. We can therefore use dispensing times, rather than volumes in defining the automated sequence.

Step No	Action	Byte Output	Time Held/s
1	Add reagent B for 3 seconds	2	3
2	Stir solution for 10 seconds	128	10
3	Dispense Reagent A for 2 seconds	1	2
4	Stir for 5 seconds	128	5
5	Add reagent C for 10 seconds and stir during the addition	132 (128 + 4)	10
6	Stir solution for 10 seconds	128	10

Fig. 3.1e. *A simple example of sequence control relating to the apparatus and interface given in Fig. 3.1d*

Now let us consider the development of a program to carry out a control sequence such as that shown in Fig. 3.1e. We could write a specific program for this particular sequence, but it is more sensible to develop a program which will allow any sequence of operations involving the apparatus in Fig. 3.1d. We must therefore generalise the problem somewhat. The approach to take is indicated by the nature of the data in Fig. 3.1e. We have a number of distinct steps: step 1, 2, 3 ... N where N is the total number of steps. It seems natural then to talk of step I where I has an integer value from 1 to N. Each step has two items of data. Firstly the byte pattern to be output and secondly the time it must be maintained at the output port. For step I, we could conveniently describe the data as byte B(I) and time T(I). We therefore need to use two arrays to represent the data in the program. Once the storage of the data is settled the program merely becomes a loop over the number of steps, and for each step, byte B(I) is output to the port and held there for T(I) seconds.

```
 10 REM SEQUENCE CONTROL PROGRAM
 20 REM DEVICES CONTROLLED VIA PORT
 30 REM WITH ADDRESS &FEOO
 40 REM DATA STATEMENT AT LINE 400, No OF STEPS (N)
 50 REM DATA STATEMENTS AT LINE 410 ONWARDS
 60 REM CONTAIN BYTE(I), TIME(I) FOR STEP I
 70 REM N DATA PAIRS REQUIRED, TIME IN SECONDS
100 READ N
110 DIM B(N), T(N)
120 FOR I = 1 TO N
130         READ B(I), N(I)
140         NEXT I
150 FOR I = 1 TO N
160         PRINT "STEP"; I;"BYTE= "; B(I);"TIME= ";T(I)
170         POKE &FEOO, B(I)
180         GOSUB 300: REM TIME DELAY
190         NEXT I
200 END
300 REM TIME DELAY SUBROUTINE
310 F=500: REM A CALIBRATION FACTOR
    (LOOPS/SECOND)
320 M=F* T(I)
330 FOR J = 1 TO M: REM DELAY LOOP
340          NEXT J
350 RETURN
400 REM DATA STATEMENT START HERE
```

Fig. 3.1f. *A program for sequence control assuming all external devices are operated by single-bit output connected to an output port with address &FE00*

Fig. 3.1f shows the listing of a program to carry out sequence control assuming that the external devices are all connected to the memory-mapped port of address &FE00.

SAQ 3.1e

(*i*) Complete the DATA statements in the program listed in Fig. 3.1f for the sequence specified in Fig. 3.1e.

(*ii*) The program in Fig. 3.1f uses a time wasting loop to produce the time delay (lines 330, 340).

(1) Is there any other way that a time delay could be produced for use in sequence control?

(2) Are there any disadvantages in using such a time wasting loop?

(*iii*) What modifications would be necessary for the program in Fig. 3.1f if,

(1) the address of the memory-mapped output port were changed to &FF00?

(2) the interface were not memory-mapped but had a port number of 205? (Some versions of BASIC which support output to a numbered port use the instruction OUT n, m where n is the port number and m is the byte to be output).

SAQ 3.1e

The example of sequence control discussed above was so simple that it was possible to gloss over a problem which sometimes arises in more complex programs. That is how to output a byte to a port, and how to change the state of a particular bit, without affecting the state of any other bits.

To be more specific consider again the steps illustrated in Fig. 3.1e. In steps 1 to 4 only one device is on at a time, so it does not matter that in step 2, for example, 128 is output to switch on the stirrer with the consequence that this byte (128) switches everything else off. In step 5 the situation is different. Here we want to switch on the valve for reagent C and not switch off the stirrer, and so we must output 128 + 4 to set both d7 and d2 to 1. The stirrer was on already and in a sense it seems silly to switch it on again in step 5 by including 128 in the byte to be output. But had we merely output 4, valve C would have been activated but the stirrer would have been switched off. If you have any doubt about this statement refer back to Fig. 3.1d and check each binary pattern (d0–d7) referred to above against the devices connected to the interface.

We need to find some way round the general problem of changing the state of one bit without affecting the state of the others. In a sense we need to 'merge' the byte we want to output (to change the state of the given bit) with the byte pattern which already exists at the output port.

This can be done with the logical OR instruction which was explained in Part 2. As a simple illustration let us consider the sequence control problem (Fig. 3.1d and Fig. 3.1e) again. Suppose that X contains the value of the byte currently output to the port. At step 4, X = 128 which corresponds to 10000000. As the most significant bit d7 is the only one set to 1, the stirrer alone would be on. We now wish to switch on the valve for reagent C without affecting the stirrer. If Y denotes the byte to switch on a given bit, then to activate valve C, Y = 4 since the valve is connected to bit d2. We need to merge these two bytes and store the result in Z say, prior to outputting Z to the port. This is done using the logical OR as follows.

	d7	d6	d5	d4	d3	d2	d1	d0
X = 128 =	1	0	0	0	0	0	0	0
Y = 4 =	0	0	0	0	0	1	0	0
Z = X OR Y =	1	0	0	0	0	1	0	0

Z = 128 + 4

Z = 132

You may recall in checking through the use of the OR operation above that it applies at the bit level comparing the ith bit (i = 0 to 7) in turn for the two bytes involved. The result is 1 if either of the two bits is 1. If both bits are 0 the result is 0. In this way we have produced a value for Z of 132 which corresponds to that originally specified for step 5 for Fig. 3.1e. The difference is that we are now using a general method which allows us to set any output bit (or bits) to 1 without affecting any other bits assuming we have an up-to-date value for the current state of the output at the port.

SAQ 3.1f

An output port is connected as follows to a series of external devices.

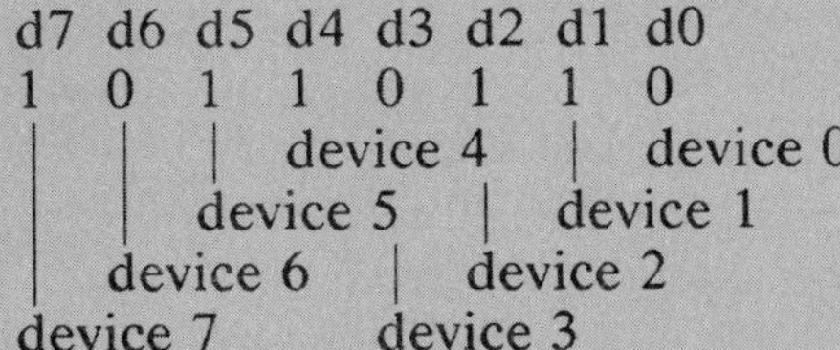

The current state of the output at the port corresponds to the byte 10110110 or 182 which is stored in a program as variable X. We wish to switch on device 3 without affecting any other device. What byte pattern would you output? In obtaining your answer use the logical OR to assign a value to the variable Z for subsequent output to the port.

The above discussion showed us how to merge two bytes to set a particular output bit to 1 without affecting other bits. But how can we reset a bit to 0, again without affecting any other output devices connected to the same port? As before we can use logical expressions except this time we need NOT and AND.

At the bit level, the NOT operator changes each bit so that if it was originally 1 it becomes 0, and if it was 0 it becomes 1. Thus if X=128, which in binary is 10000000, then NOT X would be 01111111 or 127. As you may recall from Part 2, the AND operator also works at the bit level on the two bytes to be ANDed. If Xi and Yi refer respectively to the ith bit of bytes X and Y, the result for the ith bit of X AND Y is 1 only if both Xi and Yi are 1 otherwise the result is 0.

By way of example let X = 128 and Y = 240 and the result of X AND Y is to be stored in Z. The detailed working is as follows:

```
        X = 128 = 1 0 0 0 0 0 0 0
        Y = 240 = 1 1 1 1 0 0 0 0
                  ---------------
    Z = X AND Y = 1 0 0 0 0 0 0 0
    Z = 128
```

To solve our problem we need to combine the logical operations of NOT and AND. To see how this works, suppose the current state of the output to the port is equivalent to the bit pattern 10100111, or 167, stored in the variable X. If we want to set bit 1 to 0 we first apply the NOT operator to the byte 00000010, in which there is a 1 at the bit we want to reset and the other bits are 0. Storing the result in Y gives:

Y = NOT 00000010 = 11111101

Now applying the AND operator on Y and the current output stored in X we have:

X = 10100111
Y = 11111101
\- - - - - - - - -
Z = 10100101

Comparing Z with X we see that the only difference is that bit 1 has changed from 1 to 0. We could therefore output the byte Z to the port to reset bit 1 without affecting other bits.

To generalise the procedure, the following steps are used.

(*i*) Define a byte which has 1 at each bit to be reset and 0 elsewhere.

(*ii*) Apply the NOT operator to the byte obtained in (*i*).

(*iii*) Apply the AND operator to the byte which reflects the current state of the output and that obtained in (*ii*).

(*iv*) Output the byte obtained from (*iii*) to the port.

Some versions of BASIC provide the Exclusive OR operator, EOR, which operates at the bit level and only produces a 0 result if the two bits being compared are either both 0 or both 1. If either one of the bits is 1, the result of EOR is 1. Thus, in the above example where the current output was X = 10100111 and we wished to change bit 1 to 0, all we need do is assign Z, the byte to be output, as

Z = X EOR Y

where Y = 00000010. The result is 10100101 which you may like to check for yourself.

The logical operations needed to reset a given bit to 0 can be programmed quite compactly even if EOR is not available, as indicated by the following BASIC statements which reset bit 7 assuming that the current state of the port is stored in X.

```
POKE &FE00,(X AND (NOT 128))
```

or

OUT 251,(X AND (NOT 128))

The first example assumes a memory-mapped port with address &HFE00 and the second one uses a port with port number 251.

SAQ 3.1g

An output port is connected to an external device as indicated below and the bit pattern shown represents the current state of the output.

d7	d6	d5	d4	d3	d2	d1	d0
1	0	1	1	0	1	0	1
not connected				device 3	device 2	device 1	device 0

(*i*) Use the NOT and AND operators to deduce the byte which should be output to reset bit d2 to 0 without affecting other bits.

(*ii*) Use logical operators to deduce the byte which should be output to switch off device 0 and switch on device 3 (assume that 1 = ON and 0 = OFF).

SAQ 3.1h

An output port is connected to some external devices as shown below.

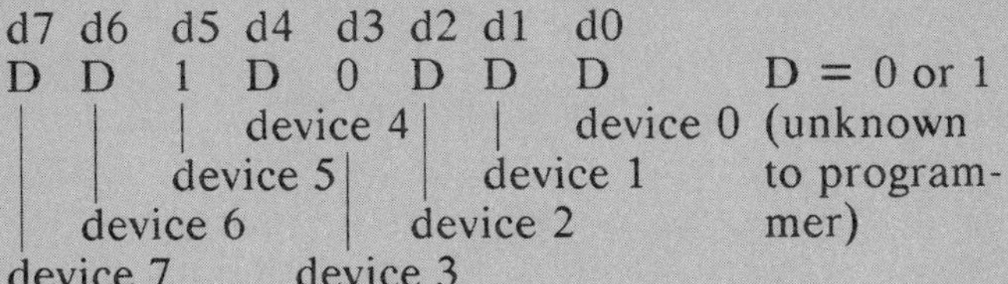

Although the programmer does not know which devices will be on or off at a particular point in the program he can assume that the last output byte is stored in a variable X.

Write the BASIC statements which will accomplish each of the following tasks (assume that 1 = ON, 0 = OFF, the port has a port number of 202 and the version of BASIC used supports the OUT instruction).

(*i*) Switch on all devices.

(*ii*) Switch off all devices.

(*iii*) Switch on device 3 without affecting other devices.

(*iv*) Switch off device 5 without altering anything else.

3.2. PROGRAMMABLE INTERFACES

3.2.1. Concept of a Programmable Interface

In the previous section we saw how specific interfaces are used to communicate with a given external device or instrument. Each application is different and requires its own pattern of input and output bits. Dedicated interfaces are designed for a specific task. If the requirements of the interface change, perhaps because of an alteration to the instrument or equipment being monitored, the interface has to be altered which usually means a soldering job.

A different approach is to use a general purpose interface which is programmed to meet the needs of a particular application. When the application changes, say from one requiring 8 bits of input to another needing 8 bits of output, we merely have to change the programming of the interface chip and no redesign of the electronic circuitry is required.

Several interfaces are available, too many to discuss each in detail here, so we will concentrate on just two which illustrate the principles.

3.2.2. The 6522 Versatile Interface Adaptor

The 6522 Versatile Interface Adaptor (VIA) has been used with a number of general purpose computers (BBC, APPLE, PET). It is a complex integrated circuit device which can be used for timing short intervals, counting input pulses, generating complicated output square-waves, as well as the simple input and output on which we shall concentrate. It also has facilities for controlling the rate of data transfer between computers, so that no information is lost, as will be explained later in Section 3.4.

The 6522 VIA has two 8 bit data ports, for input and output operations, called Port A and Port B. The signal at each pin of these ports is reflected by the byte pattern contained in two internal registers in the 6522 VIA. These are memory mapped and are called Data

Register A (DRA) for Port A, and Data Register B (DRB) for Port B. For an application using digital output, if the ith bit of DRA is set to 1 then the ith pin on Port A is at approximately 5 volts. If the ith bit is 0, then the corresponding pin on Port A is about 0 volts. Similar considerations apply for DRB and Port B.

SAQ 3.2a

A schematic pin diagram of the 6522 VIA is shown below. Pins PA0-PA7 refer to Port A, and Pins PB0-PB7 correspond to Port B. If the voltage levels given for each pin apply, choose the option which correctly describes the contents of registers DRA and DRB.

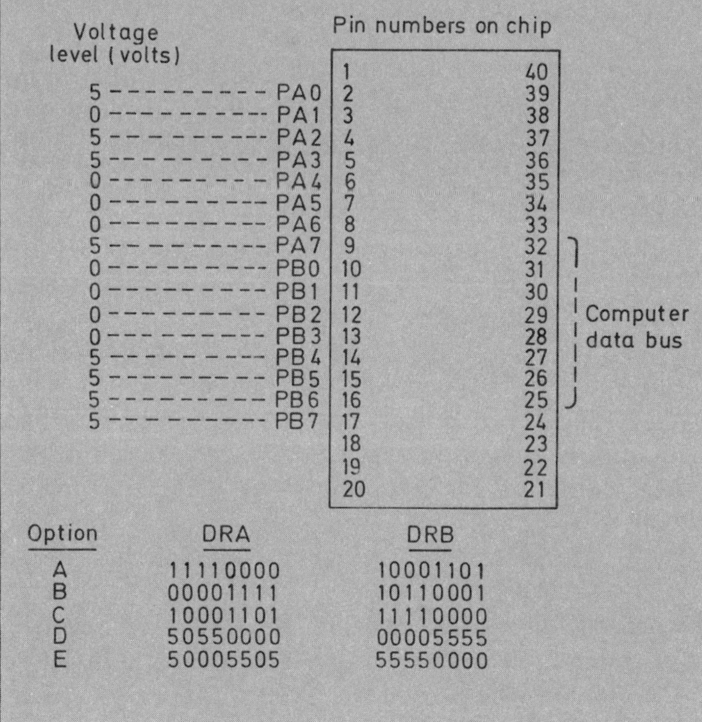

Option	DRA	DRB
A	11110000	10001101
B	00001111	10110001
C	10001101	11110000
D	50550000	00005555
E	50005505	55550000

SAQ 3.2a

Before the 6522 VIA can be used it must be programmed or 'configured'. This is done for simple input and output by writing information to two so-called Data Direction Registers. For Port A we use Data Direction Register A (DDRA) and for Port B Data Direction Register B (DDRB). For each pin of Port A which is to be used for output, the corresponding bit of DDRA must be set to 1. In contrast, setting a particular bit of DDRA to 0 configures the corresponding pin of Port A for input.

Each pin of Port B can similarly be configured for input or output by writing respectively a 0 or a 1 to the corresponding bit of DDRB. You should note that a pin configured for output should not be wired to receive an incoming signal since the possible conflict of signals (eg the 6522 VIA wants the pin at 5 volts but the external signal requires it to be 0 volts) can damage the interface chip.

By way of illustration, consider the case where we want all of Port A to be output and all of Port B to be input. To configure the 6522 VIA we would send the byte 11111111 = 255 to DDRA, and 00000000 = 0 to DDRB. Of course each port can be configured so that some bits are used for input and others for output. For example, writing 128 (or binary 1000000) to DDRA would mean that the pin corresponding to PA7 could be used for output and the rest for input.

SAQ 3.2b

An instrument provides 16 bits of data to be read as two separate bytes, and requires two control signals from the computer. One signal tells the instrument to hold or 'freeze' the existing datum in order that the computer has time to read one byte and then the other, without allowing the datum to change. We shall call this the 'RUN/HOLD' signal (1 = HOLD, 0 = RUN). The second control signal tells the instrument which byte is to be read (the most significant, or least significant one of the 16 bit number). We shall assume that 1 means the most significant, and 0 means the least significant byte. The connections with the 6522 VIA are as indicated below.

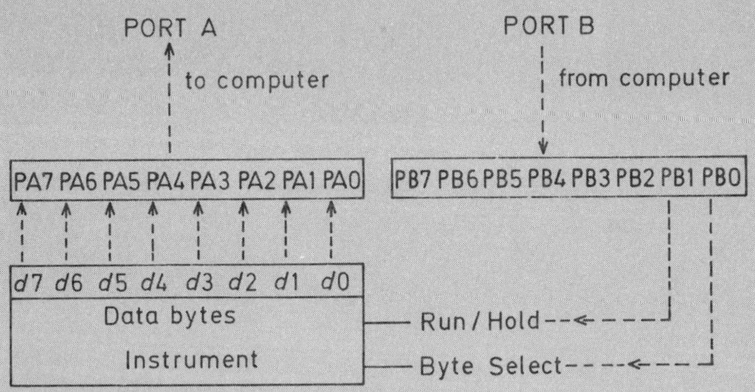

(*i*) What bytes would you write to DDRA and DDRB to configure the 6522 VIA for this application?

(*ii*) Assuming the internal registers of the 6522 VIA have the addresses given below, write the sequence of BASIC statements which would read the least significant byte into X and the most significant one into Y. The

⟶

SAQ 3.2b (cont.)

sequence of statements should then combine the bytes to give a number stored in Z, print out the value of Z, allow the instrument reading to change and repeat the whole process so that a continual stream of readings is obtained.

Assume that the version of BASIC used supports PEEK and POKE (X=PEEK(ADDRESS) and POKE X, ADDRESS where X is the byte read or written to or from location ADDRESS).

ADDRESS	REGISTER
&FE60	DRB
&FE61	DRA
&FE62	DDRB
&FE63	DDRA

3.2.3. The 8255 Programmable Peripheral Interface

The 8255 Programmable Peripheral Interface (8255 PPI) differs considerably in its operation and use from the 6522 VIA and was developed for a different family of microprocessors. Thus in the past there has been a tendency to find the 6522 and related interfaces associated with computers based on the 6502 microprocessor while the 8255 PPI has been used with Z80-based microcomputers (eg TRS-80, Sinclair ZX81, RM380Z).

The 8255 PPI provides 24 bits of digital input or output which are organised into two groups of 12 and designated as groups A and B. Each group is made up of two subgroups, one of 8 bits and another of 4 bits. Like the 6522, this interface can be used at a number of levels of sophistication. There are in fact three operational modes, Mode 0, Mode 1 and Mode 2. In Modes 1 and 2 the 8 bit subgroups are used as 8 bit ports under the control of signals placed on the 4 bit subgroups. The discussion of the detailed operation and use of these two modes is beyond the scope of this introduction, but the interested reader is recommended to consult the texts given in the Overview to Part 3. We shall restrict attention to the use of Mode 0.

The simplest way of looking at the use of the 8255 PPI in mode 0 is to regard the 24 bits of digital input and output as three 8-bit ports, labelled as Port A, Port B and Port C. Port A can be programmed as 8 bits of input, or 8 bits of output. Unlike the 6522 VIA, individual bits cannot be programmed separately. Similarly, Port B can be programmed as a byte of output or a byte of input. In contrast, the lowest 4 bits of Port C (Bits PC0-PC3) can be programmed as all input or all output independently of the other 4 bits (PC4-PC7) which can also be programmed as a nibble of input or output.

The 8255 PPI has one further port, the Control Port, to which an 8 bit control word must be written to specify precisely how ports A, B and are to be used. The significance of each bit in the control word is given in Fig. 3.2a.

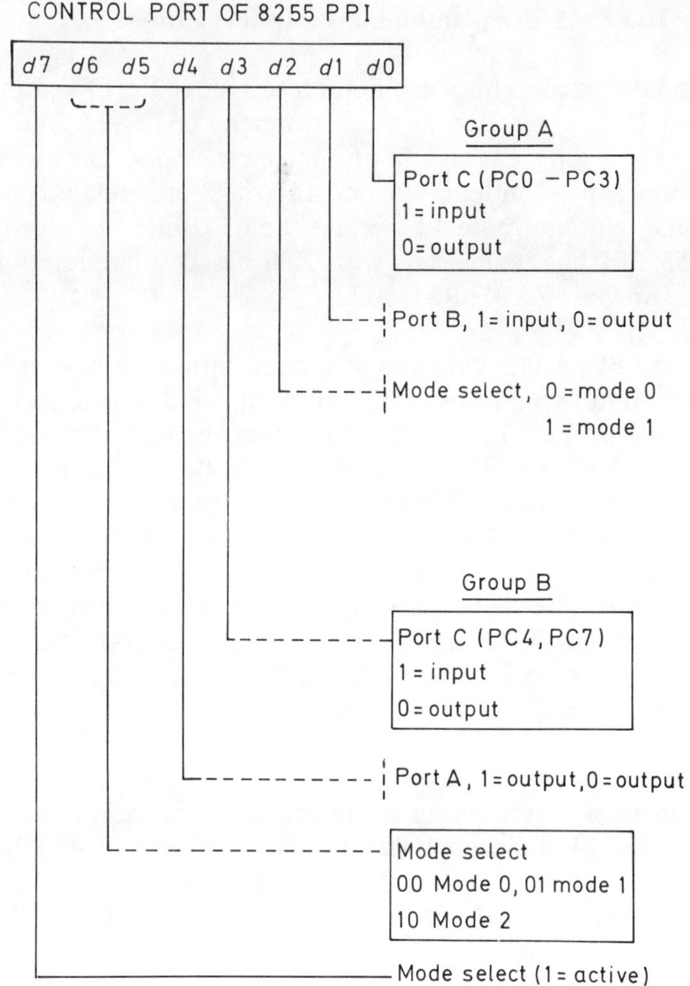

Fig. 3.2a. *The control word to configure the 8255 PPI*

To select Mode 0, Bit D7 must be at 1, and bits D6, D5 and D2 must be at 0. The pattern of input and output used for Ports A, B, and C is then determined by the bits D4, D3, D1 and D0, as shown in Fig. 3.2a. For example, to have all 24 bits configured as input the control word would be:

D7	D6	D5	D4	D3	D2	D1	D0	
1	0	0	1	1	0	1	1	= 155

SAQ 3.2c

Which one of the following control words for the 8255 PPI configures the interface for Port A as input, Port B as output and Port C with PC0-PC3 as input but PC4-PC7 as output?

	D7	D6	D5	D4	D3	D2	D1	D0
(*i*)	0	0	0	1	1	0	0	1
(*ii*)	1	1	1	1	0	1	0	1
(*iii*)	1	0	0	1	0	0	0	1
(*iv*)	0	1	0	0	1	0	0	1

SAQ 3.2d

Given the port numbers below for an 8255 PPI, write the sequence of BASIC statements which will allow a stream of bytes of information to be read from an external device, assuming the following connections between the instrument and the 8255 PPI. ⟶

SAQ 3.2d (cont.)

Port numbers
Port A = 200; Port B = 201; Port C = 202;
Control Port = 203

Connections

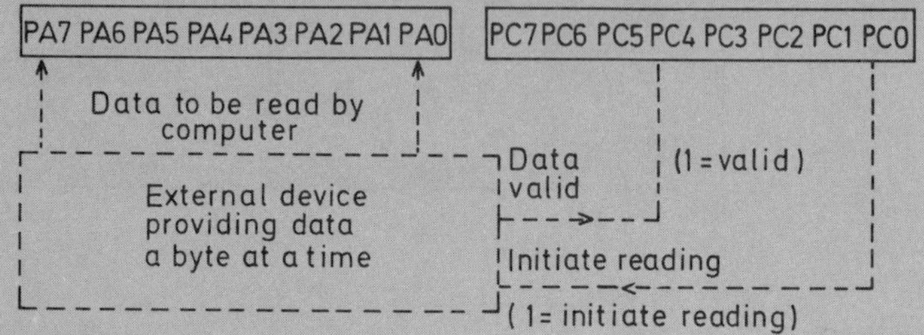

Notice that Port A is used for data input and Port C for control purposes. The upper 4 bits of Port C need to be configured for input (to test the validity of data before reading it), and the lower 4 bits for output (to indicate that a new reading from the device is required).

This concludes our brief discussion of programmable interfaces for digital input and output. Because all 8 bits of a byte are transmitted to or from the data bus together, these interfaces are described as parallel. Three points need emphasising before moving on to the next topic. Firstly, the output byte(s) presented by both the 6522 VIA and the 8255 PPI are 'latched' or caught and held at the interface by the electronic circuitry built into the interface chip. The output datum will thus remain available at the port until either the chip is reconfigured or other another byte is output. Secondly, any input signals presented to pins configured for input on either interface chip are only allowed on to the data bus when the appropriate port is being read by the computer and at other times these signals are isolated from the data bus. Finally, the current flow which can be sustained by an output pin on these interfaces is very small and the output voltages are either approximately 0 or 5 volts, therefore the interface may need additional electronics to build up the available current and voltage to the required levels for practical applications.

3.3. INTERFACES FOR ANALOGUE SIGNALS

3.3.1. Analogue and Digital Signals

Previously we have considered digital signals, or signals which are restricted to one of two values ($\approx$0 and $\approx$5 V for TTL levels). Some applications may require voltages to be output, or read by the computer, which vary continuously between certain minimum and maximum values.

These are called analogue signals because they are the electrical analogue of the physical variable being monitored, in the case of input. Fig. 3.3a shows a digital signal varying as a function of time and the time dependence of an analogue signal.

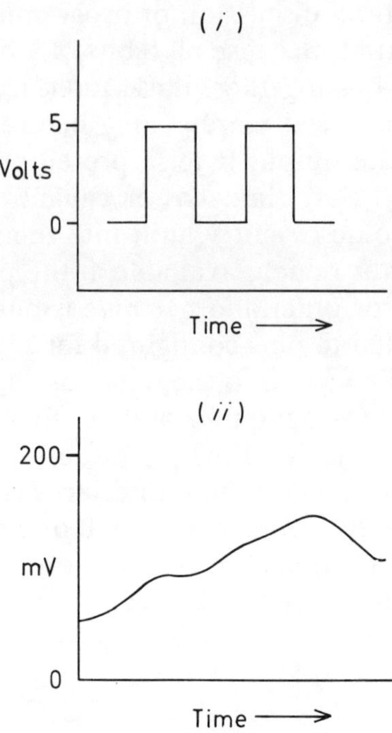

Fig. 3.3a. *Examples of (i) digital and (ii) analogue signals as a function of time*

The time dependence of the digital signal could be monitored by connecting it to a single bit of an input part and noting the value detected (0 or 1) as a function of time. This is not possible with the analogue signal since the voltage level is in general incompatible with the TTL levels expected as input at a typical computer port. There are two approaches which can be used to overcome this problem, namely voltage to frequency conversion, or analogue to digital conversion.

A voltage to frequency converter is essentially a voltage controlled oscillator. A schematic diagram showing the principles involved in using this type of device is given in Fig. 3.3b.

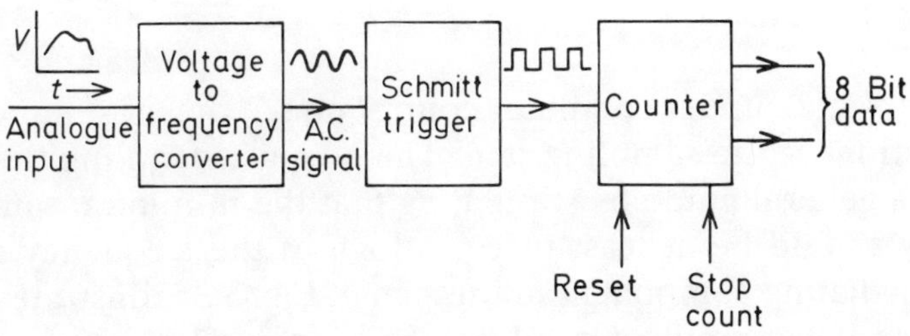

Fig. 3.3b. *Schematic diagram illustrating the production of an 8 bit data value from an analogue signal using a voltage to frequency converter*

The analogue signal is converted into an AC signal by the voltage to frequency converter. The frequency of this signal is determined by the input voltage being sampled. The AC signal can be converted to an oscillating digital signal by means of a device called a Schmitt trigger. This responds to an input signal by taking up one of the two allowed states associated with digital signals. It has the effect of 'squaring-up' the signal so that it is always clearly in one logic state or the other. Each pulse in the digital wave train can now be fed directly into a binary counter. To obtain a reading all we have to do is start the counter, wait for a specific period of time, stop the counter, and finally read it to obtain a numerical value which corresponds to the analogue voltage during the sampling period.

As an alternative approach we could dispense with the counter and simply feed the digital signal straight into a single input bit of a computer port. The computer could then be programmed to count the pulses over a given time period to obtain a value related to the magnitude of the analogue input. This would not be a successful approach using BASIC, unless a very slow stream of pulses was being monitored, and in general, monitoring would need to be done by a machine code program produced either from assembly language or a compiled high level language.

An analogue to digital converter (ADC) accepts as input an analogue signal and produces output in the form of latched parallel binary data. The method of operation will be described in the next section but, in passing, a few fundamental points should be noted.

(*a*) *Sampling Rate*

In collecting data under computer control one of the most important considerations is the sampling rate. The Nyquist sampling theorem provides a general guide by stipulating that the minimum sampling frequency should be at least twice as high as the frequency of the highest oscillating component in the signal. Fig. 3.3c illustrates what can happen if we sample at too low a frequency. The original waveform (*i*) is sampled at the points X giving the values indicated in (*ii*). The reconstructed waveform shown in (*iii*) has a much lower frequency. This effect is known as 'aliasing' and is a direct result of sampling at too low a frequency.

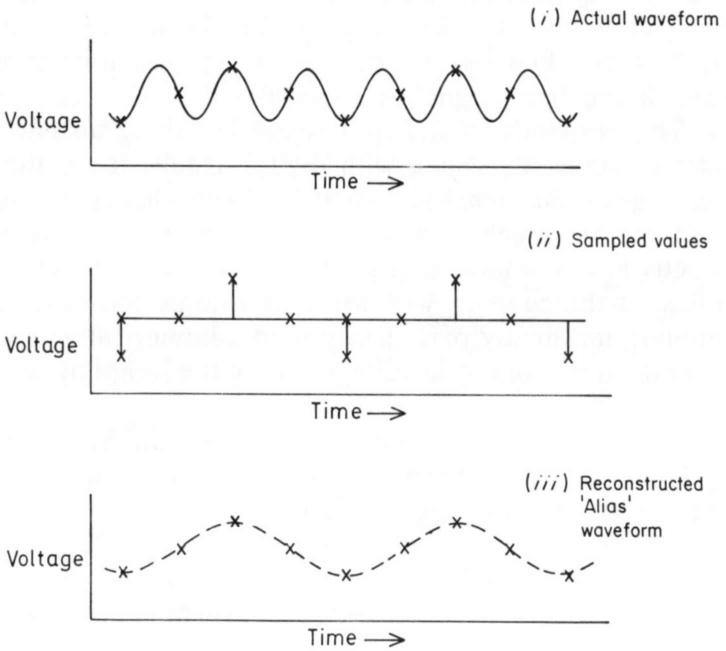

Fig. 3.3c. *The effect of too low a sampling frequency*

(*i*) Actual waveform to be sampled. Sampling points indicated by X
(*ii*) Sampled values corresponding to X on the original waveform (*i*).
(*iii*) The reconstructed waveform which has a lower frequency than the original.

To be on the safe side there is a temptation to over sample. The drawback here is that we may consume too much memory in storing essentially unwanted data. We have to use some judgement in deciding the proper sampling rate. The object is to collect sufficient data to allow the signal to be properly reconstructed later by reference to the digital values stored. To do this we have to know how the signal changes in time and choose a sufficiently rapid sampling rate so that no important information is missed. That is not to imply that we should slavishly follow the sampling theorem without taking account of the information inherent in the monitored signal. For example, the signal in Fig. 3.3d shows a base-line trace from a spectroscopic instrument being used in conditions of high gain so that a considerable amount of noise is observed. To reconstruct the precise signal, noise and all, would need very rapid sampling. However, we can use much lower sampling rates because the analytical information (the value for the baseline) is varying only slowly, if at all. Of course the measurements then taken would be subject to error by an amount determined by the level of noise. This could however be reduced by averaging a number of measurements of the baseline and the noise, being random, should average out giving a more precise figure for the baseline.

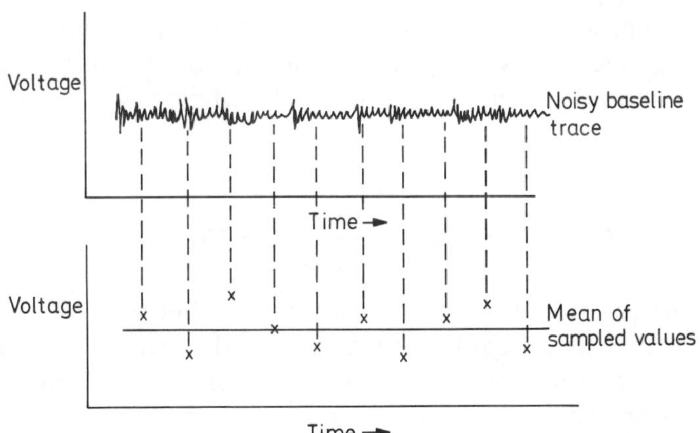

Fig. 3.3d. *The effect of sampling a randomly noisy signal at a frequency lower than that of the noise*

Finally, in considering the sampling rate one must make sure the storage medium can cope with the volume of data. For example, suppose we wish to monitor the sodium ion concentration of a river over a 24 hour period and we wanted to use a small microcomputer, with 30K bytes of memory, on-line to a pX meter and sodium electrode system. A knowledge of how the sodium ion concentration is likely to change with time indicates that we need to take a reading at least every 30 seconds.

Could we monitor for 24 hours without running out of memory given that each reading needs 10 bytes? In 24 hours we would take $24 \times 60 \times 2 = 2880$ samples which would require 2880×10 bytes, or just over 28K of memory, which just fits into the available space.

(*b*) *Conversion Time*

As should have been evident in considering the use of the voltage to frequency converter, the production of a binary number which is related to the original analogue signal takes a certain amount of time. Many devices are available to convert an analogue voltage to a digital value but they all have a finite conversion time. Some are very fast and can produce a binary value in microseconds from sampling the voltage, whereas others are much slower. Having decided the sampling rate, the appropriate device for analogue to digital conversion should be chosen to ensure a sufficiently rapid conversion speed.

(*c*) *Digital Resolution and Quantisation Error*

An analogue signal can vary continuously between specified limits whereas the digital equivalent is restricted to whole numbers in the range 0–255 (for 8 bits), 0–1023 (for 10-bits) etc. Suppose we wish to represent an analogue signal which ranges from 0 to 200 mV by an 8 bit binary number. The digital value of 0 will correspond to 0 mV and the value 255 could be made to correspond to 200 mV.

The full range of voltage 0–200 mV can then be represented by the range of numbers 0–255. Since the smallest interval in this digital range is 1, the 200 mV range is effectively divided into $2^8 - 1$ steps, or increments of 0.78 mV. If in sampling and converting the analogue voltage we agree to round to the nearest whole number in the range 0–255, we have in principle an error of up to ± 0.5 or an overall uncertainty of 1 unit. For an 8 bit representation of an analogue voltage which corresponds to the upper limit of the range, the error is 1 part in 255 or about 0.4%. For signals which are only half of the maximum voltage, the digital reading would be 127 so the error is now 1 part in 127 or 0.8%.

The above errors are all due to the discrete nature of the digital voltage which has to be fitted as closely as possible to the analogue signal. For an N bit representation, where $2^N - 1$ is used to denote the maximum analogue voltage, and 0 denotes the minimum voltage, the digital resolution is 1 part in $2^N - 1$, and the quantisation error is given by:

$$\text{Quantisation error} = \frac{V_{max} - V_{min}}{2^N - 1}$$

For a 12 bit analogue-to-digital conversion covering a range of 10 volts, the quantisation error is 10/4095 = 0.0024 V and so changes in the analogue voltage finer than this could not be reliably detected.

SAQ 3.3a

(*i*) An analyst wishes to use a computer to sample an analogue signal every second. Each reading is to be stored in the computer's main memory and takes up 5 bytes. If 10K bytes (1K = 1024) are available for data storage, what is the maximum time for which data can be collected?

What could be done to extend the time limit? ⟶

SAQ 3.3a (cont.)

(*ii*) A signal from an instrument is subject to interference from the mains supply operating at 50 Hz. It has been decided to eliminate this interference once the signal has been stored in the computer. Given that 50 Hz is the highest frequency component in the monitored signal, what is the minimum sampling frequency which is acceptable?

Would the sampling be possible using a BASIC program run by an interpreter?

If sampling is required over a 5 minute period, how much storage space would be required if each reading needed 5 bytes of memory

3.3.2. Digital to Analogue Conversion

In some applications we need to provide an analogue voltage under computer control.

Fig. 3.3e (*i*) shows a series of voltage ramps of the type which could be used in polarographic measurements. A digital to analogue converter (DAC) accepts a binary number (8, 10 or 12-bit usually) and produces a voltage proportional to its magnitude.

The output observed (Fig. 3.3e (*ii*)) in general will be made up of a number of discrete steps, due to the quantisation of the conversion, and the fineness of each step is determined by the number of bits used.

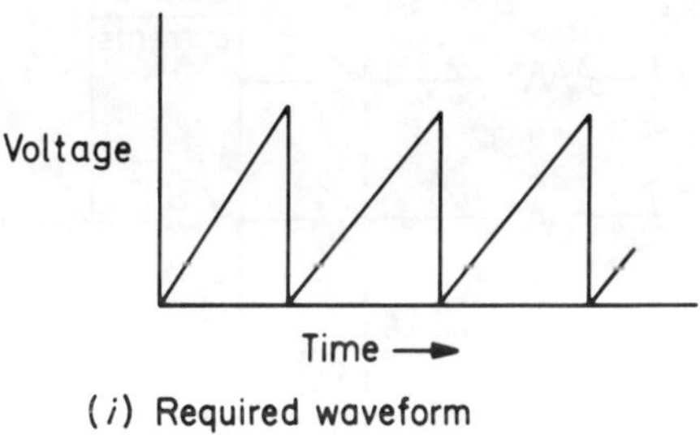

(*i*) Required waveform

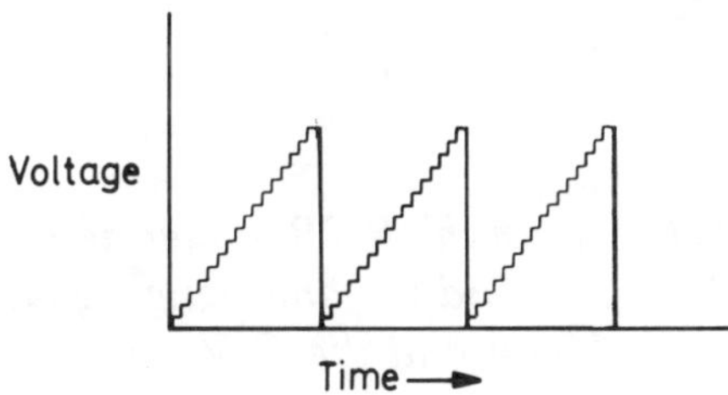

(*ii*) Output from digital-to-analogue converter

Fig. 3.3e. *Comparison of required and observed signals when analogue output is produced by a digital to analogue converter*

Fig. 3.3f shows in general forms how the DAC operates. The binary data $b_7 \ldots b_0$ control a set of switches. Each switch is closed when the corresponding bit is set to 1, and it constitutes its appropriate weight of current to a current summing amplifier.

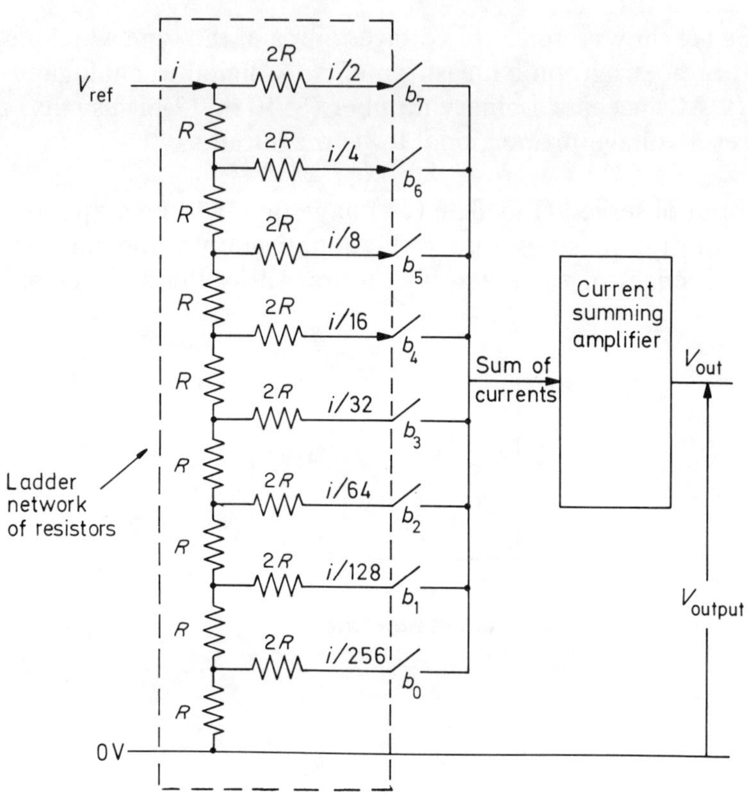

Fig. 3.3f. *A DAC based on the R-2R ladder network. The digital inputs are labelled $b_7 \ldots b_0$ and the analogue signal emerges between V out and the O V rail*

This amplifier produces an output voltage proportional to the sum of the currents and therefore proportional to the binary number expressed by $b_7 \ldots b_0$. It is not necessary for our purpose to understand the operation of the ladder network of resistors enclosed in the dashed rectangle in Fig. 3.3f. It is sufficient to note that several matched resistors are involved and the precision of matching will determine the accuracy of the DAC.

Furthermore, it does take a small amount of time for all the switches to become set and the currents so obtained to stabilise. This 'settling time' as it is called is specified by the manufacturer and is not likely to be a problem in use with BASIC programs which execute relatively slowly on the computing timescale.

Apart from the accuracy and settling time other features which ought to be considered in selecting an ADC include the output voltage range, the digital resolution as determined by the number of bits converted, and the nature of the input (binary or binary coded decimal).

SAQ 3.3b

An 8-bit digital to analogue converter (DAC) is accessed via a memory-mapped interface with address &FE01. Write a program which will produce an analogue output to resemble as closely as possible the following waveform.

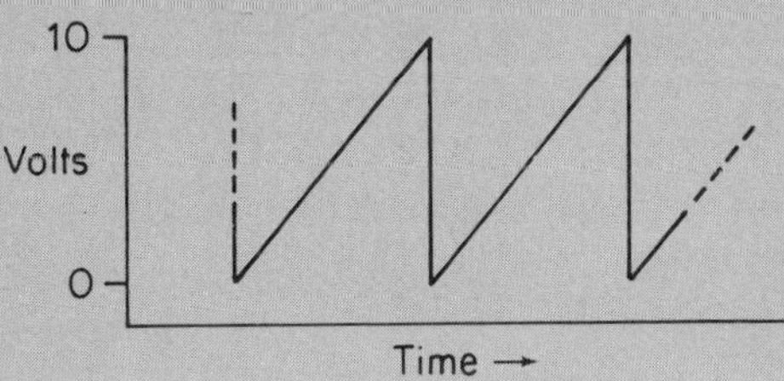

Assume that the DAC has an output which ranges from 0 to 10 V and the BASIC used supports POKE.

In what way would the analogue signal generated differ from the target waveform above?

Modify the program so that the user has control of the rate of climb, and therefore the frequency of the waveform.

SAQ 3.3b

3.3.3. Analogue to Digital Conversion

A general introduction to analogue to digital conversion was given in Section 3.3.1 in connection with quantisation error and sampling frequency. Here we examine in a little more detail the operation of a typical analogue to digital converter (ADC) and how it may be used under computer control.

Fig. 3.3g shows a schematic diagram of a so-called counter-ramp ADC. It works by generating a voltage ramp by means of an internal DAC which is driven by a binary counter. The voltage produced by the DAC is compared with the analogue input being measured and when the two voltages are the same the counter is halted, a 'data valid' signal is generated and the final count value can be read by the computer (as a digital representation of the analogue input voltage) from the binary data bits $d_7 \ldots d_0$ which are held until another analogue-to-digital conversion cycle is initiated.

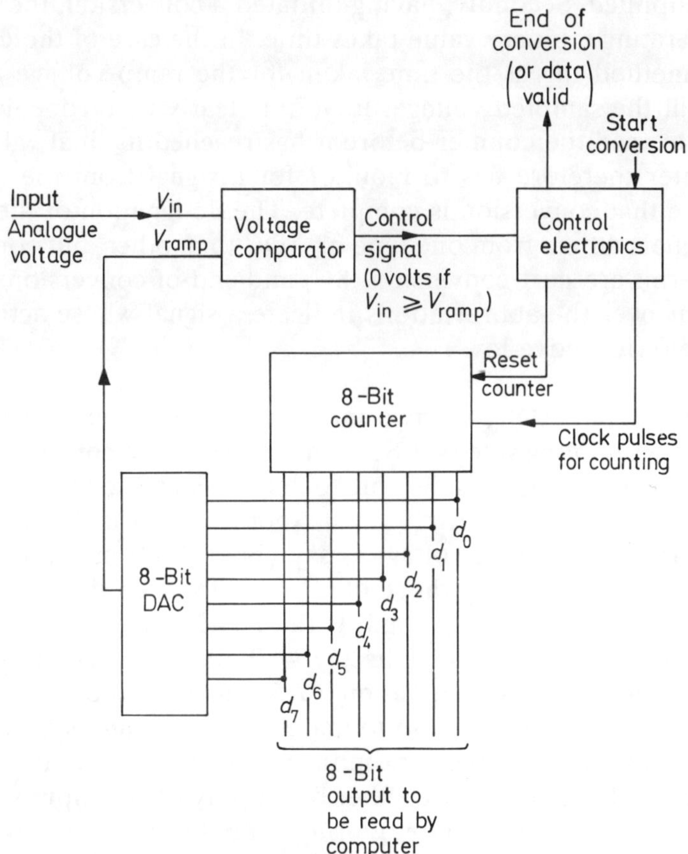

Fig. 3.3g. *A counter-ramp ADC*

The counter-ramp method of achieving analogue to digital conversion is only one of several methods in common use. For a description of other methods the reader is recommended to study one of the many specialist texts on digital electronics and computer interfacing. A number of important points emerge from the brief discussion of the counter-ramp ADC which are relevant to ADCs in general.

Firstly, to start an analogue to digital conversion a pulse, usually negative going (ie from logic 1 to logic 0 and then back to 1), has to be supplied. Secondly, having initiated a conversion, the process of generating a binary value takes time. In the case of the counter-ramp method this is the time taken for the ramp voltage to grow to equal the sampled voltage. It would clearly be erroneous to attempt to read the counter before it has reached its final value. The computer therefore has to monitor for a signal from the ADC to indicate that conversion is complete. The designation of these control signals differs from one type of ADC to another, but commonly used terms are start-conversion ($\overline{SC}$) and end-of-conversion ($\overline{EOC}$). The bar over the abbreviations indicate a signal whose active state is logic 0 (ie 'active low').

In choosing an ADC, the two most important criteria are the appropriate sampling rate (see Section 3.3.1) and the conversion time. Obviously the ADC must be able to deliver converted values at least as fast as the application requires. Another point is the input voltage range. If the input voltage is say about 2.5 V at the most, and the ADC has a working range of 0 to 5 volts, then the effect of the quantisation error is magnified. If the ADC converts to give 8 bits the quantisation error is 5 volts/255 = 0.0196 volts. On a 5 volt signal, this amounts to 0.4%, but on a 2.5 volt signal the error is 0.8%. To minimise errors it is wise to match the maximum signal voltage to be measured with the maximum value that can be accepted by the ADC. This can be done either by amplifying the input signal (or attenuating it if it is too large), or altering the working range of the ADC. Many ADCs now have the facility for a manual change of gain and offset voltage in order to aid matching with the voltage being monitored. Some ADCs even have programmable working ranges so that any necessary change from one application to another can be accomplished through software. Finally, the accuracy of values obtained from an ADC can be increased by converting to more bits.

An 8 bit ADC with a digital resolution of 1 part in 255 is limited for analytical measurements. 10 and 12 bit ADCs are more commonly used although they are slightly more difficult to handle from the computing viewpoint because each reading cannot be taken in as one byte. The solution is to use an output signal to the ADC to select which part of the 10 or 12 bit number is to be read. For example with a 12 bit ADC, the first 8 bits may be read, followed by a second read in which the next 4 bits are obtained. The two readings must then be recombined in the computer to give the digital value corresponding to the sampled voltage.

SAQ 3.3c

A computer is interfaced to a spectrophotometer via an 8-bit analogue to digital converter (ADC). The output from the spectrophotometer is linear in absorbance and 0 absorbance corresponds to 0 read from the ADC. A value of 255 corresponds to an absorbance of 1.2. If an absorbance of 0.6 is obtained, what is the quantisation error? Could an analyst expect to obtain readings in the computer which were within 0.5% of their true values?

SAQ 3.3d

An 8 bit analogue to digital converter (ADC) is interfaced to a computer via an 8255 PPI.

The 8 bit datum is read through port A and port C is used to control the ADC. Bit PCO is used to output a signal to initiate the start of conversion ($\overline{SC}$) and PC4 as an input to indicate the end-of-conversion ($\overline{EOC}$). Both SC and EOC are active low so that if EOC has a logic value of 0 the conversion is complete. Similarly conversion is started by setting SC low.

(*i*) Produce a step-wise design for a program which will read 1000 numbers from the ADC with one sample taken every 2 seconds. Assume that the computer you are using has a real-time counter which is incremental every 10 ms and the value of the counter is stored in TIME. The 8255 PPI is memory mapped with addresses:

Port A	:	&FE70
Port B	:	&FE71
Port C	:	&FE72
Control Port	:	&FE73

The control byte needed to configure the 8255 for this application is:

10011000 = 152

(*ii*) After checking your answer to (*i*), write a program to implement your design using either PEEK and POKE or the '?' equivalent of BBC BASIC.

SAQ 3.3d

Analytical applications often demand that we sample more than one analogue input.

This can be achieved by using a number of ADCs but there is a less expensive method which is often quite satisfactory. This uses a multiplexer integrated circuit device as shown schematically in Fig. 3.3h.

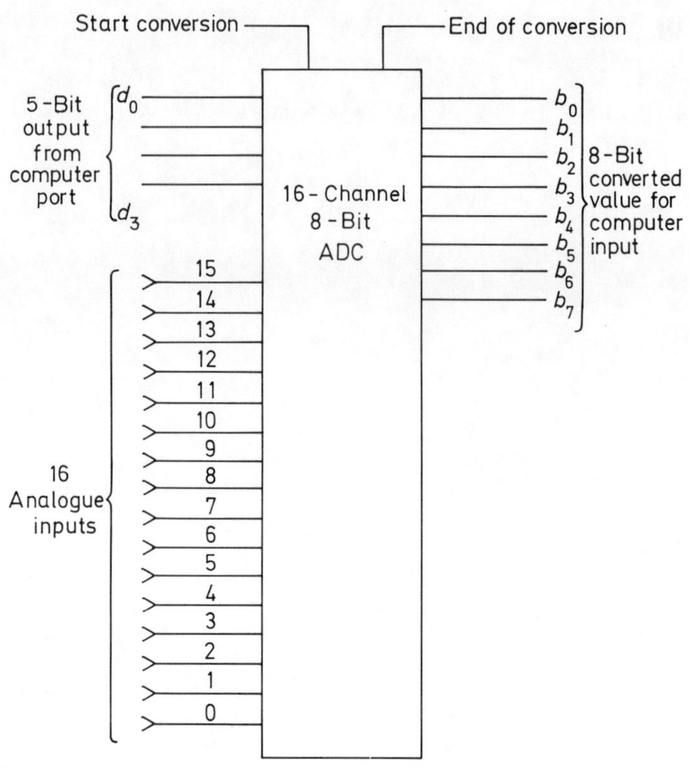

Fig. 3.3h. *Schematic representation of a 16 channel 8 bit ADC*

The channel to be sampled is selected by outputting the channel number (0–15) on the 4 bits of output labelled $d_3 \ldots d_0$. Having selected the channel, the procedure is the same as for a single channel ADC, namely initiate conversion through the SC line and then monitor the EOC signal for conversion complete. The valid data for the selected channel can then be read from the 8 bit parallel output bits $b_7 \ldots b_0$ in Fig. 3.3h.

SAQ 3.3e

Write a stepwise design for a program which will scan 500 times the first three channels of a 16-channel multiplexed ADC of the type shown in Fig. 3.3h and store the values obtained in arrays.

⟶

SAQ 3.3e (cont.)

A 2 second delay is required between successive readings of the ADC. Assume the following connections to a 6522 VIA interface through which communication with the ADC is achieved by the computer.

6522 VIA addresses:

Part A	DDRA	: &FE63
	DRA	: &FE61
Part B	DDRB	: &FE62
	DRB	: &FE60

Connections (see Fig. 3.3h):

$PA_0 \ldots PA_3$	connected to $d_0 \ldots d_3$ respectively (to select the channel number).
$PB_0 \ldots PB_7$	connected to $b_0 \ldots b_7$ (converted 8 bit data)
PA_5	start conversion
PA_7	end conversion

You should now be able to write programs which will communicate with a variety of chips involved with analogue to digital conversion. In order to consolidate your knowledge of this section, answer the following SAQ before proceeding to the final section of this Part.

SAQ 3.3f

By circling T or F, indicate whether or not each of the following statements is true or false.

(*i*) The quantisation error in a 16 bit ADC is likely to be about twice that of an 8 bit ADC because there are twice as many bits in which an error could arise. T / F

(*ii*) One must always sample analogue signals at least twice as fast as the highest frequency component, even if the analytical information is contained in a lower frequency signal, and the highest frequency component corresponds to a small amount of noise. T / F

(*iii*) Because BASIC is slow, in many applications using a fast ADC, it is sufficient to initiate a conversion by setting the $\overline{\text{SC}}$ line momentarily low and then reading the data without checking the $\overline{\text{EOC}}$ signal. T / F

(*iv*) Digital signals are restricted to one of two levels whereas analogue signals can vary continuously between 0 and 5 V. T / F

(*v*) The use of an ADC converter is the only way of reading an analogue signal by computer. T / F

SAQ 3.3f

3.4. DATA TRANSMISSION

In many applications it becomes necessary to transmit data or other information from one computer to another. A small microcomputer may be dedicated to the control of an analytical instrument. It may also be used for processing raw data to deliver individual analytical results. The operation of an analytical laboratory may then require these results to be passed on to a more powerful computer for record purposes, costings, progress chasing of samples and even the printing of invoices. The problem we consider in this section is how to transmit the data between the various computers and what are the scope and limitations of the available methods.

3.4.1. Parallel Data Transfer

In previous sections we have considered in detail digital input and output via 8 bit parallel ports. It is therefore natural to examine the use of parallel ports in communication between computers.

Superficially we may imagine the problem to be trivial. We simply write a byte of information out to a port on computer A and, assuming the proper connections are made, read the byte into an input port on computer B. This is too simple an approach since it fails to ensure synchronisation of the transfer of data between the two machines. Computer B cannot read the incoming byte unless it has already been output by computer A. Similarly computer A should not output a second byte of information unless the previous byte has been read by computer B. Clearly some control signals are needed to keep each computer informed about the state of the data being transferred. The process of controlling the data transfer is often called 'handshaking'.

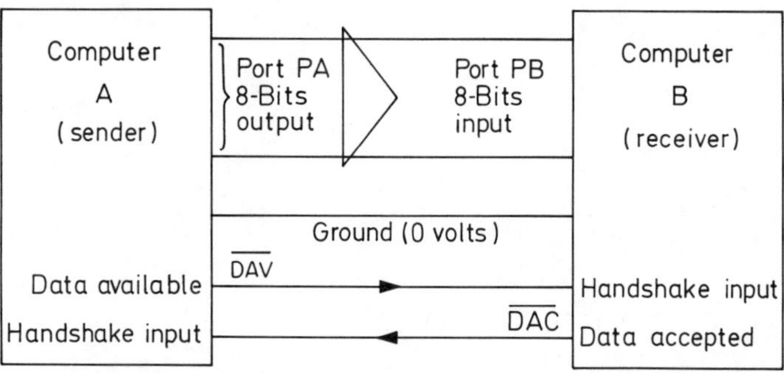

Fig. 3.4a. *An example of connections used to provide parallel 8 bit communication between computers under 'handshake' control*

To appreciate how handshaking works, let us consider two computers A and B, as shown in Fig. 3.4a. The output port for computer A, designated PA, is connected pin-for-pin to the input part on computer B, designated PB. Note the common ground line between the two machines. (If this was not present the voltage levels on one machine would not be related to those on the other and the correspondence between digital values and voltage levels would be lost in inter-computer communication). We also have in this case two handshake control lines, one labelled Data Available ($\overline{\text{DAV}}$) and the other Data Accepted ($\overline{\text{DAC}}$). Various names for handshake signals are in common use, eg Data Valid, Data Acknowledge, but they all perform essentially the same function. The Data Available signal is set to logic zero, the active state (as indicated by the bar over the symbol $\overline{DAV}$), to indicate a datum byte, currently latched on port A, is ready for reading by computer B. Computer B monitors the $\overline{DAV}$ line (it is on input at machine B) and when it goes low it can read the byte of information through its input port PB. As soon as the datum is read computer B sets the Data Accepted signal low (again active low) and returns it high after a brief time delay to provide a Data Acknowledgment pulse. Meanwhile, computer A has been monitoring the Data Accepted line, and on observing the acknowledgment pulse it can initiate a repeat of the cycle by writing the next byte for transmission to port PA.

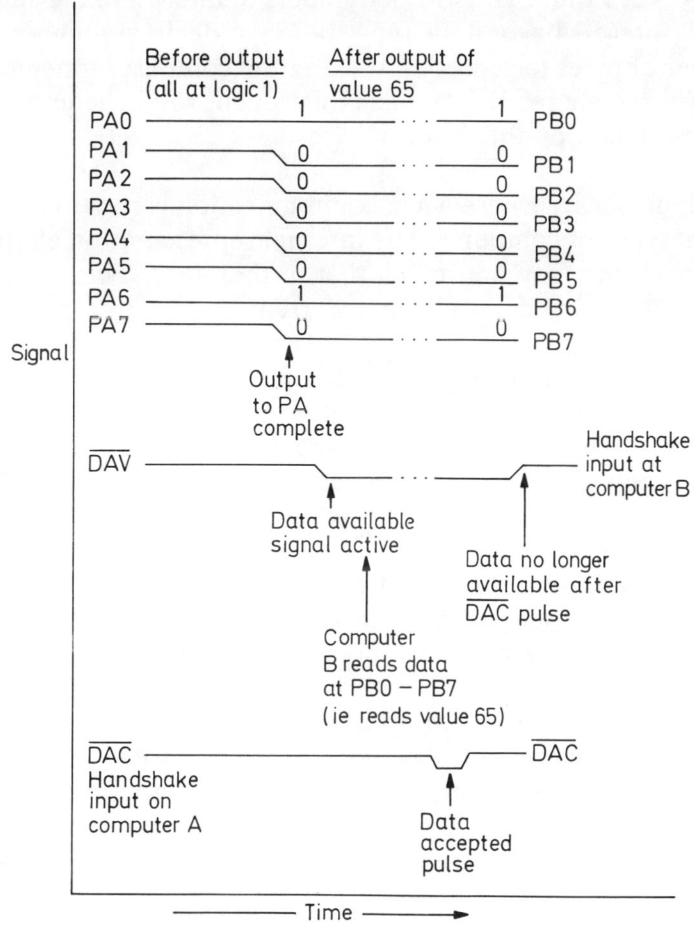

Fig 3.4b. *Sequence of signals involved in a simple handshake to pass 8 bit parallel data from a computer A to another B*

The handshake sequence is summarised in the timing diagram in Fig. 3.4b. Read this figure from the top bearing in mind that time increases from left to right. The signals correspond to a value of 65 being transferred from computer A to B. Having written 65 to port PA, so that the individual lines $PA_0 \ldots PA_7$ have the appropriate voltage levels, the $\overline{DAV}$ line is set low. Computer B then reads the datum via $PB_0 \ldots PB_7$ and sends out a pulse (negative going) on $\overline{DAC}$ to complete the transfer of data.

The above method of data transfer allows transmission of information a byte at a time. The sort of numbers of interest to the analytical chemist (integers beyond the range 0–255 or decimal numbers) need a number of bytes to represent them in the computer's memory and so to pass a single reading from an instrument will in general involve the transmission of a number of bytes.

A problem can arise here when communication is required between different types of computer. The internal representation of numbers in one machine may not match that in the other. This means that, even if data are successfully transferred, correct interpretation of the sequence of bytes as numerical data values will not be possible without the development of special software.

One way of circumventing this problem is to transmit data in a standard code. The one which is commonly used is the American Standard Code for Information Interchange, or ASCII for short. Here each keyboard character (digits, letters, punctuation marks and other symbols) is assigned a unique 7 bit code. For example capital A is represented by the code 65 or 1000001 in binary. The ASCII code for B is 66. A list of the most important ASCII codes and the corresponding characters is given in Fig. 3.4c. There are in fact 128 codes in all, 96 are used for characters and 32 are control codes used with visual display units. These control codes are of less interest in the present context and therefore are not included in Fig. 3.4c. The one display code we will need later is that for carriage return (code 13), as it is often used to terminate a string of transmitted characters.

Character	Code	Character	Code
Space	0100000	C	1000011
!	0100001	D	1000100
"	0100010	E	1000101
#	0100011	F	1000110
$	0100100	G	1000111
%	0100101	H	1001000
&	0100110	I	1001001
‘	0100111	J	1001010
(	0101000	K	1001011
)	0101001	L	1001100
*	0101010	M	1001101
+	0101011	N	1001110
,	0101100	O	1001111
-	0101101	P	1010000
.	0101110	Q	1010001
/	0101111	R	1010010
		S	1010011
0	0110000	T	1010100
1	0110001	U	1010101
2	0110010	V	1010110
3	0110011	W	1010111
4	0110100	X	1011000
5	0110101	Y	1011001
6	0110110	Z	1011010
7	0110111	[	1011011
8	0111000	\	1011100
9	0111001	]	1011101
:	0111010	^	1011110
;	0111011	_	1011111
<	0111100	‘	1100000
=	0111101	a	as for A–Z
>	0111110	:	but add 32
?	0111111	z	ie bit 5 set to 1
		{	1111010
@	1000000	|	1111100
A	1000001	}	1111101
B	1000010	≃	1111110
		DEL	1111111

Fig. 3.4c. *ASCII codes for digits, letters, punctuation marks and other display symbols*

In sending information between computers ASCII codes are very useful allowing one to transmit strings of digits as well as textual information. For example a microprocessor monitoring weight may send the message:

'1.296 grams'

This message would be dispatched as a sequence of bytes, the least significant 7 bits of which contain the ASCII code for each character in turn. Thus a '1', followed by '.', then '2', etc would be transmitted.

As all computers can accept data bytes in ASCII format, data transfer becomes independent of the internal storage of numbers in each computer. Of course, the user has to re-assemble the string of digits received into a proper number before calculations can be performed.

As an aid to this, many dialects of BASIC support the VAL function which returns a number corresponding to the value of a string. Thus if X$ contained '1.296' as a string of characters, we could convert it to a real number contained in say X by the statement:

```
X = VAL (X$)
```

There are some limitations to parallel communication of the type discussed in this section. Probably the most serious one is the requirement to keep the connecting cables short (1 metre or less is quite common). One reason is that in a noisy electrical environment transient voltages may be induced which can corrupt the data being transmitted. For high speed data transmission extra problems can arise due to the degradation of the sharpness of the transitions between the voltage levels representing logic 0 and logic 1. However, this method is quite satisfactory for coupling up a computer to an instrument or computer-controlled device immediately adjacent to the computer in a laboratory environment. In the next section we shall consider an alternative approach which allows data to be passed over greater distances and in noisy electrical environments.

SAQ 3.4a

In the context of parallel communication between computers under handshake control, indicate whether or not each statement is true (T) or false (F).

(*i*) If the sending computer halts, the receiver will wait until the appropriate handshake signals have been received and so give the appearance of halting too.

T / F

(*ii*) If the receiving computer shuts down, the sender will continue to pass data which will then be lost.

T / F

(*iii*) The ASCII character for '1' is identical with the binary representation of 1.

T / F

(*iv*) Integers in the range 0 to 255 can be passed from one computer to another as a single byte. Other integers or decimal numbers cannot be sent from one computer to another using parallel communication.

T / F

(*v*) Parallel communication is normally used on short distances.

T / F

3.4.2. Serial Transmission

Data can be transmitted from one computer to another by the use of a single line (plus a common ground). The method requires the byte to be transmitted as a series of bits. Each bit is transmitted over a specified time interval and the level of the voltage indicates the value (0 or 1) assigned to the bit.

The byte of information is therefore dispatched as a serial data packet as shown in Fig. 3.4d. The start of the data packet is introduced by a 'start bit' which is a logic 0 held for a specified time interval determined by the selected transmission speed. Then follows the datum itself.

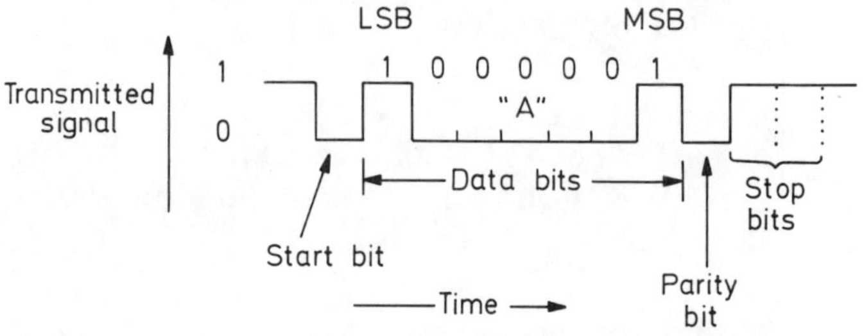

Fig. 3.4d. *An example of a serial data packet corresponding to the transmission of the character 'A' with even parity*

(LSB = least significant bit)
(MSB = most significant bit).

Normally ASCII characters are transmitted and so only 7 data bits are needed. The eighth bit is added as a check on the accuracy of transmission and is called the parity bit. For even parity, the parity bit is set to either 1 or 0 (whichever is appropriate) to make the total number of ones, in the data and parity bits, equal to an even number. Similarly if odd parity is chosen, the parity bit is set to give an odd number of ones. Once the datum is received it is a simple matter to check the total number of ones. If we have chosen even parity and an odd number of ones is found then we have a parity error due to corruption of the data during transmission.

Finally we have two stop bits (both at logic 1) to terminate the data packet. The serial transmission speed is determined by the baud rate which is the number of bits transmitted per second. Commonly used baud rates are 300, 600, 1200, 2400, 4800 and 9600. Transmission speeds at baud rates higher than 9600 can become unreliable, depending on the electrical environment. Notice that the baud rate is expressed in terms of bits/second, not characters/second. Since each character involves the transmission of 11 bits, assuming the packet structure in Fig. 3.4d, the rate of transmission of characters is slower by about an order of magnitude. When you also bear in mind that a reading from an instrument may involve of the order of 10 characters (eg '1.293 ? grams' followed by carriage return) the transmission rate for complete data values can be about 100 times slower than the baud rate. Thus at 1200 baud we could expect to transmit about 120 characters a second or about a dozen readings involving say 10 characters.

SAQ 3.4b

Which one of the following serial data packets corresponds to the transmission of the ASCII character 'A' using 300 baud even parity and two stop bits?

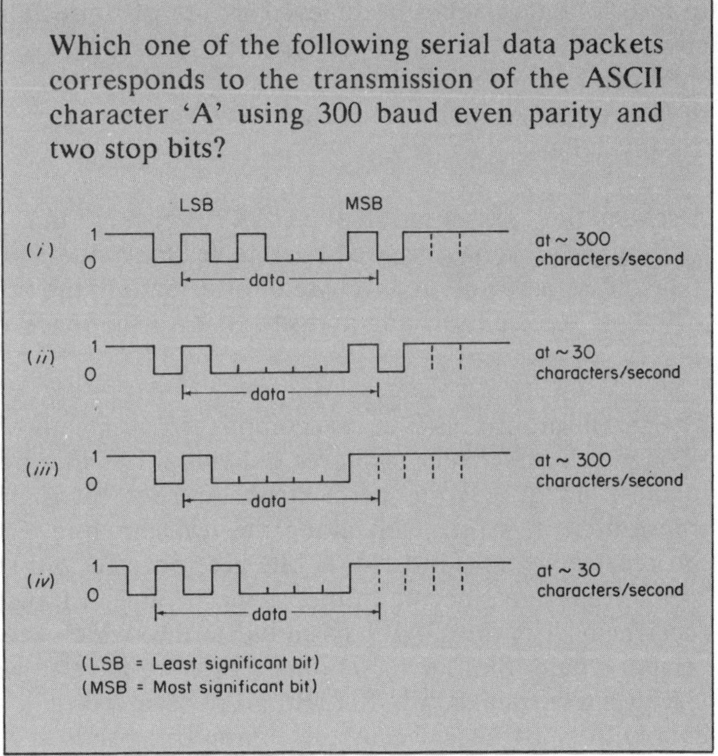

SAQ 3.4b

The conversion of the binary information from 8-bit parallel format to a serial data packet is achieved by use of a specialist integrated circuit device. A number of types of these devices exist, but we shall consider in general outline the operation of only one, namely the Universal Asynchronous Receiver/Transmitter or UART as it is usually called.

A schematic representation of a UART is given in in Fig. 3.4e. It can be thought of as having two separate sections; one to receive serial data and re-assemble it as a byte of information; the other to accept a byte of information and transmit it as a sequence of individual bits.

The serialisation of a byte is accomplished by means of a 'shift register' which effectively displaces individual bits in a byte one place to the right in a sequence of steps. As each bit is displaced it is transmitted as a serial bit along the transmit line. Conversely, in the receive section, individual bits are received and shifted into a byte format one place at a time, as the reverse of transmission. The receiver port of the UART has a buffer into which the last byte received is held. Similar to sending a byte we have to write it to the UART's transmit data buffer. Both the received data and transmit data buffers are accessed by the computer directly as ports with either port numbers or memory-mapped addresses. Additionally the

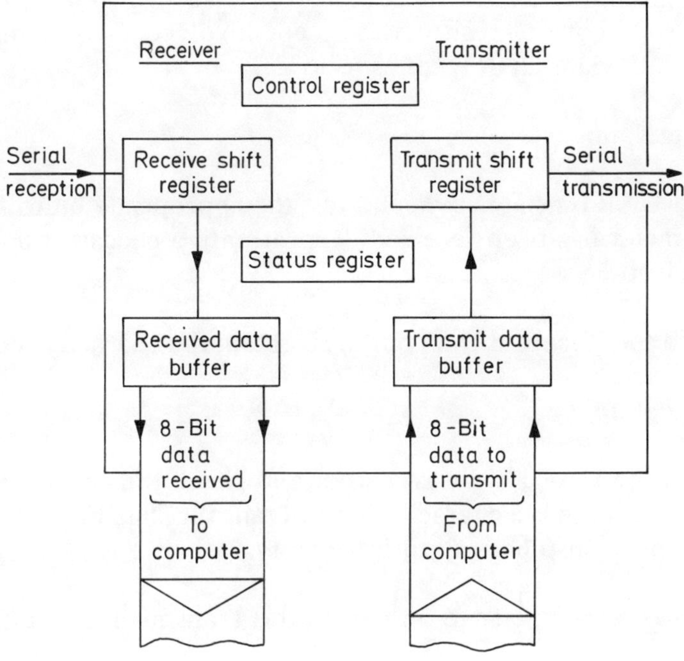

Fig. 3.4e. *A schematic representation of a UART*

UART has a status register, which again can be read by the computer to give a single byte of status information. Depending on the type of UART, particular bits of the status byte correspond to various conditions or errors. For example, if a second byte is received before the previous one has been read by the computer an over-run error is produced to warn the programmer that a byte of information (the first byte) has been overwritten and lost. By accessing specified bits of the status byte it is possible to determine whether or not a character has been received since the Received Data buffer was last read. Similar considerations apply to transmission. We can access a specified bit of the status byte to determine whether or not the Transmit Data buffer is full. If it were, it would be a mistake to write more data to it because the previous byte for dispatch would be overwritten and therefore lost.

Assuming that the UART has been properly configured for use (by writing to its control register) the steps involved in receiving and sending data via a UART are as follows:

Receiving data:

(*a*) load UART's status byte and test the appropriate bit to see if a character has been received. Repeat until a character has been received.

(*b*) Read the Received Data buffer to obtain the character received.

Sending data:

(*a*) Load the UART's status register and check the appropriate bit to see if there is a character in the Transmit Data buffer. If there is, repeat this check until it is empty.

(*b*) Write the character to be sent to the Transmit Data buffer.

It would be inappropriate in this introductory treatment to look at the details of how the UART is configured to work with the correct baud rate, number of data bits, parity and number of stop bits. The interested reader is recommended to study one of the specialist texts referred to in the Overview to Part 3.

SAQ 3.4c

> Write a stepwise design to read a sequence of characters received by a UART and assemble all the characters together in the string variable X$.
>
> Assume that the UART status register can be read from port 201 and each byte that is received can be read from port 202. Concentrate on the problem of reading in the characters one at a time and assume that the UART has been properly configured already for you. ⟶

SAQ 3.4c (cont.)

The string of characters will be terminated by a carriage return (ASCII code 13), and bit 7 of the status register will be set to 1 if a character has been received since the last read of the data port 202.

The UART accepts a byte of information from the computer using our usual 5 V, 0 V logic (TTL levels). It transmits and receives also using these levels. However, these relatively close voltages are unsatisfactory for transmission of data over anything but short distances in electrically noisy environments. For this reason the actual transmission takes place using a greater separation of voltage levels. The RS232 standard for serial communication uses −3 to −15 volts to represent a logic 1 and a logic 0 is expressed by the voltage range +3 to +15 volts. Note that the voltages used are off opposite polarity and different levels than those used in the internal logic of computers and the computer interfaces we have considered. Clearly one could damage the computer if it were connected directly to RS232 signal lines.

The minimum number of signal lines needed for serial communications is three, the common ground, the transmit and the receive lines. This does not permit any handshaking, but extra lines are included in the RS232 standard which can be used to control the transfer of data. This is particularly important when a fast computer is sending serial data to a slow device such as a printer. The printer needs to supply a 'busy' signal to inhibit the computer from sending more data until the most recent datum is printed.

A standard RS232 connector is shown in Fig. 3.4f and allows for many more than three wires to link the serial devices.

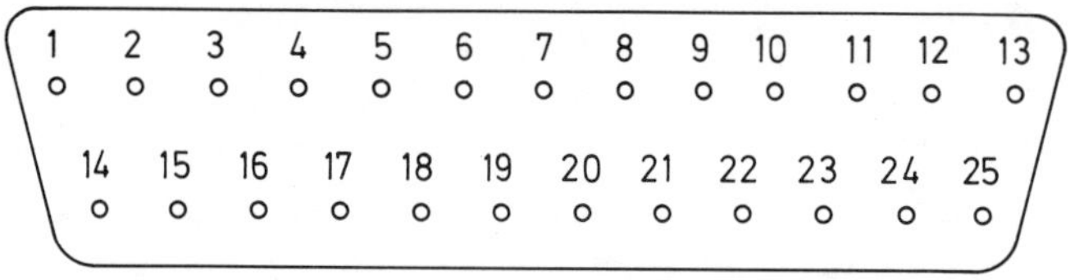

1 2 3 4 5 6 7 8 9 10 11 12 13
14 15 16 17 18 19 20 21 22 23 24 25

Fig. 3.4f. *Some designated pin connections for RS232 interfaces*

Pin No	Common Mnemonic	Function
1		Protective grand (shield)
2	TxD	Transmit (Terminal*), Receive (Comm*)
3	RxD	Receive (Terminal*), Transmit (Comm*)
4	RTS	Ready to send
5	CTS	Clear to send
6	DSR	Data set ready
7		Signal ground
8	DCR	Data carrier detect

<table>
<tr><td>9</td><td rowspan="17">Used in specialist telephone communications applications</td></tr>
<tr><td>10</td></tr>
<tr><td>11</td></tr>
<tr><td>12</td></tr>
<tr><td>13</td></tr>
<tr><td>14</td></tr>
<tr><td>15</td></tr>
<tr><td>16</td></tr>
<tr><td>17</td></tr>
<tr><td>18</td></tr>
<tr><td>19</td></tr>
<tr><td>20</td></tr>
<tr><td>21</td></tr>
<tr><td>22</td></tr>
<tr><td>23</td></tr>
<tr><td>24</td></tr>
<tr><td>25</td></tr>
</table>

* Depends whether equipment connected is classified as data terminal (such as a computer) or data communication equipment (such as a modem used for telephone transmission).

The situation is complicated somewhat by definition of a particular pin on the connector as either an input or an output depending on the equipment. For a computer, or computer terminal (described as Data Terminal Equipment), data are transmitted on pin 2, and received on pin 3. On the other hand, if the equipment is a telephone modem (a device for converting digital signals into audible tones) the serial interface receives on pin 2 and transmits on pin 3. A modem is an example of Data Communication Equipment, and all devices which use RS232 interfaces are wired either as Data Communications type or as Data Terminal Equipment. Hence the source of confusion. If your computer is wired (as it should be) as Data Terminal Equipment and you want to establish serial communication with an instrument or device which is wired as Data Communication Equipment, then you can connect pins 2, 3 and 7 on the computer's interface directly to pins 2, 3 and 7 on the other equipment.

If they are both wired as Data Terminal Equipment, which is normally the case when two computers are linked via RS232 connecting lines, then pins 2 and 3 on one interface must be connected to pins 3 and 2 respectively on the other as shown in Fig. 3.4g. Thus if you cannot establish proper RS232 communications between devices, and you are satisfied that the baud rate is correct, the first thing to try is to cross the connections linking pins 2 and 3.

Pins 4, 5, 6, 8, 17 and 20 are used for handshaking signals but it is beyond the scope of this introductory treatment to discuss how handshaking is achieved via the UART. It is sufficient to note that the software driving some serial interfaces may be using handshaking signals. It is important to make up the connecting cable correctly so that these signals can be exchanged correctly. The literature supplied with the computer or external device should specify precisely which RS232 lines need to be connected.

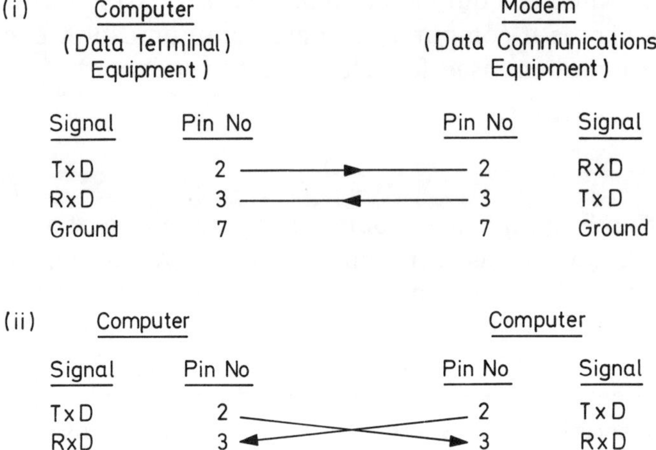

Fig. 3.4g. *3-Wire connections between a computer and* (i) *a modem, and* (ii) *another computer*

SAQ 3.4d

(*i*) What is the minimum number of connecting wires needed to establish RS232 communication between two computers? What problems can arise if this minimum linkage is used?

(*ii*) Why should RS232 signal lines never be connected directly to electronic circuits which use TTL voltage levels?

(*iii*) Which of the following are not standard baud rates?

(1) 300,	(2) 400,	(3) 600,
(4) 800,	(5) 1200,	(6) 1600,
(7) 2400,	(8) 4800,	(9) 9600

3.4.3. Instrumentation Interfaces

A general purpose instrumentation interface is designed to allow a number of instruments to be connected to a computer system at the same time, to provide fast data transfer, and to minimise the problems of replacing instruments connected to the system by others, not necessarily of the same type or function.

Ideally the interfaces and their connections should be independent of the type of computer. This attribute is particularly valuable when replacing computers, for example when they become obsolete, because no changes are necessary to the interfaces.

The RS232 interface is an example of an instrumentation interface and a range of instruments are fitted with these. However, there are drawbacks which make it less than satisfactory compared with other methods which are now available. In general we need to have a separate UART, or serial channel, for each device. For many computer systems the UART is supplied on a board to be plugged into a vacant slot on the computer's bus-system.

Often only one or two slots are available and if more serial interfaces have to be added an expansion chassis has to be used. Another problem is the data transmission rate. Most applications use 9600 baud or less and for some purposes, particularly where a lot of instruments have to be scanned, this may not be fast enough.

Data transmission rates can be improved by passing information as a sequence of bytes, with each byte being transmitted as 8 bits simultaneously. This is like the parallel data transfer discussed earlier, and results in an increase in transmission speed by an order of magnitude, at the expense of extra wiring.

The most well established instrumentation interface is the IEEE-488 standard, usually referred to as the IEEE interface. (IEEE stands for Institute of Electrical and Electronic Engineers). Its main features are summarised below:

(*a*) Uses bit-parallel, byte-serial data transfer (ie data are transmitted as a sequence of bytes).

(*b*) Handshake control is based on three control lines (NRFD, not ready for data; NDAC, not data accepted; DAV, data available).

(*c*) Up to 15 devices can be connected together on the IEEE bus at any one time.

(*d*) The maximum cable length is 20 metres.

(*e*) One device, usually a computer, is designated as bus controller.

(*f*) Each device on the bus is designated as a 'talker' or a 'listener' or both. When active as a 'talker' a device sends out information on the data lines to be picked up by another device, the 'listener'.

(*g*) Each device is assigned a primary address, and sometimes also a secondary address. Each device must have a unique primary and secondary address to identify it unambiguously on the bus.

A schematic representation of an IEEE system is shown in Fig. 3.4h, where devices are connected to a controller via 8 data lines, DIO (1-8), 3 handshake lines, and 5 bus management lines which are used for example to tell all devices how to interpret the signals on the data lines, or to initialise the the interfaces on all devices.

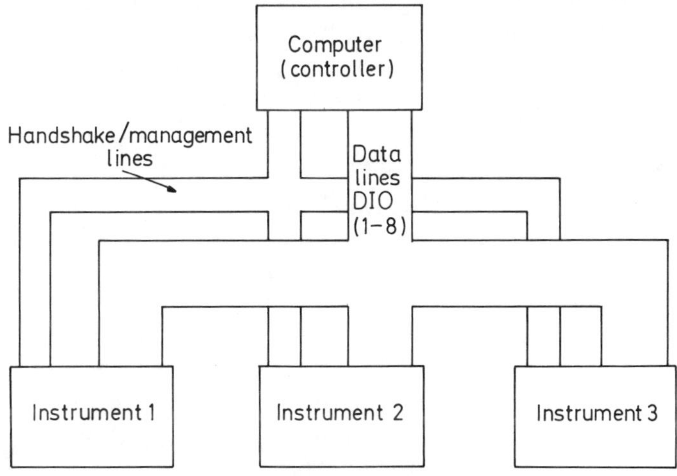

Fig. 3.4h. *IEEE-488 communication between a controlling computer and 3 instruments*

Now although the hardware aspects of the IEEE-488 are standard, so that any IEEE device can be plugged into any computer running an IEEE interface, the software to drive the interface varies from one computer to another. This software is normally supplied with the hardware as a machine code program which has to be loaded before attempting to use the interface from, say, BASIC. Unfortunately there is wide variation in the implementation of the driver software so that a program, using one computer and an IEEE interface to communicate with instruments, will need modifying if the computer is changed to a different type. This is in addition to the different driver routines needed to operate the IEEE bus system.

By way of illustration consider Fig. 3.4i which shows a BASIC program designed to send a byte of information to an external device fitted with an IEEE interface.

```
 10 NU0488 = &106
 20 PA% = 7 : REM primary address 7
 30 SA% = 17 : REM secondary address 17
 40 TIMEOUT% = 5000 : FLAGS% = 16 : ST% = 0
 50 REM Line 40 sets timeout for response (TIMEOUT%)
 60 REM from interface, certain control
 70 REM to do with message formats (FLAG%)
 80 REM and initialises a status byte (ST%)
 90 NB% = 1 : REM Number of bytes to send = 1
100 PRINT "INPUT A NUMBER (0-255)"
110 INPUT X%
120 IF X% <0 or X% >255 THEN GO to 100
130 CALL NU0488 (TIMEOUT%, ST%, FLAG%, NB%, SA%,
PA%)
140 IF ST%<>0 THEN 150 ELSE 160
160 END
```

Fig. 3.4i. *An example of a BASIC program which sends a byte of information to a device with an IEEE interface (primary address 7, secondary address 17). (A modified program based on one published by Biodata Ltd, Manchester, UK)*

The program will only run using the driver software and hardware supplied by Biodata Ltd (Manchester, UK) for an Apricot computer and a Microlink IEEE interface. Line 10 defines the memory address of the start of the appropriate driver routine supplied by Biodata Ltd for the Apricot computer. This and other IEEE driver routines have to be loaded before BASIC itself. Lines 20, 30, 40 and 90 are used to set parameters to particular values as explained in the REM statements. The data are taken in by lines 100–120 and stored in X%.

To send out the byte equal to X% over the IEEE interface to the device whose primary and secondary addresses are specified by PA% and SA%, one calls up the external machine code routine supplied by the manufacturer, as in line 130. Assuming no errors arise, the status byte ST% will remain at 0 and the program will terminate. For this particular implementation of an IEEE interface a range of driver routines is supplied to allow digital input and output, analogue input and output, as well as communication with specialist modules such as thermocouple input units, real-time clocks. Communication is also possible directly with any instrument fitted with its own IEEE interface.

The above implementation of the IEEE interface, from the point of view of software, varies considerably from that used with many other computers. For example some implementations treat each device on the IEEE bus as a data file to which data strings are written or from which strings of characters are read. Even with this approach there is variation in the instructions to initialise the interface and the format of commands and statements to read data from instruments connected to the IEEE bus, or to write data to them.

In spite of the variation in software requirements the IEEE system is probably the most successful general purpose approach to instrument interfacing.

Departing from the IEEE standard, some manufacturers have developed interfaces which suit either a particular computer, or a range of computers with similar characteristics. Again driver software is normally supplied with the interfaces.

For example, a whole range of interface boards is available for the Apple computer system. The Rexagon interface (developed by ICI), whilst not having the generality of the IEEE system, is easy to use and is applicable to a number of different types of computer.

An alternative approach to producing a standard instrumentation interface, into which a wide range of devices can be plugged without any rewiring, is to seek acceptance for a standardised computer bus system (address, control and data busses). Then, no matter which computer is used, interface cards can be plugged into an expansion chassis into which appropriate interface boards can be located. Such a system has been implemented and is called the S-100 instrumentation bus, and has been specified as an IEEE standard. The acceptance of the S-100 standard by a number of computer manufacturers has allowed a large range of S-100 interface boards to be developed. It is thus possible to buy boards for digital input and output, analogue input and output, as well as a range of specialist boards for control applications, sound generation, etc. The problem is, of course, that once committed to the use of S-100 interface boards, you must always use S-100 type computers, which may or may not be an unacceptable restriction. Readers who are interested in further details of the S-100 instrumentation bus are recommended to consult the text by S Libes and M Garetz referred to in the Overview.

SAQ 3.4e

Designate each of the following statements as either true (T) or false (F).

(*i*) An IEEE-488 instrumentation interface is correctly described as bit-parallel and byte-serial.

T / F

(*ii*) A big advantage of the IEEE-488 system is that any number and variety of instruments and devices can be connected to the controlling computer as long as each has the appropriate interface.

T / F

⟶

SAQ 3.4e (cont.)

(*iii*) Two instruments connected through IEEE-488 interfaces to the same controller can have the same primary and secondary addresses provided they are different types of instrument.

T / F

(*iv*) Although the IEEE-488 system is a high-speed instrumentation interface, rapid data capture is never possible when the application program is written in BASIC.

T / F

Summary

Section 1 introduced the idea of a computer port for 8-bit parallel input and output. The transfer of data was explained in terms of the address, data and control busses used by the computer. Data input was illustrated by means of an example where data was read from a pH meter with BCD output. Parallel data output was discussed in terms of a simple sequence control problem in which reagents were added sequentially to a reaction vessel. Logical operators such as AND and OR were also applied to this example to switch individual output bits on or off.

Section 2 was also concerned with parallel input and output, but this time programmable interfaces were used. Two examples were considered, namely the 8255 PPI and the 6522 VIA.

Section 3 dealt with analgoue input from external instrumentation. Factors which affect computer monitoring of analogue signals, such as sampling rate and digital resolution, were considered.

The final section provided an introduction to how computers can be made to communicate with each other using either serial or parallel data transfer. This led to a discussion of standard instrumentation busses illustrated by the IEEE 488 interface.

Objectives

On completing this Part of the Unit, you should now be able to:

- Explain the terms address, data and control buses and their function when a microprocessor communicates with an interface (SAQ 3.1a).

- Design a simple program to monitor the states of one or more specified bits of input from a parallel interface. (SAQs 3.1b, 3.1e and 3.1d).

- Design a simple program to switch on or off an external device connected to a specific bit of a parallel port without affecting the state of any other device connected to other output bits. (SAQs 3.1e, 3.1f, 3.1g and 3.1h).

- Configure two examples of programmable parallel interfaces for specified input or output tasks. (SAQs 3.2a, 3.2b, 3.2c, 3.2d and 3.3d).

- Explain, with reference to a BASIC program, how analogue signals can be generated under computer control. (SAQ 3.3b).

- Explain, with reference to a BASIC program, how a digital reading can be obtained from an analogue signal. (SAQs 3.3c, 3.3d and 3.3e).

- Explain the terms quantisation error, conversion speed, SC, EOC in connection with analogue input. (SAQs 3.3d and 3.3f).

- Choose an appropriate sampling frequency for a given analogue signal and relate this to the means of data storage. (SAQ 3.3a).

- Describe how handshake control of parallel data transfer between computer can be achieved. (SAQ 3.4a).

- Define the terms band rate, parity, UART and data packet as used in serial transmission of data. (SAQs 3.4b, 3.4e and 3.4d).

- State the principal communication lines and voltages used in RS232 interfaces. (SAQ 3.4d).

- Describe the main features of the IEEE-488 instrumentation interface. (SAQ 3.4e).

4. Automated Ion Selective Electrode Measurements

Overview

This case study brings together much of the material of previous parts of this Unit. It is essential that you have worked your way through Parts 1, 2 and 3 before attempting this case study.

Some new material is introduced including the use of a programmable interface to control mains power devices or other devices which use DC supplies at relatively low voltage levels. For an alternative discussion of these aspects of interfacing, relevant to chemistry, consult the text:

E. Morgan, *Laboratory Computing*, Sigma-Technical Press, distributed by J. Wiley, Chichester, 1984.

4.1. THE APPARATUS

4.1.1. Automated Standard Additions

The basic experiment involves dispensing the solution to be analysed for sodium into a vessel containing a sodium ion electrode and a reference electrode. The voltage level between these two electrodes depends on the concentration of sodium ions in solution. For the sample solution, the concentration of sodium ion, C_1, is unknown. Suppose we have a standard solution of sodium ion made up so that the concentration of Na^+ is C_2. The method of standard additions requires us to add a known volume, V_2, of this standard solution to a known volume, V_1, of the unknown solution and, after mixing, obtain a reading from the electrode system. The unknown concentration C_1 is then given by

$$C_1 = C_2.F/(F - 1 + 10^{\Delta E/S}) \qquad (4.1)$$

Where ΔE is the difference between the voltage measured using the sample with added sodium and that using sample solution, S is a constant and F is the fraction of the solution volume due to the sample ($F = V_1/(V_1 + V_2)$).

The response of the sodium ion electrode is a function of the 'activity' of the sodium ions. The activity of an ion is primarily determined by its concentration, but it is also affected by the overall concentration of ions in the solution (the 'ionic strength').

To avoid this complication we shall assume that the standard solution of sodium ions contains a high concentration of Ca^{2+} to which the sodium ion electrode is relatively unresponsive.

Similarly we shall assume that the sample solution has been made up so that it contains the same high concentration of Ca^{2+}, as the standard. This simplifies our analysis considerably because the Ca^{2+} ions effectively maintain the ionic strength, of the solution to be measured, at a constant value. At a given temperature and fixed ionic strength, the electrode response depends only on the concentration of sodium ions in the solution in which the electrodes are immersed. The ionic strength does not change significantly following an addition of standard solution because the concentration of Ca^{2+} is equally high in both the unknown and standard solutions.

To a first approximation the voltage of the ion selective electrode relative to the reference electrode is given by

$$E = B + S \log_{10}[Na^+] \qquad (4.2)$$

where B is a constant and S is the same quantity which appeared in Eq. 4.1 and is a factor which is related to the response of the sodium electrode to sodium ions.

Ideally S should equal 59 mV at 25 °C but in practice it may have a somewhat different value and can vary between electrodes. The constant S needs to be determined for a given electrode system by taking measurements on two standard solutions of sodium ions with different concentrations but the same overall ionic strength.

Π Suppose we obtain a voltage of −126 mV for a solution containing 10^{-3} molar Na^{2+} and −70 mV for a solution with 10^{-2} molar Na^{2+}. What is the value of S in Eq. 4.2?

We have two equations and two unknowns:

$$E_1 = -126 = B + S \log_{10}(10^{-3}) \quad (4.3)$$

$$E_2 = -70 = B + S \log_{10}(10^{-2}) \quad (4.4)$$

Subtracting Eq. 4.3 from Eq. 4.4 gives,

$$56 = S \log_{10}(10^{-2}/10^{-3}) = S \times 1$$

$$\therefore \quad S = 56 \text{ mV}$$

Having determined S once for the electrode system, we can use Eq. 4.1 to deduce the concentration of the unknown C_1. All we have to do is measure the voltage produced when the sample alone is measured, then note the voltage after the addition of a volume of standard, V_2. The difference in voltage, ΔE, can then be inserted into Eq. 4.1 to obtain a value for C_1.

A better method is to carry out a number of additions, noting the response of the electrode system after each addition (and proper mixing!). We could then take the mean of a number of results for C_1.

We now have enough information to specify the instrumental requirements needed for this application. The apparatus must allow:

(*i*) the sample solution to be dispensed accurately into the measuring vessel. (The pre-treatment of the sample by the addition of the Ca^{2+} ions as swamping agent is not included in the automation for the sake of simplicity);

(*ii*) a method of dispensing accurate yet small volumes of the standard solution of Na^{2+};

(*iii*) a method of mixing the solutions after addition of the standard to the unknown solution;

(*iv*) a method of draining the vessel which holds the mixture of unknown solution and added standard in contact with the electrodes;

(*v*) a method of rinsing the electrodes, and the vessel containing the mixture, between analyses;

(*vi*) a method of measuring the response of the electrode system and passing on the data to the computer system.

4.1.2. Dispensing Methods

Before choosing the most appropriate methods for our application, a brief review of the methods available for dispensing liquids under computer control is in order.

Fig. 4.1a shows four methods. The peristaltic pump is perhaps the most familiar to chemists. A definite volume can be delivered by switching on this pump under computer control and then waiting for a specified time before switching it off. The longer the pump is switched on, the greater the volume delivered. Clearly this apparatus has to be calibrated to establish the relationship between the pumping time and the volume delivered.

The pump works by displacing liquid through the tube in contact with the triangular cam as illustrated in Fig. 4.1a(*i*). Problems can arise due to wear and tear in the tube. The method is usually unsuitable for non-aqueous solvents since the tubing degrades. For aqueous solutions, and provided that small volumes (say less than a cm^3)

are *not* required, this method is quite good and relatively inexpensive (typically £40 for a single channel mains operated peristaltic pump).

The gas displacement method (Fig. 4.1a(*ii*)) has been used with success to dispense liquids. It is particularly good for agressive solvents because the working parts (the gas valve here) are not in contact with the liquid being dispensed (it should be noted however, that the vapour pressures of some agressive solvents are sufficient to cause swelling of the '0' ring fittings on gas valves).

The time required to pressurise the storage bottle depends on the volume of free space in it and so the bottle is often permanently pressurised. In this case the gas valve is not required, but to control the flow of liquid and thus the amount dispensed, we would need to insert a control valve directly in the liquid line. This can be expensive; a PTFE engineered valve is required. Even high grade stainless steel valves are between five and ten times more expensive than gas valves.

In Fig. 4.1a(*iii*) we illustrate an alternative method for accurately dispensing liquids. A motor drives a threaded rod. The coupling between the syringe and the rod is free to move along the rod as it is rotated. The motor could be an ordinary DC motor which can be switched on and off under computer control. The problem here is obtaining some feedback to indicate when the motor has rotated the rod a sufficient number of times for a particular volume to be delivered. We could fit an optical sensor, for example, which would give a signal each time the motor does one revolution. An alternative method is to use a stepper motor. This is a DC motor which rotates in either direction by executing a number of discrete steps. This type of motor has several independently wound excitation coils (A, B and C say for a three-coil, or 'three-phase', motor). To produce a rotation in one direction each coil is excited (by passing current through it) in turn. Thus excitation of the coils in the sequence A,B,C,A,B will produce a rotation in one direction, but the sequence C,B,A,C,B,A will cause the opposite rotation. Electronic interface circuiting can be provided to drive a stepper motor quite simply. In some cases we need only send out a bit-pattern to indicate the direction of rotation required and the number of steps to be taken. Some motors have quite large step size (eg 4 per revolution), whereas others are much finer (eg 24 per revolution). The main advantage of using a stepper motor is that no feedback is required to ascertain how far the motor has rotated the shaft. If you use a 24-step motor and specify 24 steps, a single revolution will be obtained.

Fig. 4.1a(*iv*) shows a single shot method of delivery. Under computer control, the nitrogen gas valve can be opened to pressurise the cylinder and drive back the piston to actuate the syringe. This is only useful for fixed volumes, although repeated shots could be used to increase the volume dispensed, but only as multiples of the single shot volume.

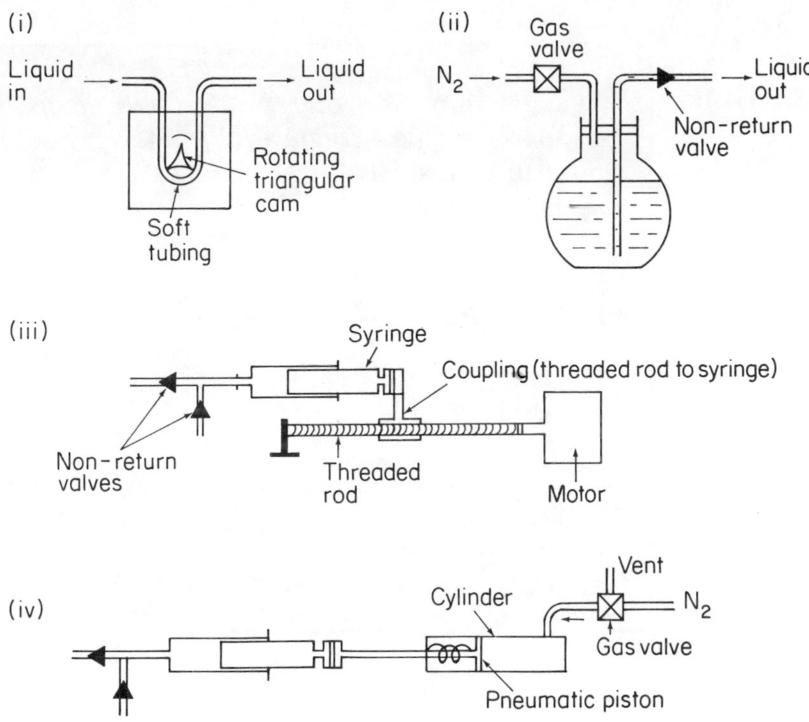

Fig. 4.1a. *Methods for dispensing liquids under computer control*

(i) A peristaltic pump *(ii) Gas displacement*
(iii) Motor driven syringe *(iv) Pneumatically operated syringe*

Finally we should be aware of the possibilities of using a balance to monitor the amount of liquid dispensed. The principle here is that the vessel into which liquid flows is weighed before and after dispensing. Problems can sometimes arise if the balance fails to 'settle' during the addition of liquid and no valid weighing is obtained. One way round this is to dispense roughly say 70% of the required volume (for example by time-based gas displacement of liquid) and then obtain the accurate weight dispensed as a means of predicting how much more liquid needs to be added. Quite precise amounts of liquid can be dispensed in this way but only at the expense of programming complexity. Another disadvantage is that the cost of the balance interfaced to a computer may be too high compared with the cost of the whole apparatus.

SAQ 4.1a

The determination of Na^+ by the method of standard additions requires the following solutions to be dispensed into the measuring vessel under computer control.

(*A*) the sample,

(*B*) standard Na^+ solution,

(*C*) rinse solution to prepare the measuring vessel for the next sample.

For each of the above solutions (*A*), (*B*) and (*C*) choose the most appropriate method of dispensing the solution, bearing in mind both the accuracy required and cost. ⟶

SAQ 4.1a (cont.)

(*i*) A peristaltic pump operating at 240 V AC.

(*ii*) A 24 volt stepper motor driving a threaded rod which in turn drives a syringe.

(*iii*) Weigh the measuring vessel as a means of monitoring the amount dispensed by gas displacement of the solution from its storage bottle.

(*iv*) A syringe capable of delivering a single shot, equal to the capacity of the syringe, at one stroke.

4.1.3. The Design of the Apparatus

First let us consider the measuring instrument. As is often the case, choice is influenced by the equipment already available. For this reason we adopt a high quality pX meter with BCD readout of either pH or the mV reading of the voltage produced by an ion selective electrode relative to the reference electrode. To achieve the required accuracy, we need to read three significant digits from the pX meter, in BCD format. The implications for the computer interface will be considered in the next section.

Another point to consider in the design is the method of draining the measuring vessel under computer control. As mentioned earlier, in-line liquid control valves are more expensive than gas valves. Also corrosion can be a problem with aqueous solutions unless the in-line liquid valve has wetted parts of high grade stainless steel or PTFE.

SAQ 4.1b

Which one of the following experimental arrangements would you use to empty the measuring vessel between measurements on different samples?

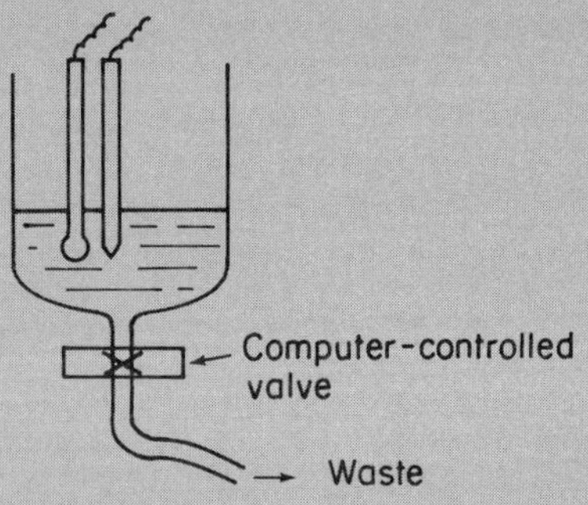

(*i*) In-line liquid control valve with gravity feed. ⟶

SAQ 4.1b (cont.)

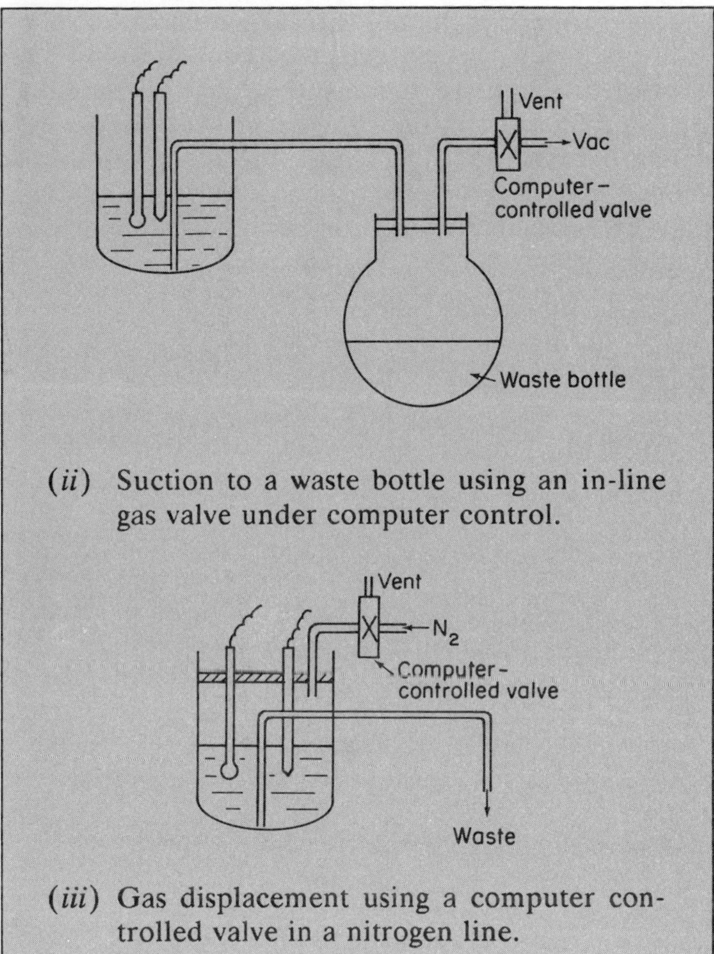

(*ii*) Suction to a waste bottle using an in-line gas valve under computer control.

(*iii*) Gas displacement using a computer controlled valve in a nitrogen line.

Another aspect of the design is how to stir the solution automatically before each measurement. It is quite possible to operate motor driven stirrers, or magnetic stirrers, under computer control. The main problem here though is the risk of damage to the electrode system. To avoid this we choose to stir the solution using a stream of nitrogen gas or air as a gas bubbler. The gas flow is stopped before each measurement and a brief period of bubbling is used after each addition of standard sodium solution.

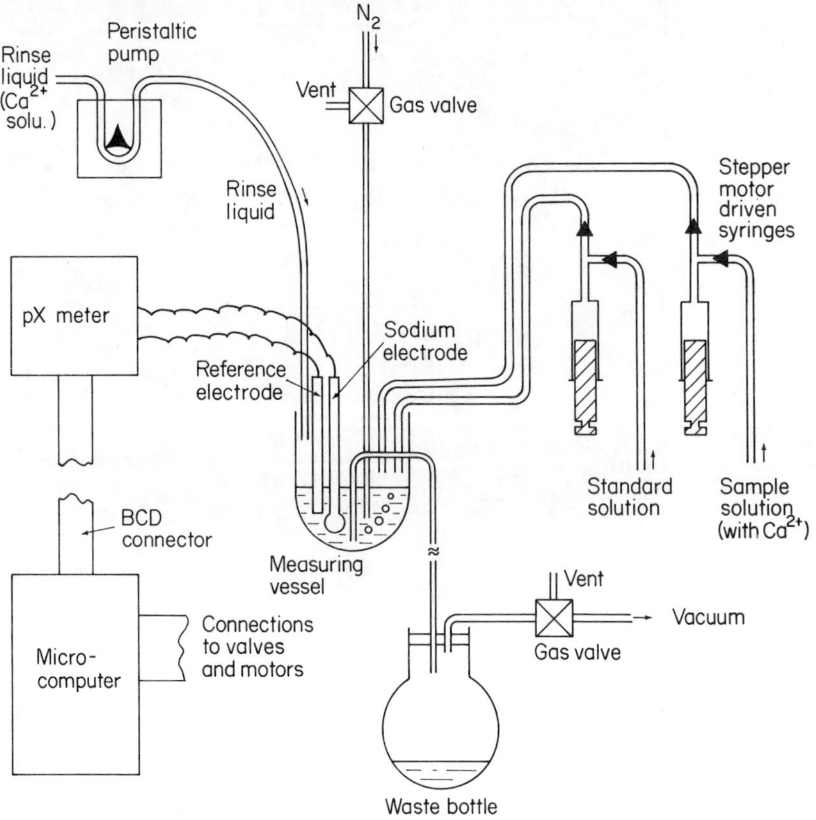

Fig. 4.1b. *Apparatus for automated standard additions to determine Na^{2+} in aqueous solutions*

We are now in a position to provide an outline design of a prototype of the whole apparatus. The experimental arrangement is shown in Fig. 4.1b. The measuring vessel has entry tubes for the sample solution, the standard solution, nitrogen gas as a stirrer, and rinse liquid pumped in from a peristaltic pump. The liquid level during the measurement must always be sufficient to cover the electrodes and during measurement the gas bubbler should be turned off. The measuring vessel has no level device fitted. Accidental flooding is unlikely however, as the waste pipe acts as a siphon if the liquid level in the vessel becomes too high.

The operational sequence using this apparatus is as follows:

(*i*) drain the previous sample solution;
(*ii*) dispense rinse liquid into the measuring vessel and drain;
(*iii*) dispense the sample;
(*iv*) measure the voltage generated between the sodium and reference electrodes;
(*v*) add a small amount of standard solution;
(*vi*) bubble air to mix the solution;
(*vii*) cease bubbling, measure the voltage produced and calculate the concentration of thc unknown solution;
(*viii*) repeat steps (*v*), (*vi*) and (*vii*) until the required number of measurements has been obtained;
(*ix*) compute the mean answer for the sodium ion concentration in the sample.

Before going on to consider how this sequence of operation may be programmed, we need to examine details of the computer interface and the selection of the electrical equipment.

4.2. THE INTERFACE

4.2.1. Devices Requiring Mains Power Supplies

Referring to Fig. 4.1b there are two valves and a peristaltic pump which could be powered from the mains (240 V, AC). It is possible to obtain gas valves which operate on 24 V DC which in general are to be preferred from the point of view of safety, particularly if the valve is likely to come into contact with aqueous solutions. To show a variety of interfacing techniques, we choose a 24 V (DC) valve to control the vacuum line, and a 240 V (AC) valve for the nitrogen supply for mixing.

The peristaltic pump can only be obtained as a mains operated device because the steady pumping action is determined by the AC frequency of the mains, by use of a so-called synchronous motor.

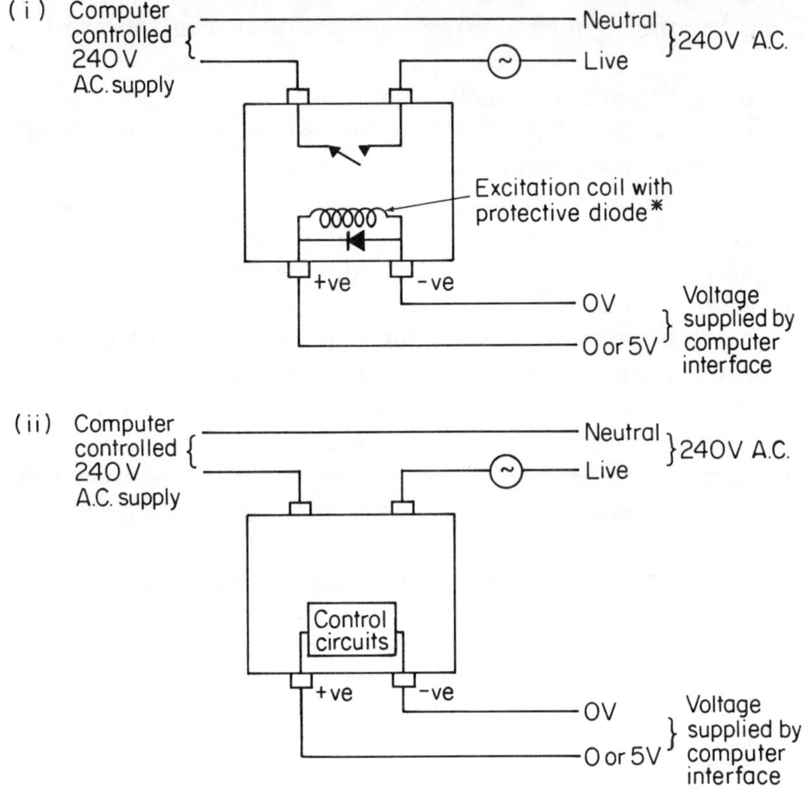

Fig. 4.2a. *Relays for switching mains operated devices on or off*

(*i*) Electromechanical type (single pole switch over).
(*ii*) Solid-state relay.

* A diode is a solid-state device which allows current to flow in one direction only.

Two possibilities arise for switching mains devices on or off under computer control.

We could use a mains relay which is schematically represented in Fig. 4.2a(*i*), or a solid-state relay as depicted in Fig. 4.2a(*ii*).

A solid-state relay accepts digital voltage levels to switch on or off the mains supply. It usually has a so-called 'zero-state crossing' principle of operation. This means that when the signal arrives from the computer to switch on the relay, the control circuitry in the relay waits until the voltage level between the live and neutral lines of the mains supply is at zero volts as the AC voltage swings between the positive and the negative parts of the mains cycle (hence the term 'zero-crossing'). This provides very smooth switching with no sudden load being placed on the mains. In contrast the electromechanical type (Fig 4.2a(*i*)) switches on the mains no matter what is the voltage level between the live and neutral lines. This can produce a spark as a sudden surge of current flows. In many applications, the risk of a spark would be unacceptable and so a solid-state relay would have to be used.On the other hand solid-state relays often suffer a leakage current so that, when they are switched off, a low power device may keep running! A final point in favour of a solid-state relay is that usually the high voltage and low voltage sides of the relay are optically isolated. This means that current flowing on the low voltage (computer) side stimulates the emission of light (from a light emitting diode*) and this is used to trigger the control circuitry. There is therefore little danger of high voltages reaching the computer's circuitry, which would of course lead to irretrievable damage.

Referring to Fig. 4.1b, we choose to operate the peristaltic pump using an electromechanical relay (to avoid the problem with a solid-state relay of the leakage current driving the motor even when the relay is switched off). To illustrate the alternative approach, we control the nitrogen gas valve, operated by 240 V AC, by means of a solid-state relay. The drain valve used operates at 24 V from the same power source as that used for the stepper motors, as will be discussed in the next section.

* A light emitting diode is a device which allows current to flow in one direction only and which emits light while conducting.

SAQ 4.2a

(*i*) Is the following statement true?

Electromechanical relays can only be used for mains voltage switching.

(*ii*) A solid-state relay can be switched on by supplying a 5 V signal with a current loading of 8 mA. Which of the following connections to a single pin of a computer output port, capable of providing 1 mA at 5 V, will allow computer control of the relay?

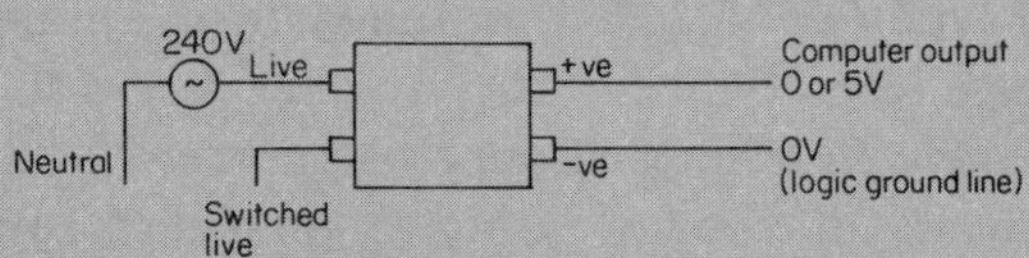

(*a*) Direct connection

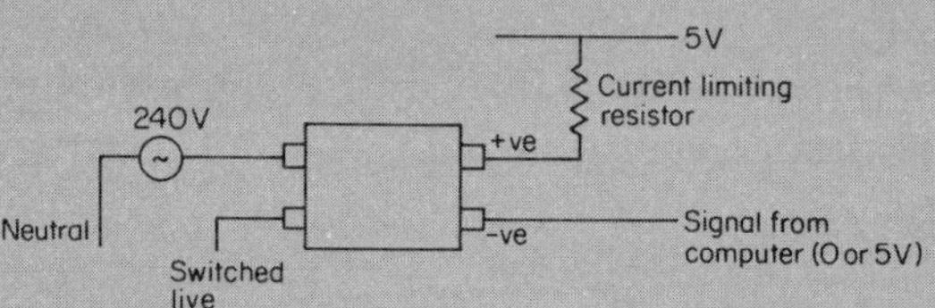

(*b*) Use of external power to provide the current to drive the input side of the solid-state relay.

SAQ 4.2a

4.2.2. Control of the DC Valve and Motors

The basic problem here is to increase the current and voltage levels produced at an output pin at the computer interface to suit the application. Let us first consider the DC valve which typically operates at 24 V DC with a current of 100 mA. If we use a programmable interface we are unlikely to provide more than about 1 mA at roughly 4.5 V. The current and voltage can be amplified using a transistor as a switch, as shown in Fig. 4.2b(*i*) and the equivalent schematic diagram in Fig. 4.2b(*ii*). Readers with some knowledge of electronics may wish to examine the circuit (*i*) in Fig. 4.2b whereas others may wish to follow the explanation given below with reference to Fig. 4.2b(*ii*).

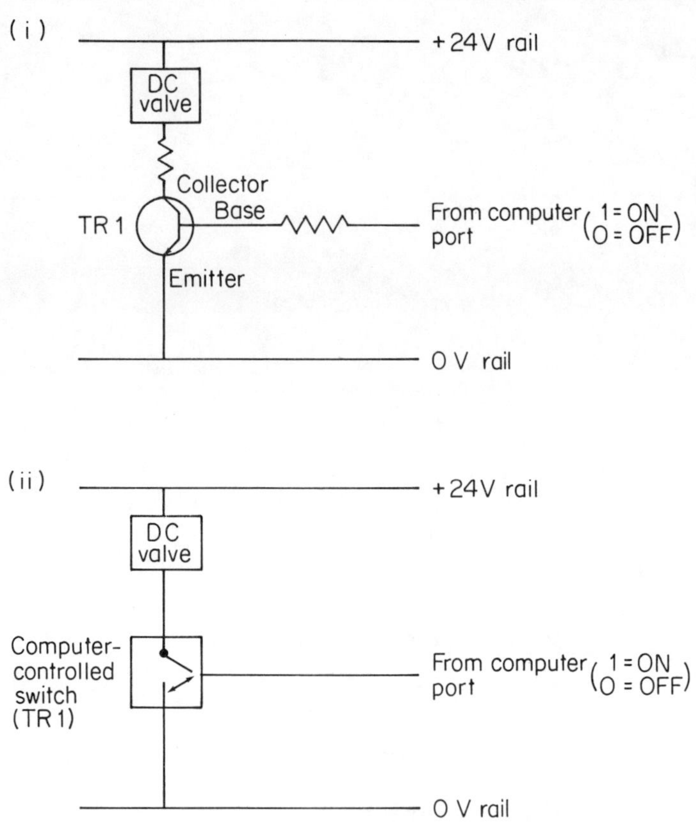

Fig. 4.2b. *(i) Circuit for control of DC valve (ii) Equivalent schematic circuit with transistor TR1 represented as a computer controlled switch*

The transistor, TR1, can be regarded as an electronic switch. If a 5 V signal, corresponding to digital '1', is output from the computer interface (for example from a single output pin of a programmable interface) to the transistor at the appropriate connector (the so-called base), then the transistor is switched to its conducting state. This means that current will flow from the 24 V supply rail, through the load which in this case is our DC valve, then through the transistor to the ground rail. If the computer interface outputs 0 V, corresponding to digital '0', the transistor is switched off and does not allow current to flow from the 24 V rail down to the 0 V rail. Thus no current flows through the excitation coil of the valve and the valve closes.

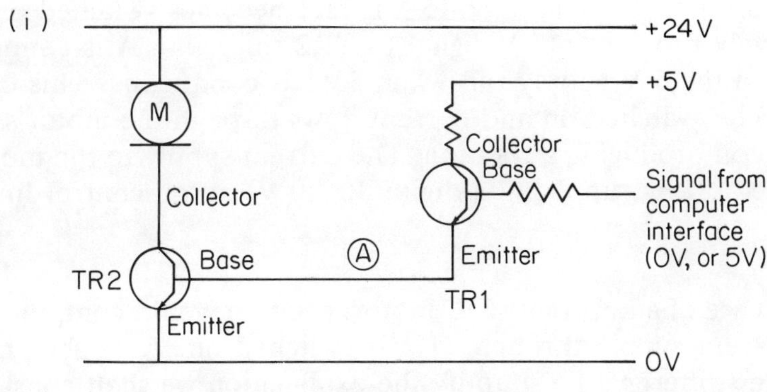

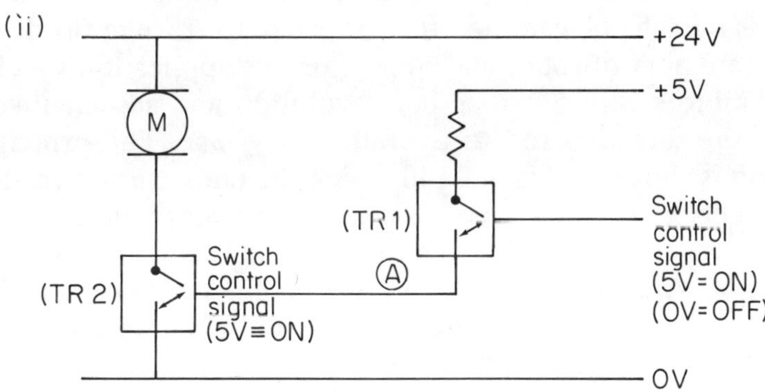

Fig. 4.2c. *(i) Circuit for supplying current under computer control to drive a small DC motor (M) (ii) Schematic diagram equivalent to (i)*

The control of the stepper motors can similarly be achieved by using transistors as computer controlled switches. The problem is that the current drawn by each excitation coil may be of the order of a few amperes and so two stages of current amplification are necessary, as shown in Fig. 4.2c. The digital electronic signal from one of the output pins on the computer interface is fed to the base of the first transistor TR1 which acts as a switch.

When a digital '1' is applied (5 V), TR1 becomes switched on and effectively provides a 5 V signal to TR2 since line A is connected direct to the 5 V supply rail when TR1 is conducting. This causes TR2 to be switched on and current flows through the motor's excitation coil producing a rotation. The current supply to the motor is switched off by supplying a digital '0' (0 V) at the control line for TR1.

In the case of an ordinary DC motor, rotation would continue until the current supply through TR2 is switched off. A stepper motor is rather different. To simplify the explanation we shall consider a stepper motor with 3 excitation coils A, B and C. (Stepper motors with 4 excitation coils are also readily available). Each of the three coils is controlled separately through a circuit similar to that given in Fig 4.2c(*i*). Each coil can therefore be excited in turn to produce the required number of rotational steps. For this application we choose stepper motors with 24 steps per revolution and sufficient torque to drive the threaded rod to operate the syringe. The principle of operation is shown in Fig. 4.2d in which the coils are shown excited in sequence. Each time a coil is excited the magnetic core of the motor moves to the position required by the excitation coil. In Fig. 4.2d(*i*), coil A is excited, in (*ii*) coil B is the only one excited, and in (*iii*) C alone is excited. The arrow in the diagram indicates the orientation of the motor at these stages. Notice that excitation of coil A following C results in continuous rotational motion.

All the computer has to do to drive the stepper motor is provide the appropriate sequence of coil excitations. Remember that we have to allow time for the rotor to move, so that we would need a sequence such as:

(*i*) excite coil A and wait for 0.5 s,
(*ii*) excite coil B and wait for 0.5 s,
(*iii*) excite coil C and wait for 0.5 s,
(*iv*) excite coil A and wait for 0.5 s.

In this case study we shall actually drive the stepper motors through such a sequence, directly under computer control. It is important to realise that stepper motor driver circuits, which accept data such

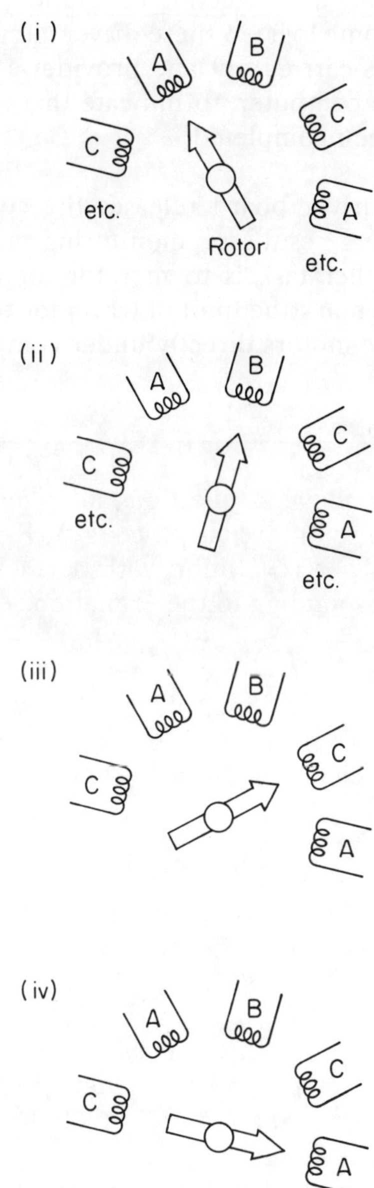

Fig 4.2d. *Principle of operation of a 3-coil stepper motor illustrated by the sequence: (i) Coil A excited (ii) Coil B excited (iii) Coil C excited (iv) Coil A excited*

as the number of pulses and direction of rotation, are commercially available. The electronic logic in these driver circuits keeps track of the number of pulses carried out and provides a signal, which can be monitored by the computer, to indicate that the task (eg rotate left 48 pulses) has been completed.

The use of such a driver board releases the computer for other things, like calculating results or monitoring another process. In our case the only other task is to monitor for a signal from the microswitches which mark the limit of travel for the syringes and so we opted to drive the motors directly under computer control.

SAQ 4.2b

The following questions all refer to a stepper motor with excitation coils A, B and C, and 24 steps per revolution with a clockwise rotation corresponding to the excitation sequence A,B,C etc.

(*i*) Which one of the following sequences would provide one quarter of a turn anticlockwise from a point where coil C is energised?

(*a*) ABC ABC ABC ABC
(*b*) CBA CBA CBA
(*c*) BAC BAC
(*d*) CAB CAB

(*ii*) What time delay is required between changes in coil excitation to give a clockwise rotation of 2 revolutions per minute?

(*iii*) What problem could arise if the time delay between excitation of stepper motor coils, in a rotation sequence, is too short?

SAQ 4.2b

Given the appropriate digital electronic signals you should now be able to understand how the apparatus can operate under computer control. In the following sections we shall examine the feedback signals from the apparatus to the computer needed to complete the automated operation of the apparatus.

4.2.3. Microswitches

When driving the syringes under computer control we need some signal to feed back to the computer to indicate whether or not an end stop has been reached.

The design we choose for driving the syringes using the stepper motor is shown in Fig. 4.2e.

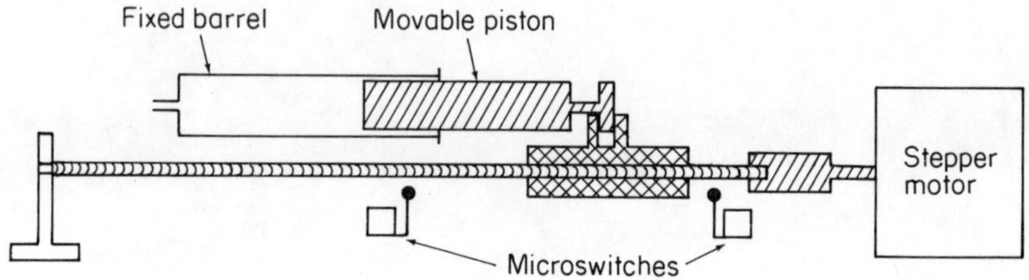

Fig. 4.2e. *Arrangement of microswitches to mark limit of travel of the syringe piston*

The piston moves backwards and forwards whilst the barrel is fixed. A microswitch is fitted to the trigger when the piston is withdrawn to the point where the syringe is full, and another one is fitted to indicate when the syringe is empty. These microswitches provide digital electronic signals that are at logic '1' (5 V) if the switch is open and logic '0' (0 V) if it is closed, as shown in Fig. 4.2f.

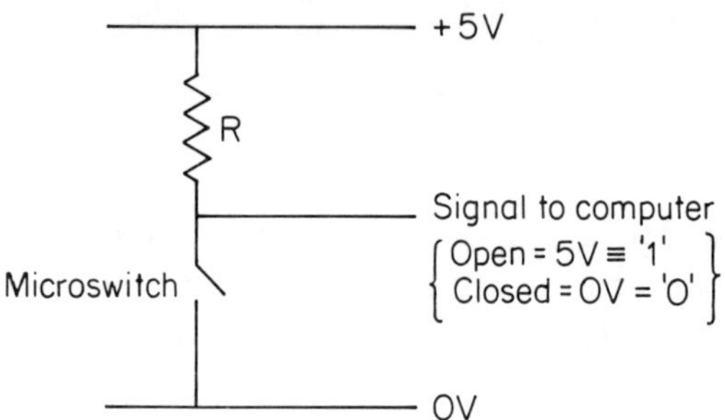

Fig 4.2f. *The use of a microswitch to generate a digital signal ('1' or '0')*

When the switch is open the computer signal is taken straight from the 5 V supply rail via the resistor R. Negligible current flows and so there is little voltage drop across the resistor.

The voltage available as a signal to the computer interface is therefore 5 V which is read as a digital '1'. When the switch is closed, the signal line to the computer is directly connected to the 0 V rail and therefore produces a logic '0' signal.

We have four microswitches, two for each syringe, which means we need to plan for four lines of input from the microswitches alone. The next two sections will examine the total requirements of input and output bits in connection with the choice of interface to handle the digital signals needed to drive the apparatus and read the data.

SAQ 4.2c

Which one of the following arrangements would generate an electrical signal at X which would be equivalent to a digital '0' at the limit of travel of the syringe barrel (not the piston) when the syringe is full.

(*i*)

(*ii*)

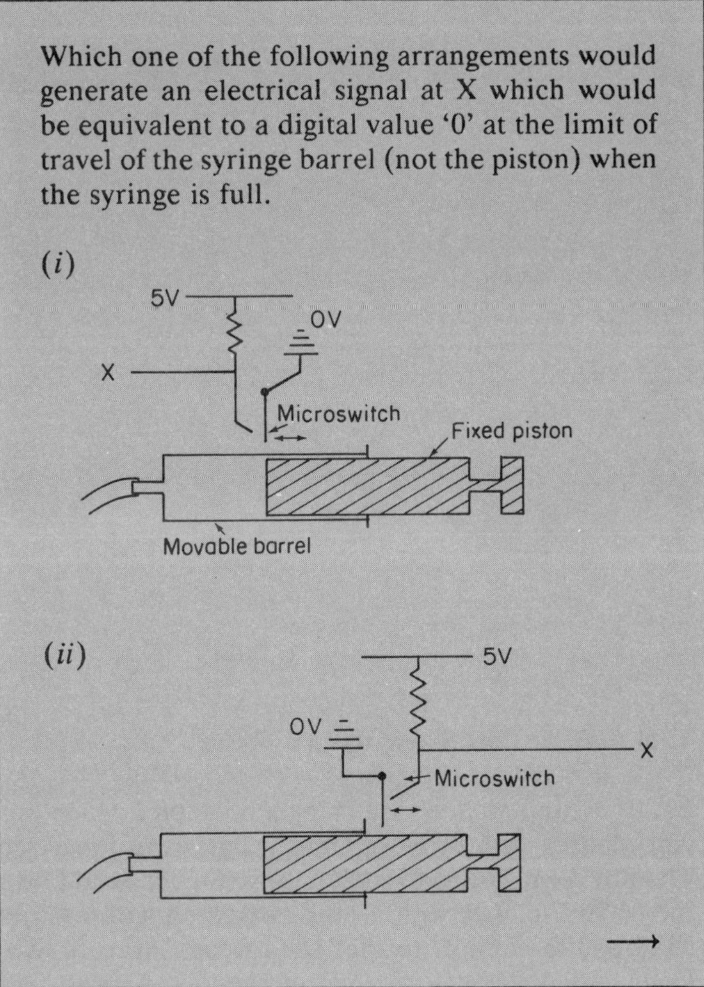

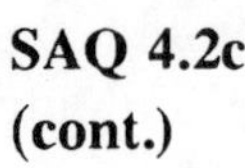

SAQ 4.2c (cont.)

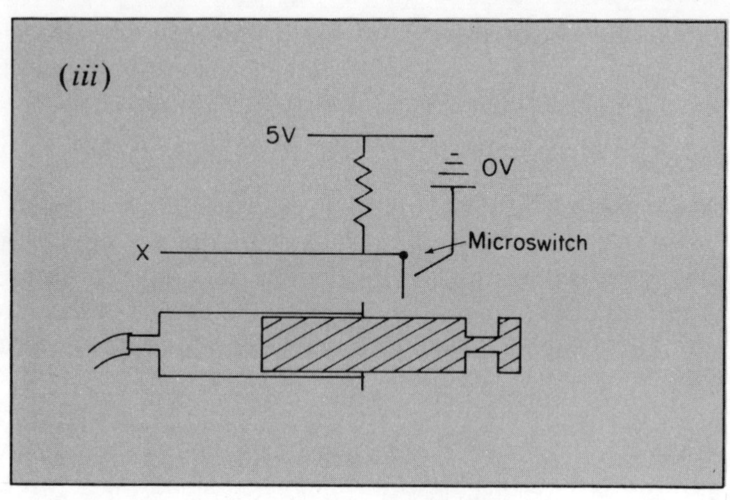

4.2.4. BCD Data from the pX Meter

Let us assume that we are using a pX meter which provides a millivolt reading with three significant figures in binary coded decimal. The pin connections to such an instrument would be similar to that shown in Fig. 4.2g with 4 pins associated with each BCD digit. We will need to connect to the 'Data Ready' line, to test that the analogue to digital conversion is completed and the BCD datum is valid.

All of the voltages measured will be negative as the pNa electrodes is negative with respect to the standard calomel electrode used as reference for the concentration of sodium ions of interest. The negative sign of the measured voltage is indicated by a logic '1' at pin 15 on the pX meter interface.

millivolt reading	digit 3	digit 2	digit 1
	hundreds	tens	units

Pin connection at pX meter	*Assignment*
1, 2, 3, 4	Digit 1 (pin 4 is most significant bit)
5, 6, 7, 8	Digit 2 (pin 8 is most significant bit)
9, 10, 11, 12	Digit 3 (pin 12 is most significant bit)
13	Digital ground
14	Data Ready
15	negative reading
16	millivolt reading

Fig. 4.2g. *BCD connections to a typical pX meter with 3 significant digits (Most instruments would provide additional output appropriate to pH measurements)*

To obtain a millivolt reading using the information output by the pX meter through the pin-out shown in Fig. 4.2g the following sequence is necessary.

(*i*) The computer scans the 'Data Ready' signal until a logic '1' is detected.

(*ii*) Read digits 1 and 2 as one byte (8 bits) of computer input.

(*iii*) Read digit 3 and the individual 'bits' corresponding to the sign of the signal and the indicator for a mv reading.

Ion selective electrode measurements tend to be slow on the scale of time associated with computers, so that there is no need to insert a 'Hold-the-data' step between (*i*) and (*ii*) above (in fact there is no 'Data Hold' control line available in the pin-out given in Fig. 4.2g).

SAQ 4.2d

If the pX meter gives a reading of −128 mV, which one of the following sets (A,B,C,D) of signals at the output pins of the meter (see Fig. 4.2g) is correct?

Pin number at pX meter	*Digital values* A	B	C	D
1	1	0	0	0
2	0	0	0	0
3	0	0	0	0
4	0	1	1	1
5	0	0	0	0
6	0	1	1	1
7	1	0	0	0
8	0	0	0	0
9	0	1	1	1
10	0	0	0	0
11	0	0	0	0
12	1	0	0	0
13	0	0	0	0
14	0	0	0	1
15	1	0	1	1
16	1	1	1	1

4.2.5. Use of a Programmable Interface

Interfaces are required for automated operations to allow a microcomputer to control the motors and valves, to monitor the state of the syringes, and to collect data from the electrode system:

Output

(*i*) Three lines for each of two stepper motors
(*ii*) One line for the drain valve
(*iii*) One line for the air valve
(*iv*) One line for the rinse pump

Total number of output lines = 9

Input

(*i*) One input line from each of the four microswitches which monitor the syringes

(*ii*) Sixteen input lines from the digital meter:
- Four for each of three BCD digits
- One to indicate 'data valid'
- One to indicate sign (+ or −)
- One to indicate mode of instrument (pH or mV)
- One for the fourth digit (0 or 1) when in the pH mode.

Total number of input lines = 20

The total requirement of lines could be meet by two 8255 PPIs or two 6522 VIAs but to broaden our discussion we take one 8255 and one 6522 and assign the ports and bits as shown in Fig. 4.2h. Two ports, A and B of the 8255, have all pins configured for input while port C is configured part for input and part for output. The 6522 has two ports, normally designated as A and B. We shall use the second of these and refer to it in the program as Port D.

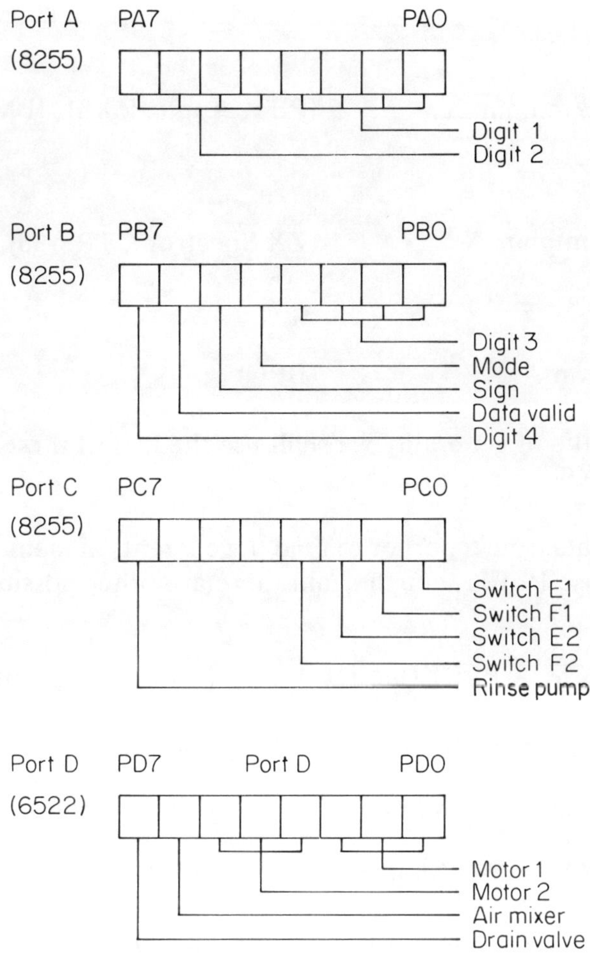

Fig 4.2h. *Port and pin assignments*

As far as a BASIC program is concerned sending control data through a port is similar to placing bytes in memory locations. The program statement to place an integer X% in location mmmm depends on the particular computer, typical examples being:

```
LET X% = 1

POKE mmmm,X%           (PET, Apple, ZX81, IBM)

LET X% = 2

OUT mmmm,X%            (ZX Spectrum, TRS-80)

LET X% = 4

?mmmm = X%             (BBC)
```

For the sake of generality we shall use the first of these, *viz*, POKE mmmm, X% in our programs.

To read data from a port or to read the contents of memory locations we shall use PEEK though again there are other possibilities:

```
LET X% = PEEK(mmmm)        (PET, Apple, IBM)

LET X% = PEEK mmmm         (ZX81)

LET X% = IN mmmm           (ZX Spectrum)

LET X% = INP(mmmm)         (TRS-80)

LET X% = ?mmmm             (BBC)
```

For our purposes we want an 8255 to be configured with Ports A and B as input, pins PC0–PC3 of Port C as input and pins PC4–PC7 as output. As shown in Section 3.2 this requires the control word 147 to be sent to the *control port* of the 8255. Similarly, to configure a 6522 VIA so that all its port B (our Port D) pins carry output signals the control word 255 must be sent to the *Data Direction Register B* (DDRB) of the 6522. Let us represent the addresses of the Ports A–D by PA, PB, etc., the address of the 8255 control port by P1 and the address of the 6522 DDRB by P2. The correspondence between these symbolic addresses and the interface chips may be summarised:

Address	*Chip Port or Register*
PA	8255 Port A
PB	8255 Port B
PC	8255 Port C
PD	6522 Output register B
P1	8255 Control Port
P2	6522 Data Direction register B

At the start of a program (initialisation) the actual address for PA, PB, etc must be assigned and the ports configured by writing to the appropriate addresses. The method of doing this depends on the particular computer and though we are keeping our program as general as possible, two examples might be helpful.

Using the syntax of TRS-80 BASIC and assuming that P1 corresponds to address 195, the statement for configuration is:

```
OUT 195,147    or    OUT P1,147
```

Once configured, the ports may be addressed; eg, to switch on the peristaltic pump by setting bit 7 of Port C (see Fig. 4.2h):

```
OUT 194,128    or    OUT PC,128
```

or to read Port B:

```
X% = INP(193)    or    X% = INP(PB)
```

(Here 193 and 194 are taken to be equivalent to PB and PC).

The BBC machine has a resident 6522 VIA and though the manual advises the use of certain operating system subroutines to communicate with external devices this may not always be necessary. As an example it may be possible to configure the 6522 for output by taking &FE62 as address P2 and writing 255:

```
?&FE62=255    or    ?P2=255
```

Control signals may then be sent to Output Register B at address &FE60 (our address PD). Thus, to operate the air mixer valve by setting pin PD6:

```
?&FE60=64    or    ?PD=64
```

It must be emphasised that these examples are only applicable to particular machines though the principles involved are the same for all computers and interface chips.

4.3. PROGRAM FOR STANDARD ADDITION

4.3.1. Outline of Program

In this section we develop a program for automated analysis by the method of standard addition. In this method a measured volume of sample is taken, a *reading* (eg emf, absorbance, polarographic current) is made, and then a standard solution is added to the sample one or more times, *readings* being taken after each addition. The calculation depends on the fact that there exists a known relationship between the *reading* and the concentration of substance being analysed.

In our case concentration is indicated by the potential of an electrode and so the readings are of electromotive force. A supporting electrolyte is to be added by means of a peristaltic pump and the sample and standard are to be added by syringes driven by stepper motors. The logical structure of a suitable program can be summarised:

Initialisation

(*i*) Assign addresses
(*ii*) Set motor controls
(*iii*) Set meter to read millivolts
(*iv*) Charge syringes

Determination

(*i*) Drain previous solution
(*ii*) Rinse vessel with supporting electrolyte
(*iii*) Dispense sample of unknown solution
(*iv*) Mix
(*v*) Measure emf

Repeat several times:

Add standard solution
Mix
Measure emf
Calculate unknown concentration

Terminate

Calculate and report mean concentration

The actual volumes or the number of motor steps made in dispensing the sample and standard solution must be known because the calculation depends on these volumes.

The emf values must be measured accurately because the emf is related linearly to the logarithm of the concentration. The program will measure each emf four times and take a mean.

In constructing a complete program it is usually best to write procedures or subroutines for each element or operation. It is a matter of choice whether one uses procedures or subroutines and we shall use the latter. Writing separate subroutines for each different element of the overall program makes for a legible program and usually yields a bonus in that the subroutines can be used virtually unchanged in other programs.

4.3.2. The Syringes

Two 3-phase stepper motors are employed to drive syringes, the direction of rotation depending on the switching sequence. Forward motion is achieved by applying high or low voltage signals to the three drive transistors in the following sequence:

Drive Transistor:	C	B	A
Step 1	LOW	LOW	HIGH
Step 2	LOW	HIGH	LOW
Step 3	HIGH	LOW	LOW

The coils are switched on in the order A ... B ... C ... A ... B ... for forward motion and a reversal of this order (ie C ... B ... A ... C ...) produces reverse motion.

In the present application the signals to activate the coils are sent by a microcomputer through an interface but it should be noted that these signals only *control* the motor current. Since the motor requires a higher voltage (eg 18–24 V) and draws more current (*ca* 0.5 A) than is available from a computer directly, the power supply for the motor and the transistor switching arrangement or *driver* are separate from the computer. Some stepper motor drivers include a unit which provides the proper switching sequence so that a step results when the potential of a single terminal is changed from 0 to 5 volts, the direction of rotation depending on the potential of another terminal. This kind of device is useful if there is a shortage of pins on the interface port.

In general terms, the microcomputer selects a stepper motor and then sends a signal through the interface to the driver of that motor; the driver switches the coils of the selected motor:

```
                          ┌──→ Driver ──→ Motor 1
Micro ──→ Interface ──→ ──┤
                          └──→ Driver ──→ Motor 2
```

(Note the direction of the signals)

If pins PD0 to PD2 of the 6522 VIA (see Fig. 4.2h) are connecting to drive transistors A, B and C of motor 1 then forward rotation of this motor is achieved by placing high (1) or low (0) signals on the pins in the following sequence:

	PD2	PD1	PD0
Step 1	0	0	1
Step 2	0	1	0
Step 3	1	0	0

Now, a port can be regarded as a memory location of the computer and placing signals on pins corresponds to writing the correct byte to the location. Provided we are only concerned with pins 0 to 2 the sequence of steps shown above requires the bytes to have decimal values 1, 2 and 4 for steps 1, 2 and 3 respectively. The actual program statements depend on the computer but, as discussed above, we shall use POKE and refer to the memory location associated with Port D as PD. Forward rotation of motor 1 can therefore be obtained by repeating the following program segment as long as rotation is required:

```
POKE PD,1

POKE PD,2

POKE PD,4
```

(Line numbers have been omitted meantime.)

The reverse sequence produces reverse motion.

In actual practice it is usually necessary to move the motor a small number of steps at a time and to monitor another line or pin after each movement. It is therefore convenient to define a subroutine or procedure which produces a number of steps and which keeps track of the position of the motor. A method of doing this for a single step of motor 1 employs an integer variable, M1%, as a control variable:

```
M1%=2*M1%

IF M1%>4 THEN M1%=1

POKE PD,M1%
```

Before these statements are executed for the first time M1% has the value 4. Then, when the program reaches this segment, the value 1 is output to Port D. When next called, M1% changes to 2 and this is output. Next, M1% becomes 4, is output, and the sequence re-commences. At any time the value of M1% is equal to the last value output. To produce more than one step the three statements are executed the appropriate number of times in a subroutine.

In a similar subroutine for the reverse operation, M1% is divided by 2 and a conditional statement makes value 4 follow value 1:

```
M1%=M1%/2

IF M1%<1 THEN M1%=4

POKE PD,M1%
```

For the motor controlled by pins 3 to 5 of Port D the subroutines are similar, the only difference being that the control variable, M2%, must change between the limits 8 and 32 (corresponding to bits 3 to 5). M2% must therefore be given the value 32 initially and the conditional statements changed to:

IF M2%>32 THEN M2%=8 (forward)

IF M2%<8 THEN M2%=32 (reverse)

The same effect could be achieved by using the previous two subroutines with M1% replaced by M2% and POKE PD,M1% replaced by POKE PD,8*M2%.

The state of each syringe is monitored by signals supplied by two microswitches fitted to the supports of each syringe and interfaced through Port C. One of the switches (E1) is closed when the plunger of syringe 1 is fully forward (empty), the other (F1) when it is fully withdrawn (filled). Switches E2 and F2 monitor the second syringe. These switches are connected to the 8255 interface Port C as shown in Fig. 4.2h. If the memory location associated with this port is designated PC, the extreme positions of the syringes are indicated by the contents of PC:

PC contents	*Significance*
1	E1 on. Syringe 1 empty
2	F1 on. Syringe 1 filled
4	E2 on. Syringe 2 empty
8	F2 on. Syringe 2 filled

Monitoring of these switches is simplified by the fact that we only need to check the one we are interested in at a particular time. Thus, if the contents of syringe 1 are being dispensed it is only necessary to monitor bit 0 of the byte at location PC. The best way of doing this is by means of the AND operator. Memory location PC is read by a PEEK or similar statement and the contents ANDed with 1:

LET X%=PEEK(PC)

LET Y%=X% AND 1

A value 1 for Y% indicates that syringe 1 is empty. Similarly, when syringe 1 is being refilled we AND with 2; a result Y% = 2 shows that PC1 is set, switch F1 is on and therefore refilling is complete. Similarly, for syringe 2 we AND with 4 and 8. Communication between the monitoring switches and the computer can be represented by the diagram:

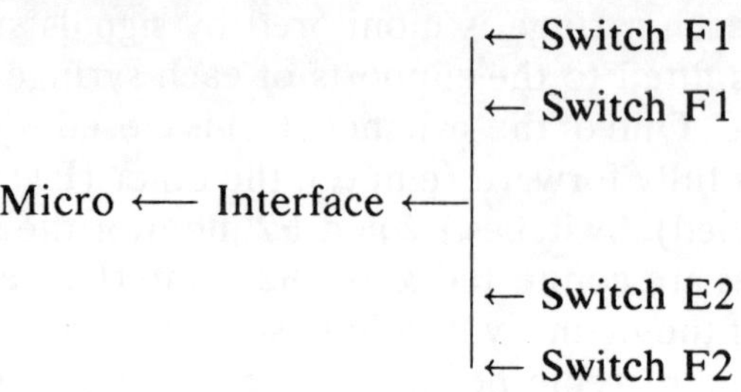

(Again note the direction of the signals)

We are now in a position to design a proper subroutine. Let us outline part of a flow diagram for dispensing from syringe 1:

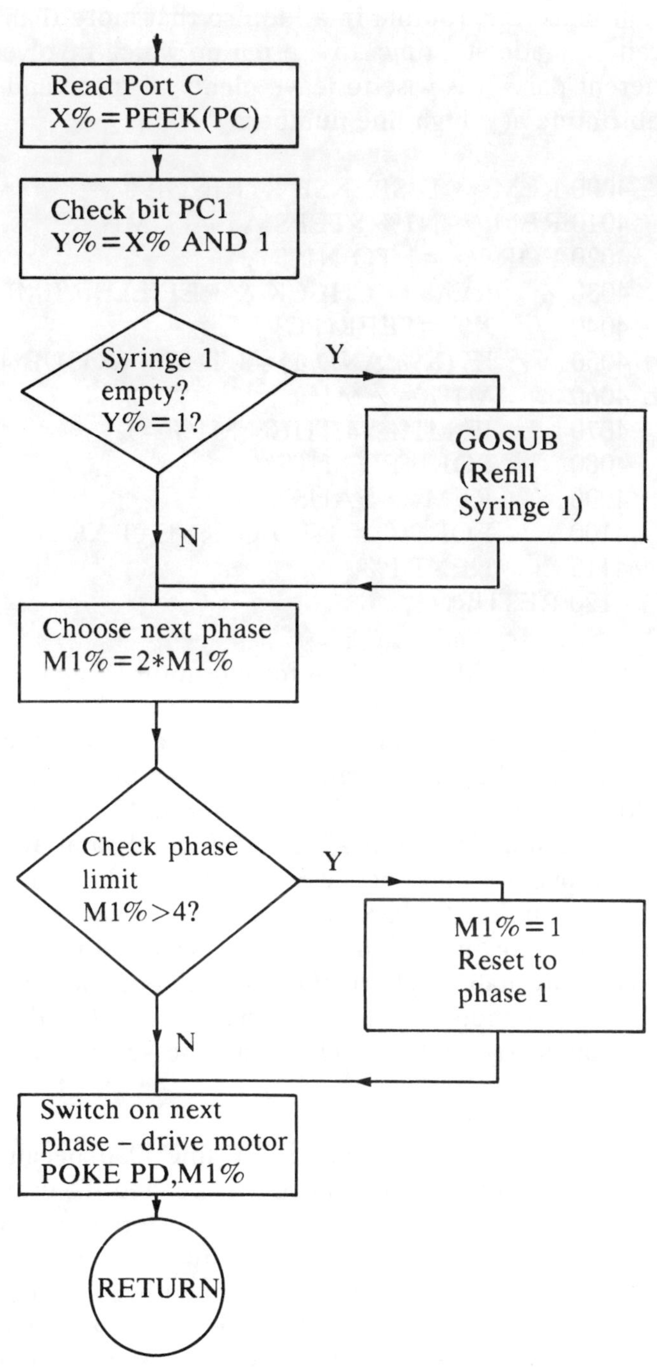
Read Port C
X%=PEEK(PC)
Check bit PC1
Y%=X% AND 1
Syringe 1
empty?
Y%=1?
Y
GOSUB
(Refill
Syringe 1)
N
Choose next phase
M1%=2*M1%
Check phase
limit
M1%>4?
Y
M1%=1
Reset to
phase 1
N
Switch on next
phase – drive motor
POKE PD,M1%
RETURN

We shall place this routine in a loop so that more than one motor step can be made at a time. In a program which involves a number of different parts it is wise to leave plenty of space and so we start the subroutine at a high line number:

```
4000 REM ** DISPENSE SYRINGE 1 ***************
4010 REM ** N1% STEPS AT A TIME
4020 FOR I%=1 TO N1%
4030      REM ** CHECK & REFILL IF EMPTY
4040      X%=PEEK(PC)
4050      IF (X% AND 1)=1 THEN GOSUB 4200
4060      M1%=2*M1%
4070      IF M1%>4 THEN M1%=1
4080      POKE PD,M1%
4090      REM ** PAUSE
4100      FOR X%=1 TO Q%:NEXT X%
4110      NEXT I%
4120 RETURN
```

(NB Make M1% = 4 on initialisation.)

The variable N1% determines the number of single steps which the motor will make before returning to the main program. This number is decided in another part of the program. The loop at line 4090 introduces a short pause of about 0.5 second to allow the stepper motor to move; without this the motor may not have turned through the step angle before the next signal arrives. Some versions of BASIC have a facility for introducing a timed pause within a program but if this is not available a short FOR-NEXT loop like that use here is quite a good method of passing some time. The length of the pause depends on integer Q% which should be set in the initialisation procedure.

When pin 0 of Port C is found to be set (line 4050) because switch E1 has been closed the refilling subroutine is called. To draw liquid into the syringe it is only necessary to drive the stepper motor in reverse because non-return valves between syringe and and reaction vessel and between syringe and liquid reservoir ensure that the liquid flows in the correct direction. The subroutine starts with a warning that the syringe is being refilled. After every step of the motor the byte at address PC is monitored since pin 1 of Port C goes high when the 'syringe full' microswitch is closed:

```
4200 REM ** REFILL SYRINGE 1 ***************
4210 PRINT "SYRINGE 1 REFILLING"
4220 M1%=M1%/2
4230 IF M1%<1 THEN M1%=4
4240 POKE PD,M1%
4250 REM ** PAUSE
4260 FOR X%=1 TO Q%:NEXT X%
4270 REM ** CHECK & REPEAT UNTIL FULL
4280 X%=PEEK(PC)
4290 IF (X% AND 2)<>2 THEN 4220
4300 PRINT "SYRINGE 1 REFILLED"
4310 RETURN
```

This subroutine is effectively a loop in which Port C is checked in each pass. A REPEAT-UNTIL structure is ideal for this kind of loop, lines 4220–4270 being repeated until X% has the value 2.

At the commencement of the operation old solution often has to be driven out and the syringes filled with fresh solution. A subroutine for this is similar to that for dispensing liquid but without the number (N1%) of steps:

```
4600 REM ** CHARGING SYRINGE 1 **************
4610 PRINT "EMPTYING SYRINGE 1 BEFORE CHARGING"
4620 M1%=2*M1%
4630 IF M1%>4 THEN M1%=1
4640 POKE PD,M1%
4650 REM ** PAUSE
4660 FOR X%=1 TO Q%:NEXT X%
4670 REM ** CHECK & REPEAT UNTIL EMPTY
4680 X%=PEEK(PC)
4690 IF (X% AND 1)<>1 THEN 4620
4700 GOSUB 4200:REM REFILL
4710 RETURN
```

SAQ 4.3a

Write the dispensing, refilling and initial charging subroutines for syringe 2 starting at line number 5000.

4.3.3. Rinsing, Mixing and Draining

The peristaltic pump for filling and rinsing the vessel, the solenoid valve to drain the vessel, and the air pump for mixing the contents, are controlled by the computer through pins of Port C and D:

	Pin	*Decimal to set*
Rinse pump	PC7	128
Drain valve	PD7	128
Air mixer	PD6	64

The rinse pump is started by sending 128 to location PC and stopped by sending zero or a number less than 128:

```
POKE PC,128:REM START RINSING

POKE PC,0:REM STOP RINSE PUMP
```

We must ensure that the pump operates long enough but not too long. The time necessary for filling can be found by experiment and a suitable pause built into the program. If the controlling computer has a timing facility the correct number of seconds can be programmed but otherwise a dummy FOR-NEXT loop can be used. The following is a filling subroutine which uses a loop:

```
6000 REM ** PERISTALTIC PUMP ************
6010 REM ** START THE PUMP
6020 PRINT "RINSING"
6030 POKE PC,128
6040 REM ** WAIT
6050 FOR X%=1 TO F%:NEXT X%
6060 POKE PC,0
6070 PRINT "RINSING COMPLETE"
6080 RETURN
```

The length of delay introduced by the dummy loop at line 6050 depends of F%. This value is assigned at the start of the program (eg F%=73160) but it must be determined by experiment *before* the program is used in earnest.

Writing a 128 to Port D operates the solenoid which opens the drain valve. Again, a delay loop is necessary to allow sufficient time for the liquid to drain away:

```
6200 REM ** DRAINING VESSEL **************
6210 PRINT "DRAINING LIQUID"
6220 POKE PD,128
6230 REM ** WAIT
6240 FOR X%=1 TO D%:NEXT X%
```

```
6250 POKE PC,0
6260 PRINT "DRAIN VALVE CLOSED"
6270 RETURN
```

The value of D% is set at initialisation (eg D% = 12000).

A short subroutine for mixing the contents of the vessel is convenient. The air pump is activated by writing 64 to Port D and a time delay can be introduced by a variable M% which, like F% and D%, is set on initialisation:

```
6400 REM ** MIXING
6410 POKE PD,64
6420 FOR X%=1 TO M%:NEXT X%
6430 POKE PD,0
6440 RETURN
```

4.3.4. Measuring Electromotive Force

As shown in Fig. 4.2h the data from the meter are received by Ports A and B as two 8-bit integers. The program can access the information by reading memory locations PA and PB. The byte at PA is read as two digits in BCD (binary coded decimal) form and a third BCD digit is taken from the four low bits of PB. Before reading these data, however, it is necessary to ensure that conversion of the emf to digital form has been completed and this is done by noting the state of pin 6 of Port B. The potential of this pin remains low while the meter is actually measuring emf and becomes high when the bytes are ready to be read (*data valid*). The state of this pin can be established by finding whether PB6 is set (1) or not (0) and this involves an AND with decimal 64. The program is made to stay in a loop until PB6 is set:

```
(line x) P%=PEEK(PB)

(line y) IF (P% AND 64)=0 THEN (line x)
```

This little loop can be written as a single line and followed immediately by the routine for reading the digits.

We shall number the digits of an emf display from 1 to 3 reading from the right; eg for an emf of 467 the digits and their values are as follows:

Digit:	3	2	1
Value:	4	6	7

Let us suppose that the digits are represented by variables X, Y and Z. Variable X (digit 3) has value 4, Y has value 6 and Z value 7. The emf could then be displayed as the *string* XYZ.

The values of X, Y and Z are found by reading memory locations PA and PB and then separating the 'nibbles' by means of the AND operator. Digit 1 (units of millivolts), represented by variable Z, is obtained from the lower four bits of the byte in location PA:

```
P%=PEEK(PA)

Z=P% AND 15
```

Digit 2 (terms of millivolts) is obtained from the same byte but we must AND with 240 and divide by 16:

```
P%=PEEK(PA)

Q%=P% AND 240

Y=Q%/16
```

Division by 16 is necessary because the digit is held in the higher four bits of the byte.

Digit 3 (hundreds of millivolts) is obtained from Port B in the same way as digit 1 from Port A.

When the three digits have been extracted from the port addresses the numerical value of the emf is calculated from the individual contributions:

```
V=100*X+10*Y+Z
```

We have still to take account of the sign of the emf. The polarity of the indicator electrode determines the state of pin 5 of Port B. This pin goes high if the indicator is negative with respect to the reference electrode and so the final step in measuring the emf involves another AND of the byte at PB:

```
P%=PEEK(PB)

Q%=P% AND 32

IF Q%=32 THEN V=-V
```

When brought together in a subroutine these program segments return the emf of the electrode combination in variable V though the operations need not be carried out exactly as we have discussed.

SAQ 4.3b

Write in correct order, but without line numbers, the statements necessary to obtain an emf value from BCD data input through Ports A, B and C.

To obtain reasonably accurate emf values the subroutine is written to take the mean of four readings. A variable, E, is used to hold the sum of four values of V and is then divided by 4 to give the mean emf:

```
3000 REM ** MEASURE EMF (4 times) **********
3010 E=0:REM SUMS READINGS & BECOMES MEAN
3020 FOR I%=1 TO 4
3030      REM ** FIRST TEST VALID FLAG
3040      IF (PEEK(PB) AND 64)=0 THEN 3040
3050      REM ** READ THE NIBBLES
3060      P%=PEEK(PA)
3070      Z=(P% AND 15)
3080      Y=(P% AND 240)/16
3090      P%=PEEK(PB)
3100      X=(P% AND 15)
3110      REM ** COMPUTE EMF NUMBER
3120      V=100*X+10*Y+Z
3130      REM ** FIND SIGN
3140      IF (P% AND 32)=32 THEN V=-V
3150      PRINT INT(V+0.5)
3160      E=E+V
3170      NEXT I%
3180 REM ** TAKE MEAN OF 4 VALUES & ROUND
3190 E=INT(E/4+0.5)
3200 RETURN
```

The value of E is the emf to the nearest millivolt.

4.3.5. Calculation

The basis of the calculation has already been discussed (4.1.1). For programming purposes it is convenient to introduce two variables, F and D:

F Fraction of the total volume due to the sample. If N1% motor steps of sample have been taken and N2% steps of standard have been added I% times then the total number of steps is N1%+I%*N2% and the value of F is given by:

F=N1%/(N1%+I%*N2%)

D Difference between the emf (E) with sample plus standard and that (E0) with sample alone divided by a constant (S) which depends on the particular electrodes and on the experimental temperature:

D=(E–E0)/S

The factor S is usually in the range 56–59 mV for a univalent ion but since it is temperature dependent we shall make allowance for the value to be supplied in the initialisation procedure. Taking C1 and C2 as the sample and standard concentrations respectively, a short calculation subroutine may be written:

```
3400 REM ** CALCULATION ***************
3410 F=N1%/(N1%+I%*N2%):REM SAMPLE FRACTION
3420 D=(E-E0)/S
3430 C1=C2*F/(F-1+10^D)
3440 C1=INT(1000*C1+0.5)/1000
3450 RETURN
```

If user-defined functions are allowed by the computer, C1 could be defined as a function instead of being calculated in a subroutine.

4.3.6. Initialisation

The full initialisation procedure must:

(*i*) Alert the operator to see that the meter is in the proper mode to read millivolts (not pH).

(*ii*) Check the times for dispensing supporting electrolyte and draining the vessel.

(*iii*) Charge the syringe.

(*iv*) Assign values to the port addresses, PA, PB, etc. and to certain variables like the number of motor steps, the number of standard additions to be made, the concentration of the standard.

(*v*) Configure the various ports.

(*vi*) Initialise the motor control variables.

(*vii*) Supply the factor S.

The way in which the ports are addressed and the means of configuration have been discussed in 4.2.5 and in Section 3.2. For the required configuration control word 147 must be sent to the *control port* of the 8255 and 255 must be sent to the *Data Direction Register* B (DDRB) of the 6522. We represent the address of the 8255 control port by P1 and the address of the 6522 DDRB by P2. At the start of the program (initialisation) the actual addresses for P1, P2, PA, PB, etc. must be assigned and the ports configured by writing to the appropriate addresses.

The initialisation procedure starts with statements to do this and we insert question marks where the assignment depends on the computer. Then follows the configuration statements and the assignment of values to other variables. After filling syringes if necessary, the program should be ready to run.

A subroutine may be used for initialisation or, since the operations involved must be carried out early, the necessary statements may be placed near the beginning of the program. Since there is a certain logical satisfaction in placing initialisation routines at the start of a program, we shall take the latter course, starting at line 1000:

```
1000 REM ** INITIALISATION *****************
1010 REM ** ASSIGN ADDRESSES (INCOMPLETE)
1020 PA=?:PB=?:PC=?:PD=?
1030 P1=?:P2=?
1040 REM ** CONFIGURE PORTS
1050 POKE P1,147:POKE P2,255
1060 REM ** DELAYS FOR FILL, DRAIN, MIX, MOTORS
1070 F%=73160:D%=12000:M%=10000:Q%=2000
1080 REM ** MOTOR CONTROLS
1090 M1%=4:M2%=32
1100 PRINT "AUTOMATIC DETERMINATION OF SODIUM ION"
```

```
1110 PRINT "CONCENTRATION BY STANDARD ADDITION METHOD."
1120 PRINT
1130 REM ** CHECK METER, ETC.
1140 PRINT "SET METER FOR millivolts."
1150 PRINT "SEE THAT SUPPORTING ELECTROLYTE"
1160 PRINT "RESERVOIR IS FULL."
1170 PRINT
1180 INPUT "CHARGE SYRINGE 1 (SAMPLE) ? (Y/N)", A$
1190 IF A$="Y"THEN GOSUB 4600
1200 PRINT
1210 INPUT "CHARGE SYRINGE 2 (STANDARD) ? (Y/N)", A$
1220 IF A$="Y"THEN GOSUB 5600
1230 REM ** VOLUMES & CONCENTRATION
1240 PRINT
1250 INPUT "HOW MUCH SAMPLE? (NO. OF STEPS)", N1%
1260 INPUT "HOW MUCH STANDARD ? (NO. OF STEPS)", N2%
1270 PRINT
1280 INPUT "CONCENTRATION OF STANDARD ? ", C2
1290 REM ** DECIDE NO. OF ADDITIONS OF STANDARD
1300 PRINT "HOW MANY STANDARD ADDITIONS"
1310 INPUT "AND MEASUREMENTS ? ", A%
1320 PRINT
1330 INPUT "ENTER FACTOR S FOR ELECTRODES ", S
1340 PRINT
1350 PRINT "NOTE SETTINGS OF DELAYS:-"
1360 PRINT
1370 PRINT "RINSE PUMP "; F%
1380 INPUT "ENTER 0 IF OK OR ENTER NEW VALUE ", X%
1390 IF X% <>0 THEN F%=X%
1400 PRINT
1410 PRINT "DRAIN VALVE "; D%
1420 INPUT "ENTER 0 IF OK OR ENTER NEW VALUE ", X%
1430 IF X%<>0 THEN D%=X%
1440 PRINT
1450 PRINT "R  RUN DETERMINATION"
1460 PRINT "I  REPEAT INITIALISATION"
1470 PRINT "Q  QUIT"
1480 PRINT
1490 INPUT "ENTER OPTION "; A$
1500 IF A$="Q" THEN STOP
1510 IF A$="I" THEN 1100
1520 IF A$<>"R" THEN 1440
```

While the program as written assigns values to F%, D%, M% and Q% the numbers used must be regarded as examples only. Actual values for a particular apparatus and computer are best found by experiment. An escape route has been built in at lines 1450–1520 in case problems are encountered during setting-up. This is always a good idea, especially when peripheral devices are involved.

Instead of printing a message asking the operator to check the mode of the meter(line 1140) the state of the meter could be found by the program reading the state of pin 4 of Port B (AND PB with 16) and staying within a loop until PB4 has the proper value to indicate a mV reading.

4.3.7. Main Routine

After initialisation the following operations take place in sequence:

(*i*) Drain previous solution
(*ii*) Rinse vessel with supporting electrolyte
(*iii*) Dispense sample of unknown solution
(*iv*) Mix
(*v*) Measure emf

Repeat several times:

- Add standard solution
- Mix
- Measure emf
- Calculate unknown concentration

Calculate and report mean concentration

To a large extent the main routine simply calls the various subroutines as required. Some other statements are included to display the progress of the operations on the screen and to report the final result:

```
2000 REM ** MAIN ROUTINE*************
2010 REM ** RINSE VESSEL
2020 GOSUB 6200:REM DRAIN
2030 GOSUB 6000:REM ADD SUPPORT
2040 GOSUB 6200:REM DRAIN
2050 REM ** TAKE SAMPLE & MEASURE
2060 GOSUB 4000:REM ADD SAMPLE
2070 GOSUB 6400:REM MIX
2080 GOSUB 3000:REM MEASURE
```

```
2090 EO=E:REM EO IS EMF WITH SAMPLE
2100 REM ** ADD STANDARD & MEASURE
2110 S=0:REM S SUMS CONCENTRATIONS
2120 FOR J%=1 TO A%
2130        GOSUB 5000:REM ADD STANDARD
2140        GOSUB 6400:REM MIX
2150        GOSUB 3000:REM MEASURE
2160        GOSUB 3400:REM CALCULATE
2170        PRINT "ADDITION "; J%
2180        PRINT "EMF="; E;
2190        PRINT "CONCENTRATION="; C1
2200        S=S+C1
2210        NEXT J%
2220 REM ** AVERAGE AND REPORT
2230 C=S/A%
2240 PRINT
2250 PRINT:PRINT "MEAN CONCENTRATION=";
2260 PRINT INT (10000*C+0.5)/10000
2270 INPUT "RUN THE PROGRAM AGAIN ? (Y/N)", A$
2280 IF A$="Y"THEN 1100
2290 END
```

The program is so written that all of the initialisation procedure need not be repeated when the program is re-run. We have chosen to return to line 1100 but any convenient point could be chosen. If the jump from line 2280 to 1100 seems rather large or if greater flexibility is required initialisation could be done by means of three subroutines, one for the addresses, ports and constants, one for the initial charging of syringes, and one for variables like the standard concentration and the number of additions to be made. This would allow the main routine to be reduced in length and also permit some latitude in selecting setting-up procedures.

SAQ 4.3c

Outline a procedure and write a program to determine the factor S of the equation

$$E = B + S*\log[Na^+]$$

if the syringes of the standard addition apparatus contain two different standard solutions.

4.4. PROGRAM FOR AUTOMATED NEUTRALISATION

4.4.1. Outline of Operations

The apparatus used for automated analysis by the standard addition technique is suitable for other automated operations. In this Section, we shall develop a program for the neutralisation and disposal of a liquid under the control of a microcomputer. Acid or alkali will be added to a liquid by the syringes and as far as possible we shall make use of the material of the previous Section.

We first look at the overall operation. The sequence of steps (after initialisation) is as follows:

(*i*) Fill the vessel with liquid by means of the peristaltic pump.

(*ii*) Measure pH.

(*iii*) Decide if adjustment is necessary and whether to add acid or alkali.

(*iv*) Operate appropriate syringes (acid or alkali).

(*v*) Repeat the measurement and adjustment routines until the pH is within prescribed limits.

(*vi*) Drain the liquid through the solenoid-operated valve.

It will be obvious that the subroutines developed in Section 4.3 for driving the peristaltic pump and draining the vessel and for operating the syringes can be used as they stand. Filling and draining times may require adjustment but this is done during initialisation. Mixing of the solutions need not be timed, the air pump being switched off and on as desired. The principal changes are in measuring pH and in making decisions about syringe operations. In the ensuing discussion we shall assume that the subroutines previously written are available.

4.4.2. Measuring pH

If pH is to be read to the second decimal place four digits are required but the most significant fourth digit will be either 0 or 1. The meter supplies three of the digits in BCD form and indicates the fourth digit through pin 7 of Port B (see Fig. 4.2h). If a pH is 10.58 the digits and their values are as follows:

Digit	4	3	2	1
Value	1	0	5	8

Provided the meter is in the pH mode the decimal point comes between digit 3 and digit 2 when the meter displays a pH value.

As in reading emf the potential of pin 6 of Port B remains low while the meter is actually measuring pH and then becomes high when the data are ready to be read but there should be no problem about sign when using the meter in pH mode. We shall represent the 4 digits of a pH number by the string WXYZ; that is, a pH (P) would be displayed in the form

P=WX.YZ

Digit 1 (1/100 of pH unit), represented by variable Z, is obtained from Port A:

```
P%=PEEK(PA)

Z=P% AND 15
```

Digit 2 (1/10 of pH unit) is also obtained from Port A but we must AND with 240 and, since the digit is in the high nibble of PA, we must divide by 16:

```
Y=(P% AND 240)/16
```

P% is the same byte as read from PA and used in evaluating variable X.

Digit 3 (units of pH) is obtained from Port B but the procedure for digit 4 is a little different because this variable (W) has only two possible values. Pin 7 of Port B is set high if W is 1 and set low if W is zero. Consequently, an AND with decimal 128 is a sufficient test:

```
P%=PEEK(PB)
Q%=P% AND 128
W=Q%/128
```

In pH units, W makes a contribution of 10 or 0.

When all four digits have been obtained from the input addresses the pH value is calculated by combining the individual contributions:

```
P=10*W+X+Y/10+Z/100
```

In the following subroutine the air pump used for mixing is stopped for a short time before a reading is attempted:

```
3000 REM *** MEASURE PH **********************
3010 POKE PC,0:REM STOP AIR MIXER
3020 FOR X%=1 TO M%:NEXT X%:REM WAIT
3030 REM ** TEST VALID FLAG
3040 IF (PEEK(PB) AND 64)=0 THEN 3040
3050 REM ** READ THE NIBBLES
3060 P%=PEEK(PA)
3070 Z=(P% AND 15)
3080 Y=(P% AND 240)/16
3090 P%=PEEK(PB)
3100 X=(P% AND 15)
3110 REM ** GET DIGIT 4
3120 W=(P% AND 128)/128
3130 REM ** COMPUTE THE PH VALUE
3140 P=10*W+X+Y/10+Z/100
3150 POKE PC,1:REM START MIXER
3160 RETURN
```

At line 3020 we have used variable M% to introduce a short pause after stopping the mixer and before reading the BCD data. This variable is not needed to control the mixing time as in the standard addition program. It must of course be given a value at initialisation.

If it is necessary or desirable to take an average of several pH readings, the contents of this subroutine (except the RETURN) may be included in a FOR-NEXT loop so that a number of values of P can be acquired, added, and averaged. For our purposes at present this is not necessary.

4.4.3. Adjusting pH

It is convenient to write separate subroutines to control the addition of acid and alkali. Suppose that we know by the dimensions of the syringes and the concentrations of the solutions that acid can be added in 10 motor steps at a time when the pH is higher than 8 but in only 2 steps at a time when the pH is lower than 8. The corresponding figures for alkali might be different, say 25 steps for a pH less than 6 but 5 steps for higher pH values.

```
3200 REM ** PH TOO HIGH **********************
3210 REM ** DECIDE ADDITIONS (N1%)
3220 IF P>8 THEN N1%=10 ELSE N1%=2
3230 GOSUB 4000:REM ADD ACID
3240 GOSUB 3000:REM MEASURE pH
3250 REM ** PH LOW ENOUGH?
3260 IF P>7.2 THEN 3220
3270 RETURN
```

This kind of procedure is suited to a REPEAT-UNTIL structure. The values of N1% (10 or 2 here) are chosen to allow as close control as is desired. If these values are likely to change often it may be convenient to use variables whose values can be assigned in the initialisation routine.

The subroutine for adding alkali or base is very similar.

SAQ 4.4a

> Write a subroutine starting at line 3400 for adding alkali when the pH is too low.

SAQ 4.4a

4.4.4. Initialisation

That part of the initialisation procedure which assigns addresses and configures the ports is the same as in the standard addition program. The remainder of the setting-up is simpler than in that program because no calculations are involved and therefore concentrations are not required. It is assumed also that no allowance need be made for temperature fluctuations.

```
1000 REM ** INITIALISATION *********************
1010 REM ** ASSIGN ADDRESSES
1010 PA=?:PB=?:PC=?:PD=?
1030 P1=?:P2=?
1040 REM ** CONFIGURE PORTS
1050 POKE P1,147:POKE P2,255
1060 REM ** DELAYS FOR FILL, DRAIN, MOTORS, PAUSE
1070 F%=73160:D%=12000:Q%=2000:M%=5000
1080 REM ** MOTOR CONTROLS
1090 M1%=4:M2%=32
1100 PRINT "AUTOMATIC NEUTRALISATION AND DISCHARGE"
1110 PRINT
1120 REM ** CHECK METER, ETC
1130 PRINT "STANDARDISE pH METER AND"
1140 PRINT "CHECK SYRINGES"
1150 PRINT
1160 INPUT "CHARGE SYRINGE 1 (ACID) ? (Y/N)",A$
1170 IF A$="Y"THEN GOSUB 4600
```

```
1180 PRINT
1190 INPUT "CHARGE SYRINGE 2 (ALKALI) ? (Y/N)",A$
1200 IF A$="Y"THEN GOSUB 5600
1210 PRINT
1220 PRINT "NOTE SETTING OF DELAYS:-"
1230 PRINT "SAMPLING PUMP";F%
1240 INPUT "ENTER 0 IF OK OR ENTER NEW VALUE",X%
1250 IF X% <>0 THEN F%=X%
1260 PRINT
1270 PRINT "DRAIN VALVE";D%
1280 INPUT 0 IF OK OR ENTER NEW VALUE",X%
1290 IF X%<>0 THEN D%=X%
1300 PRINT
1310 PRINT "R RUN NEUTRALISATION"
1320 PRINT "I REPEAT INITIALISATION"
1330 PRINT "Q QUIT"
1340 PRINT
1350 INPUT "ENTER OPTION";A$
1360 IF A$="Q" THEN STOP
1370 IF A$="I" THEN 1100
1380 IF A$<>"R" THEN 1340
```

As mentioned in 4.3 it is possible to check the mode setting of the meter by reading PB and finding if PB4 is on or off but since a pH meter must always be standardised before use this should be unnecessary.

4.4.5. Main Routine

The central part of the program is concerned with deciding whether the pH is to be changed up or down and when the liquid is ready for discharge. Let us specify that the solution pH must be adjusted to a value between 6.8 and 7.2 before it is discharged. After setting up the system and filling the reaction vessel the first decision to be made is whether or not to take any action. All the subroutines are available and the main program therefore measures the pH, decides what to do and calls the appropriate subroutines:

```
2000 REM ** MAIN ROUTINE *********************
2020 GOSUB 6000:REM FILL VESSEL
```

```
2040 POKE PC,1:REM START MIXER
2050 FOR X%=1 TO 2*Q%:NEXT:REM WAIT
2060 GOSUB 3000:REM MEASURE pH
2070 REM ** NEUTRALISE IF NECESSARY
2080 IF P>7.2 THEN GOSUB 3200:REM ADD ACID
2090 IF P<6.8 THEN GOSUB 3400:REM ADD ALKALI
2100 IF ABS(P-7)>0.2 THEN 2070
2110 POKE PC,0:REM STOP MIXER
2120 REM ** DISCHARGE TO WASTE
2130 GOSUB 6200:REM OPEN DRAIN VALVE
2140 REM ** ROUND pH FOR REPORT
2150 P=INT(100*P+0.5)/100
2160 PRINT "ALL DISCHARGED AT pH ";P
2170 INPUT "REPEAT THE PROCESS? (Y/N)",A$
2180 IF A$="Y"THEN 1100
2190 END
```

At 2180 the program is directed to line 1100 if the operator wishes to repeat the operation. This line is in the initialisation routine.

The section of program from 2070 to 2100 is ideally suited to a REPEAT-UNTIL structure since it must be repeated until the pH lies between the specified limits. Actually, there can be danger here if some of the parameters are incorrectly chosen. If the acid and base strengths are too high or the number of motor steps taken at a time too large, the pH may oscillate from below 6.8 to above 7.2 and never be discharged. Of course, the syringes will empty and the operator be alerted but it might be a good idea to build in a safeguard such as limiting the number of times the 2070–2100 loop can be traversed. This could be done by including a 'counter' at say 2072:

```
2072 N%=N%+1

2074 IF N%>5 THEN GOSUB ...
```

The new subroutine would be designed to stop the process or sound an alarm or to alter the size of the steps N1% or N2%. Of course N% should be zero when the neutralisation routine starts.

SAQ 4.4b

Write statements which keep an operator informed of the pH and the current operation. (eg the VDU displays the message 'pH=7.8. Adding acid 2 steps at a time'.)

It is not difficult to make this program more flexible by allowing the pH limits to be specified during the setting-up procedures or to make the rates of acid or alkali addition depend on how far the measured pH is outside the set limits. If you have studied both programs, standard addition and pH control, you have probably yourself formed opinions on how they might be improved, eg by making a single statement do the work of several or by introducing other subroutines for common tasks. Provided the vessel is not too large and a third pump is available to introduce a buffer solution you might arrange for the standardisation of the pH meter to be done under computer control.

Objectives

Given a procedure for automated determination of Na^+ in aqueous solutions, using the method of standard additions, the reader should be able to achieve the following on completion of this case study:

- choose appropriate methods of dispensing liquids automatically for the standard additions method used; (SAQ 4.1a)

- choose a suitable method of draining the measurement vessel under computer control; (SAQ 4.1b)

- discuss the overall design of the apparatus for automated standard additions; (SAQs 4.1a, 4.1b)

- describe how a mains voltage device can be controlled by a computer; (SAQ 4.2a)

- describe how a DC device such as a stepper motor with three excitation coils can be controlled by a computer; (SAQ 4.2b)

- describe a method of interfacing a microswitch to a computer to act as a limit-of-movement indicator; (SAQ 4.2c)

- explain how the signals at a BCD output connector for a pX meter are related to the digital mV display; (SAQ 4.2d)

- write program subroutines for controlling devices through interface chips; (SAQs 4.3a, 4.4a)

- write program subroutines to read BCD data from a digital meter interfaced to a microcomputer; (SAQ 4.3b)

- write a program which involves subroutines for controlling devices, reading BCD data and performing calculations. (SAQs 4.3c, 4.4b)

5. Simple Programs for Curves and Peaks

Overview

In many analytical methods important measurements are often presented in graphical form and interest attaches to the shape of the graph, to the positions and heights of peaks, and to the area enclosed between a plotted curve and an axis. In Parts 3 and 4 of this unit we saw how data could be acquired and transferred to a computer and we now look at how the data may be studied by means of relatively simple BASIC programs. In the first section we consider how random fluctuations in measurements can be eliminated or their effect reduced. Subsequent sections deal with the shape of curves, with measurement of area and, finally, with a program which locates peaks and measure areas at the same time.

5.1. SMOOTHING DIGITAL DATA

5.1.1. Smoothing and Averaging

One of the first things we must do is to ensure that the points making up a plotted curve are reasonable representations of the experimental measurements and not subject to too many errors. Experimental data are always subject to errors and data collected by microprocessors are no exception. The electrical signals produced by the detection systems of instruments like spectrophotometers and chromatographs always exhibit random fluctuations or *noise* which results in a display (spectrum, chromatogram) with a rough trace. One effect of noise in analysis is to limit the detection of species present in low concentration.

There are several sources of noise but for our purposes it is convenient to think in terms of two general categories. One of these, *environmental noise*, arises through interference from external sources such as power lines and electrical apparatus in the vicinity of the instrument being used. This type can often be reduced if not eliminated by careful instrumental design and good working practice. On the other hand the second type, *random noise*, is inherent in the system and must be tolerated to some extent because it has its origin in the thermal motion of electrons or other charge carriers in the components of the instrument. One way of reducing this type of noise is to lower the temperature but this is seldom a practicable method of achieving a significant improvement. More usually a combination of hardware and software methods is employed to enhance the signal to noise ratio. Hardware methods involve the incorporation of electrical smoothing circuits which act on the noise but have little effect on the analytical signal. Software methods normally employ a computer to process the signal after it has been converted to digital form and these are the methods which we shall discuss here. We shall concentrate on relatively simple methods which can be applied to digital data collected by a microcomputer either through an input port or by means of an analogue–digital converter (ADC) connected to the recorder output of an instrument.

One of the simplest methods of improving the ratio of analytical signal to instrument noise is by making a series of measurements and taking the average. Because of its random nature the noise component of a signal tends to cancel itself out in a series of repetitive measurements. Provided noise is truly random the standard deviation of the measured signal may be taken as a measure of the noise and statistical theory shows that the standard deviation is inversely proportional to the square root of the number of measurements. Let us suppose that N measurements of the same datum are made and the N results averaged (eg the absorbance at a fixed wavelength is measured N times). When the N measurements are added the component of the sum which is due to the pure signal increases in proportion to N while the component due to the noise increases in proportion to the square root of N. It follows that the average signal/noise ratio is proportional to the square root of N. This is the theoretical justification for taking an average of several readings. Provided the signal itself does not change with time, or changes very slowly, several readings of the same datum are made and the average calculated. Of course the time interval between readings must be chosen to allow time for the reading to be taken and to avoid *aliasing* (see 3.3.1 and Fig. 3.3c). If the number of data points to be measured is large this procedure requires a lot of instrument time.

5.1.2. Ensemble Average

In principle, this method is similar to taking the average of a number of readings but, instead of sampling an amplitude several times before moving on to the next, a number of runs is made and a complete set of amplitudes is read in each run. For example, a spectrum is scanned several times and absorbances are measured at the same wavelengths each time. The results are summed in an array R(1), R(2), ..., one element for each wavelength and averages S(1), S(2), S(3), ... obtained by dividing the sum for each wavelength by the number of runs. This method yields a signal to noise improvement in proportion to the square root of the number of runs but it is very demanding of instrument time and requires that both the signals and the instrument remain very stable over successive runs.

5.1.3. Boxcar Average

Another way of smoothing data subject to random fluctuations is to replace one datum point by the average of a number of points. Take for example a spectrum represented by a set of absorbances at 2 nm intervals starting at 400 nm. The measured (raw) absorbances may be represented by R(1), R(2), R(3), ... over the range of wavelengths. To achieve some smoothing we might replace R(2) by a new point which is the average of three raw points:

$$S(2) = \frac{R(1) + R(2) + R(3)}{3} \qquad (5.1)$$

Instead of plotting the three raw points at 400, 402 and 404 nm we plot only S(2) at 402 nm.

The procedure is repeated for R(4), R(5) and R(6) to produce a new S(5) and so on. Obviously, the more points used in each average the greater is the degree of smoothing and in fact the signal to noise ratio increases in proportion to the square root of the number of points used in the averaging. It will be obvious that there must be some loss of detail in an averaging process in which one point replaces a number of points and more detail is lost when more points are used. A balance must therefore be struck between obtaining a smooth curve and being able to observe small changes.

The *boxcar* or *moving window* method can be used for real-time processing when the signal varies slowly with time or for processing signals after collection.

SAQ 5.1a

An analogue–digital converter needs 50 microseconds to complete a conversion and store the result.

How much computer time per item is required by the time-average method if the signal-to-noise ratio is to be enhanced by a factor of four over that for a single measurement?

Would ensemble averaging effect any saving in time?

SAQ 5.1a

5.1.4. The Moving Average

This method is similar to the boxcar but the set of points used in finding the average moves along only one point at a time.

You will probably find it convenient to think of a spectrum or a chromatographic peak which has been produced by sampling an amplitude (absorbance, transmittance, wave height) at regular, known intervals of wavelength or frequency or time. The amplitude is plotted as ordinate (Y) against the regularly changing variable (X) as abscissa.

Consider again the process of averaging three signals R(1), R(2) and R(3). We use Eq. 5.1 above to produce a new point S(2) but instead of then discarding R(2) we use it in calculating S(3):

$$S(3) = \frac{R(2) + R(3) + R(4)}{3} \tag{5.2}$$

The original R(3) is replaced by this new S(3). Similarly, the original

R(4) is replaced by S(4) which is the average of R(3), R(4) and R(5).

In this *moving average* method every point is influenced by the points on either side. If we want to increase the influence of the other points we average over more than three points. Similarly, we can use a weighting factor to give more weight to a particular point as in:

$$S(3) = \frac{R(2) + 2 \times R(3) + R(4)}{4} \tag{5.3}$$

The effect of this is to give double weight to the central point. The procedure has two effects apart from smoothing the data: at least one point will be lost at each end of the data set and some fine detail of the true curve may be lost. Such effects are inherent in any system of smoothing and cannot be avoided.

The moving average is relatively simple to apply and as it can provide us with some useful practice in programming techniques we shall write a program to apply the method to a set of numbers. In many cases the raw data will be held in memory locations of a computer and retrieved by a PEEK or equivalent statement but because the syntax for writing to and reading from memory is machine-dependent our examples will make use of DATA statements and usually read the data into an array. We shall always assume that graphs are produced by plotting Y values against X values and that the latter are spaced at regular intervals.

Three arrays are used. One, R(), holds the original raw data, the second, S(), holds the smoothed data, and the third, W(), is the working array which holds the data currently being smoothed. In the following program which averages over 3 points we dimension the working array for 3 data items and the other two for up to 50 data items each.

```
100 REM ** 3-POINT SMOOTHING. EVEN WEIGHTING
110 REM ** DIMENSION ARRAYS
120 DIM R(50),S(50),W(3)
130 REM ** GET RAW DATA INTO ARRAY
140 READ N%
150 FOR I%=1 TO N%
```

```
160       READ R(I%)
170       NEXT I%
180 REM ** INITIALISE WORKING ARRAY
190 W(1)=R(1)
200 W(2)=R(2)
210 REM ** NOTE W(3) STILL TO COME
220 REM ** MAIN LOOP FOR NEW ARRAY
230 REM ** ONE ITEM LOST AT EACH END
240 FOR I%=2 TO N%-1
250       REM ** BRING IN W(3) AND AVERAGE
260       W(3)=R(I%+1)
270       S(I%)=(W(1)+W(2)+W(3))/3
280       REM ** S(I%) IS A 3-POINT AVERAGE
290       REM ** NOW MOVE ALONG
300       W(1)=W(2)
310       W(2)=W(3)
320       NEXT I%
330 REM ** NEW ARRAY NOW COMPLETE
340 REM ** LET US SEE OLD AND NEW
350 FOR I%=2 TO N%-1
360       R(I%)=INT(R(I%)+0.5)
370       S(I%)=INT(S(I%)+0.5)
380       PRINT R(I%), S(I%)
390       NEXT I%
400 END
500 REM ** DATA STATEMENTS
510 DATA 47
520 DATA 4,4,7,8,8,12,18,21,22,31,36
530 DATA 40,50,57,55,69,73,75,81,91,97
540 DATA 99,97,95,94,93,95,87,85,82,78
550 DATA 67,58,56,47,45,42,35,26,24,23
560 DATA 15,9,5,6,5,2
```

The results from this program are tabulated below in a slightly different format from that obtained on running the program. The numbers are given to the nearest integer for convenience in making comparisons but the output should really be plotted. When plotting you will have to multiply the new S() values by a scale factor and this should be done before rounding the values. Alterations to lines 360–380 should be sufficient for plotting. Fig. 5.1a compares both sets of data and shows that the 3-point average has produced some smoothing of the original data.

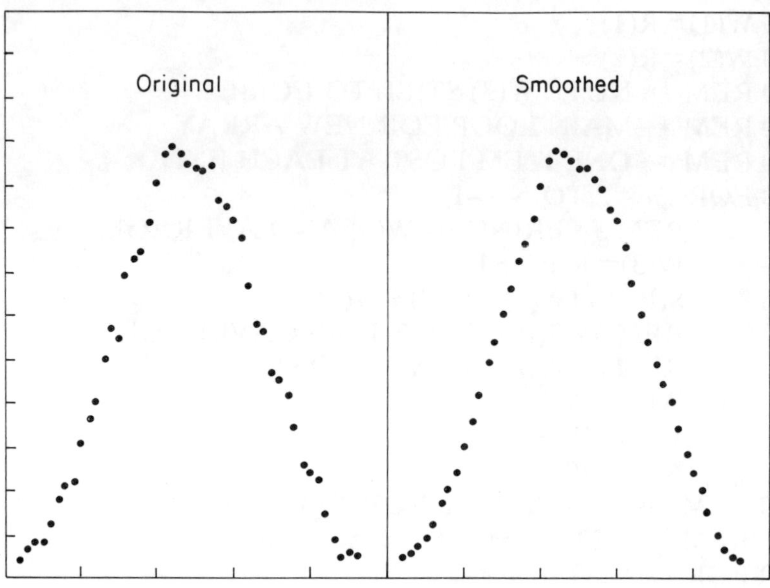

Fig. 5.1a. *3-Point average*

Original:	(4)	4	7	8	8	12	18	21
Smoothed:	–	5	6	8	9	13	17	20
Original:	22	31	36	40	50	57	55	69
Smoothed:	25	30	36	42	49	54	60	66
Original:	73	75	81	91	97	99	97	95
Smoothed:	72	76	82	90	96	98	97	95
Original:	94	93	95	87	85	82	78	67
Smoothed:	94	94	92	89	85	82	76	68
Original:	58	56	47	45	42	35	26	24
Smoothed:	60	54	49	45	41	34	28	24
Original:	23	15	9	5	6	5	(2)	
Smoothed:	21	16	10	7	5	4	–	

A word of warning may be in order here. Smoothed points are derived from raw points which are obtained by some kind of device and often the device will set limits to the accuracy with which a particular measurement can be made. Thus, if an 8-bit ADC is used to collect data the *best* that can ever be expected is an accuracy of 1 in 256. An 8-bit ADC works on a scale from 0 to 255 and a voltage which should give a reading of 57.8 on such a scale will produce a digitised reading of either 57 or 58 depending on the type of ADC employed.

When actually applying any smoothing procedure it will often be more convenient or satisfactory to send smoothed data to a recorder via a digital–analogue converter (DAC) so that a proper graph is drawn on paper. From the point of view of the computer the actual drawing of a graph through a DAC usually means placing the output data in certain memory locations. As with plotting on the screen we shall not include details of such operations in our programs because the procedures for setting up or configuring the external equipment are very device-dependent. In any event, when a program is being developed you usually want to see the data printed out or plotted on the screen before connecting up external devices.

SAQ 5.1b

Smooth the data given below by the 5-point moving average method but give double weight to the central point.

Data (as in example program)

4	4	7	8	8	12	18	21	22	31	36	40
50	57	55	69	73	75	81	91	97	99	97	95
94	93	95	87	85	82	78	67	58	56	47	45
42	35	26	24	23	15	9	5	6	5	2	

SAQ 5.1b

Before moving on let us introduce a few terms. The processes we are discussing are known generally as *digital smoothing*. We have all at one time plotted points on graph paper and then either placed a ruler on the paper to draw the 'best' straight line or, if we expected a curve, simply drew a free-hand curve through the points. Digital smoothing is simply a more scientific approach to this activity. A very readable introduction to the subject will be found in Binkley and Dessy (1979).

In the operations described in the moving average method we performed a *convolution* of the R(I%) numbers by the following steps:

A set of 3 R(I%) numbers were multiplied by a set of *convoluting integers*. These integers were 1 1 1.

The products were added and the sum divided by a *normalising factor*. This factor was 3.

To give double weight to the central point of the three we used Eq. 5.3 instead of 5.2 when computing the average. In this case the set of convoluting integers, or the convolution function, is 1 2 1 and the normalising factor 4.

5.1.5. The Least-squares Method

You have probably used the method of least squares at some time to find the best straight line through a set of points. This is a method of smoothing which may be described in the following terms:

For each value of a variable x let D be the difference between the actual Y value plotted and the Y value read from the best straight line. Then the best line is the line for which the sum of all the D^2 values is a minimum.

The line obtained by this method may be represented by an equation of the type:

$$Y = A + Bx \tag{5.4}$$

Here A and B are constants. We use the lower case x here deliberately because it is not the same as X used in plotting the points. The difference is best seen by considering the case in which we smooth by replacing 5 points spaced at regular intervals along the X axis by a single average point. In effect we take the middle point as being at $x = 0$, the two on the lower side as being at $x = -2$ and $x = -1$, and the two on the higher side as at $x = +1$ and $x = +2$. In other words x represents an *offset* from the central point. It follows from this that the constant term, A, is equivalent to the Y value of the central point since $Y = A$ when $x = 0$. For future reference you should also note that the constant B represents the gradient of the straight line defined by Eq. 5.4.

When the plotted points cannot possibly be represented by a straight line a more general application of the least-squares principle is still possible but the equation used may have to be a polynomial of higher degree. This principle can be applied to smoothing a set of data points quite effectively. The method is similar to the moving average method described above in that a group of points is taken, but the convoluting function (ie the set of integers) and the normalising factor depend on the number of points used and on the degree of the polynomial to which they are to be fitted. Tables published by Savitzky and Golay (1964) give the integer sets and normalising factors for polynomials up to degree 6 and for smoothing based on from 5 to 25 points. The derivations are outside the scope of this unit but the results are relatively simple to apply. We first decide how many points are to be used in the averaging and the degree of the polynomial. In Fig. 5.1b a small extract from the published tables for a quadratic or cubic function is shown. The integer set and normalising factor are the same for a quadratic and a cubic.

No. of points	Integer set	Factor
5	−3 12 17 12 −3	35
7	−2 3 6 7 6 3 −2	21
9	−21 14 39 54 59 54 39 14 −21	231

From Savitsky and Golay, *Anal. Chem.*, **36**, 1627 (1964)

Fig. 5.1b. *Smoothing by quadratic or cubic functions*

These numbers are used in the same way as in the program of 5.1.4. If five points are to be used the working array of the program must be dimensioned as W(5) and line 270 replaced by:

```
270 S(I%)=(-3×(W(1)+W(5)) +12×(W(2)+W(4))+17×W(3))/35
```

A greater degree of smoothing results from the use of a larger number of data points though when five points are used the smoothed data set will have four points less than the original.

SAQ 5.1c

Write and run a program to smooth the following data by the Savitsky–Golay method for a 7-point cubic smooth. If possible plot the raw and smoothed points.

Data (47 points as before):

4	4	7	8	8	12	18	21	22	31	36	40
50	57	55	69	73	75	81	91	97	99	97	95
94	93	95	87	85	82	78	67	58	56	47	45
42	35	26	24	23	15	9	5	6	5	2	

5.2. PEAK HEIGHT AND SHAPE

5.2.1. Locating Maxima

In spectroscopy, chromatography and voltammetry it is often necessary to find the position of a peak maximum. Provided the signal amplitudes are digitised the exercise is equivalent to finding which number of a series has the highest value and this can be done easily when there is only one peak. In 5.1.4 we applied the moving average method to smooth a single peak and a few extra lines in that program are sufficient to locate the highest point. Insert the following lines:

```
345 M%=1
355 IF S(I%)>S(M%) THEN M%=I%
395 PRINT
396 PRINT "MAXIMUM IS ";S(M%);" AT POINT ";M%
```

This results in the printing of a final message saying that the maximum is 98 at point 22. It should be fairly obvious how this works but only the highest peak will be noted if the points pass through more than one maximum. When an array of points can include several peaks the strategy must be a little more complex because we need to detect when one peak has been passed and the next is being approached. We shall address this problem in stages, starting with a straightforward program to locate maxima in a series of points.

It is usually best to smooth the points before testing for peaks and so the program below employs a Savitsky–Golay 5-point smoothing procedure. As each point is smoothed several tests are made:

First, the new amplitude is compared with the previous value (stored in variable H) and if the new point is found to be higher a flag, F%, is set to 1. If the new point is lower than or equal to the previous then F% is set to 2.

Flag F% therefore shows whether the points are rising (F% = 1) or not rising (F% = 2).

If F% = 1 (rising) then the index, I%, of the point being tested is stored in an array of peak points, P(). Thus, if we are expecting peak 3 then P(3) is given the value I%. The value of P(3) indicates the highest point since we started looking for peak 3.

To find if a peak has just been passed the value of F% is compared with the value found when the previous point was tested. This, the previous F%, is held in another flag, G%. If G% = 1 and F% = 2 then the points have just stopped rising and therefore the index of the peak array is increased by one to prepare for the next peak.

At the conclusion of this series of tests the current amplitude is stored in variable M and the current F% stored in G%.

As before we must leave you to insert your own procedure for actually plotting points.

```
100 REM *** PEAK LOCATION - 5 POINT SMOOTH
120 REM ** DIMENSION ARRAYS
130 DIM R(100),S(100),W(9),P%(10)
140 REM ** RAW DATA INTO ARRAY
150 READ N%
160 K% = 1:REM REFER TO PEAKS AS P(K%)
170 H = 0:REM AMPLITUDE OF LAST POINT
180 F% = 1:REM UP/DOWN FLAG. ASSUME UP TO START
190 G% = 1:REM G% REMEMBERS LAST VALUE OF F%
200 FOR I% = 1 TO N%
210       READ R(I%)
220       NEXT I%
230 REM ** INITIALISE WORKING ARRAY
240 FOR I% = 1 TO 4
250       W(I%) = R(I%)
260       NEXT I%
270 REM ** W(5) STILL TO COME
280 REM ** MAIN LOOP
290 FOR I% = 3 TO N%-2
300       REM ** FIRST BRING IN W(5)
310       W(5) = R(I%+2)
320       REM ** USE INTEGERS FOR SMOOTHING
330       S = -3*(W(1)+W(5))+12*(W(2)+W(4))+17*W(3)
340       S(I%) = S/35:REM S() IS SMOOTHED ARRAY
350       REM ** TESTING PROCEDURES
```

```
360       IF S(I%)>H THEN F% = 1 ELSE F% = 2
370       IF F% = 1 THEN P%(K%) = I%:REM STORE IF HIGHER
380       REM ** CHANGE K% IF STOPPED RISING
390       IF F% = 2 AND G% = 1 THEN K% = K%+1
400       REM ** MOVE ALONG
410       FOR J% = 1 TO 4
420       W(J%) = W(J%+1)
430       NEXT J%
440       REM ** STORE FOR NEXT CYCLE
450       H = S(I%)
460       G% = F%
470       REM ** (PLOT RAW & SMOOTH POINT IF DESIRED)
480       NEXT I%
490 REM ** K% IS TOO BIG IF WE STOP ON DESCENT
500 IF F% = 2 THEN K% = K%-1
510 REM ** LIST PEAKS DETECTED
520 FOR I% = 1 TO K%
530       PRINT P%(I%),S(P%(I%))
540       NEXT I%
550 END
600 REM ** DATA STATEMENTS
610 DATA 79
620 DATA 20,20,21,22,23,23,24,25,26,27
630 DATA 28,31,35,38,43,51,59,70,80,95
640 DATA 109,128,135,137,128,115,106,89,75,56
650 DATA 48,43,40,38,37,36,37,42,46,56
660 DATA 63,73,84,92,99,105,105,93,85,81
670 DATA 83,88,94,100,109,115,105,86,72,60
680 DATA 55,52,48,45,42,40,39,38,36,35
690 DATA 34,33,32,32,31,31,30,30,30
```

The graphical output from this program is shown in Fig. 5.2a. The text output lists the peaks and smoothed heights as:

```
24 136.1
46 105.1
56 113.3
```

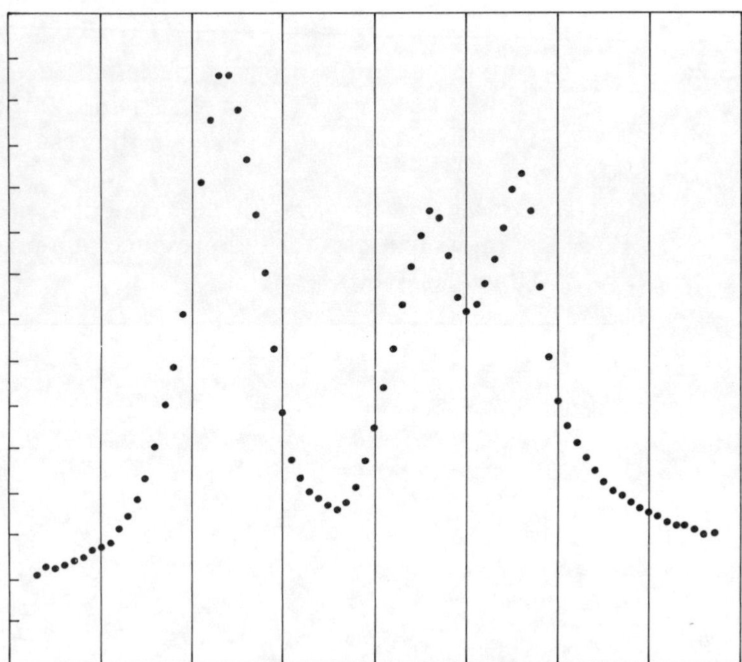

Fig 5.2a. *Peak location*

If you run this program with other data or if you use more than five points in the smoothing process you may find that the point at which a peak occurs is not always the same as the highest raw point. For example, if the same data are run in a nine-point smoothing program it will be found that the third peak occurs at point number 55 instead of at point 56 and the height is 108.2. The differences are, of course, consequences of the smoothing process and demonstrate how important it is to standardise data handling methods in analysis.

SAQ 5.2a

(*i*) Run the example program but alter the test IF F% = 2 AND G% = 1 ... (line 390) to IF F%<>G% ... and explain the result.

(*ii*) Draw a flow diagram for the segment of program which tests for the presence of a peak (lines 360–480).

5.2.2. Finding Maxima Using Derivatives

You will be aware that the first derivative of an X–Y plot indicates the gradient or rate of change of the curve while the second derivative gives the rate of change of the gradient. An array of digitised data lends itself very readily to the production of derivative plots and such plots can yield interesting information in many fields of analysis (eg spectroscopy, chromatography, potentiometric titrations, voltammetry, thermal analysis). In some cases the fourth derivative of signal amplitude with respect to time or wavelength can be profitably utilised.

We shall continue to look at peaks though the principles we shall discuss are applicable to any kind of curve and to any derivative. The only restrictions are that signal amplitudes must be measured at regular intervals and the curves be sufficiently smooth to allow the changes sought to be distinguished from noise. As far as peaks are concerned a maximum occurs where the first derivative of Y with respect to X passes through zero. This means that the gradient of the curve is zero (ie the curve is horizontal) at a maximum. The same criterion holds also for a minimum but the sign of the gradient changes from positive to negative at a maximum and from negative to positive at a minimum. Mathematically, this means that the second derivative is negative at a maximum. The last program employed these criteria simply by noting the difference in height between successive points and testing to see if the change was in the same direction (up or down) as in the previous cycle.

In that program smoothing was achieved by finding the quadratic equation which best fitted five points at a time as the program scanned through the points. The same principles can be used not only to smooth the points but also to find the derivatives and to locate accurately the position of a peak maximum. Convoluting integers and normalising factors have been tabulated for derivatives up to order 5 using polynomials up to degree 6. We shall restrict ourselves to a quadratic (ie a polynomial of degree 2) and look at the first and second derivatives only. The relevant equation is then:

$$Y = A + Bx + Cx^2 \qquad (5.5)$$

The integer sets and factors for 5, 7 and 9 points are given in Fig. 5.2b. The original papers give tables for up to 25 points.

First derivative by quadratic function

No. of points	Integer set									Factor
5			−2	−1	0	1	2			10
7		−3	−2	−1	0	1	2	3		28
9	−4	−3	−2	−1	0	1	2	3	4	60

Second derivative by quadratic function

No. of points	Integer set									Factor
5			2	−1	−2	−1	2			7
7		5	0	−3	−4	−3	0	5		42
9	28	7	−8	−17	−20	−17	−8	7	28	462

From Savitsky and Golay, *Anal. Chem.*, **36**, 1627 (1964)
Steiner *et al*, *Anal. Chem.*, **44**, 1906 (1972)

Fig. 5.2b.

As it is applied to the location and measurement of a single peak the principle of the method may be summarised:

The raw points are smoothed and the peak located using the appropriate convoluting integers as discussed in 5.1.6. Let us assume that we use five points and that point M% is found to have the largest amplitude.

We now focus attention on the five points centred on point M%. A quadratic equation is fitted to these five points and the constants A, B and C of Eq. 5.5 are calculated. The constant term, A, is the same as the smoothed height of point M%. Coefficients B and C are obtained using the first and second derivative functions respectively and the position of the curve maximum is calculated from the properties of the equation. The maximum occurs at $-B/2C$ *relative to* point M% since this is the value of x at which the first derivative of Eq. 5.5 is zero. (Differentiate 5.5 with respect to x, equate to zero, and then rearrange to $x = -B/2C$.)

Once A, B, C and the position of the curve maximum are known the height at the maximum is calculated from Eq. 5.5.

The mathematical derivation of the convoluting functions is somewhat involved but there is no great difficulty in applying the procedure. The program below applies it to the data used in the program of 5.1.4 (Fig. 5.1a) though, when we only need to locate the maximum of a peak derived from instrumental data, it may be much easier, noise and time permitting, simply to plot points more frequently. A good, fast analogue–digital converter can save much computation. That said, numerical methods have other uses and location of a peak maximum is a good starting exercise.

```
100 REM ** PEAK POSITION & HEIGHT (5 POINTS)
110 REM ** DIMENSION ARRAYS
120 DIM R(50),S(50),W(5)
130 REM ** GET RAW DATA INTO ARRAY
140 READ N%
150 FOR I% = 1 TO N%
160       READ R(I%)
170       NEXT I%
180 REM ** INITIALISE WORKING ARRAY
190 FOR I% = 1 TO 4
200       W(I%) = R(I%)
210       NEXT I%
220 REM ** W(5) STILL TO COME
230 REM ** LET M% BE INDEX OF HIGHEST POINT
240 REM ** LET M% BE 1 AT START
250 M% = 1
260 REM ** MAIN LOOP. 4 ITEMS LESS
270 FOR I% = 3 TO N%-2
280       REM ** BRING IN W(5) AND CONVOLUTE
290       W(5) = R(I%+2)
300       S = -3*(W(1)+W(5))+12*(W(2)+W(4))+17*W(3)
310       S(I%) = S/35
320       REM ** CHECK HEIGHT & CHANGE M%
330       IF S(I%)>S(M%) THEN M% = I%
340       REM ** (PLOT S(I%) AGAINST I%)
350       REM ** MOVE ALONG
360       FOR J% = 1 TO 4
370       W(J%) = W(J%+1)
380       NEXT J%
390 NEXT I%
```

```
400 REM ** NOW WORK ON SMOOTHED POINTS
410 REM ** TAKE 5 POINTS CENTRED ON POINT M%
420 J% = 1
430 FOR I% = M%-2 TO M%+2
440      W(J%) = S(I%)
450      J% = J%+1
460      NEXT I%
470 REM ** USE SMOOTHING INTEGERS FOR A
480 S = -3*(W(1)+W(5))+12*(W(2)+W(4))+17*W(3)
490 A = S/35
500 REM ** USE 1ST DERIV. INTEGERS FOR B
510 S = 2*(W(5)-W(1))+W(4)-W(2)
520 B = S/10
530 REM ** USE 2ND DERIV. INTEGERS FOR C
540 S = 2*(W(1)+W(5))-W(2)-W(4)-2*W(3)
550 C = S/14
560 REM ** SEE NOTE BELOW ABOUT THIS DIV'N
570 REM ** CALCULATE x RELATIVE TO POINT M%
580 X = -B/2/C
590 REM ** TRUE MAXIMUM IS AT P
600 P = M%+x
610 REM ** CALCULATE HEIGHT FROM EQUATION
620 Y = A+B*x+C*(x↑2)
630 REM ** LET CALCULATED HEIGHT BE H
640 H = Y
650 REM ** PRINT OUT DATA POINT & HEIGHT
660 PRINT "POINT "; M%;" SMOOTHS TO "; S(M%)
670 PRINT
680 REM ** PRINT OUT CALCULATED PEAK & HEIGHT
690 PRINT "CALCULATED PEAK IS AT ";P;" HEIGHT"; H
700 END
800 REM ** DATA STATEMENTS
810 DATA 47
820 DATA 4,4,7,8,8,12,18,21,22,31
830 DATA 36,40,50,57,55,69,73,75,81,91
840 DATA 97,99,97,95,94,93,95,87,85,82
850 DATA 78,67,58,56,47,45,42,35,26,24
860 DATA 23,15,9,5,6,5,2
```

At line 550 the constant C was obtained by dividing by 14 though the normalising factor is given as 7. In fact, the division is by 7 and by factorial 2. If we were calculating the third derivative it would be necessary to divide by factorial 3 as well as by the normalising factor.

The program reports the curve maximum as 98.8 at position 22.3. This is slightly to the right of point 22 which has a raw amplitude of 99. Plotting the data gives an output similar to Fig. 5.1a.

When we want to find a peak height accurately it is of course necessary to make proper allowance for the base line. One way of doing this is to set up an equation to represent the base line. When the position of the peak has been established it is then only necessary to calculate the amplitude of the base line at this position and subtract from the total height of the peak. A linear equation of amplitude (Y) against position (X or I%) may be all that is required for the base line equation but this will depend very much on the behaviour of the curve on either side of the peak.

SAQ 5.2b

Suggest how the position of a selected peak of a spectrum like that of Fig. 5.2a might be located precisely.

5.2.3. Plotting Derivatives

When a trace includes a number of peaks the methods just described can be combined to find the true positions and heights if these need be known precisely. Similar methods can be applied to produce graphs of the derivatives of a curve and this can yield valuable information when two or more peaks overlap. We shall consider only the first and second derivatives but you should have no difficulty in extending the method.

We use the same data as in the program of 5.2.1 but, for a change, employ 9 points instead of 5 (see Fig. 5.2b). The steps follow those of the program above very closely but the constants A, B and C are calculated for *every* point and the values of these constants are plotted as they are calculated. You will recall that B represents the first and C the second derivative at the point where $x = 0$. With 9-point smoothing there are more terms to be added when finding the sums (S) and so we evaluate the sum in stages.

```
100 REM ** 9 POINT SMOOTH AND 1ST & 2ND DERIVS
110 REM ** DIMENSION ARRAYS
120 DIM R(100),W(9)
130 REM ** GET RAW DATA INTO ARRAY
140 READ N%
150 FOR I% = 1 TO N%
160       READ R(I%)
170       NEXT I%
180 REM ** INITIALISE WORKING ARRAY
190 FOR I% = 1 TO 8
200       W(I%) = R(I%)
210       NEXT I%
220 REM ** NOTE W(9) STILL TO COME
230 REM ** MAIN LOOP
240 FOR I% = 5 TO N%-4
250       REM ** FIRST BRING IN W(9)
260       W(9) = R(I%+4)
270       REM ** USE INTEGERS FOR SMOOTHING
280       REM ** DO ADDITIONS IN 3 STEPS
290       S = -21*(W(1)+W(9))+14*(W(2)+W(8))
300       S = S+39*(W(3)+W(7))+54*(W(4)+W(6))
```

```
310     S = S+59*W(5)
320     A = S/231
330     REM ** A IS 0th DERIV (SMOOTHED)
340     REM ** USE INTEGERS FOR 1st DERIV
350     S = 4*(W(9)-W(1))+3*(W(8)-W(2))
360     S = S+2*(W(7)-W(3))+W(6)-W(4)
370     B = S/60
380     REM ** B IS 1st DERIV AT POINT I%
390     REM ** USE INTEGERS FOR 2nd DERIV
400     S = 28*(W(1)+W(9))+7*(W(2)+W(8))
410     S = S-8*(W(3)+W(7))-17*(W(4)+W(6))
420     S = S-20*W(5)
430     C = S/924
440     REM ** C IS 2nd DERIV AT POINT I%
450     REM ** NOW MOVE ALONG
460     FOR J% = 1 TO 8
470     W(J%) = W(J%+1)
480     NEXT J%
490     REM ** PLOT A, B AND C AGAINST I%
500     REM ** NEXT 4 LINES APPLY FOR BBC MICRO
510     X = 200+12*I%
520     PLOT 69,X,200+2*A
530     PLOT 69,X,600+4*B
540     PLOT 69,X,800+16*C
550     REM ** AND DO IT ALL AGAIN
560 NEXT I%
570 END
600 REM ** DATA STATEMENTS
610 DATA 79
620 DATA 20,20,21,22,23,23,24,25,26,27
630 DATA 28,31,35,38,43,51,59,70,80,95
640 DATA 109,128,135,137,128,115,106,89,75,56
650 DATA 48,43,40,38,37,36,37,42,46,56
660 DATA 63,73,84,92,99,105,105,93,85,81
670 DATA 83,88,94,100,109,115,105,86,72,60
680 DATA 55,52,48,45,42,40,39,38,36,35
690 DATA 34,33,32,32,31,31,30,30,30
```

On plotting the output a graph like Fig. 5.2c is displayed showing the smoothed data points and the first and second derivatives. Included in the program are statements for plotting the points using BBC syntax to give an idea of where they should come and to show how the scaling factors have to be adjusted for the different derivatives (we use factors of 2 for A, 4 for B and 16 for C). You will of course have to modify these statements for your own circumstances as well as insert statements to clear the screen before plotting. You will find it instructive to draw vertical lines at, say, every ten points and to draw lines showing the zeros of the three Y axes. We did not collect the calculated data in arrays but this could easily be done if any further computation is envisaged. The output could of course be directed through a DAC to a recorder if continuous curves on paper are required.

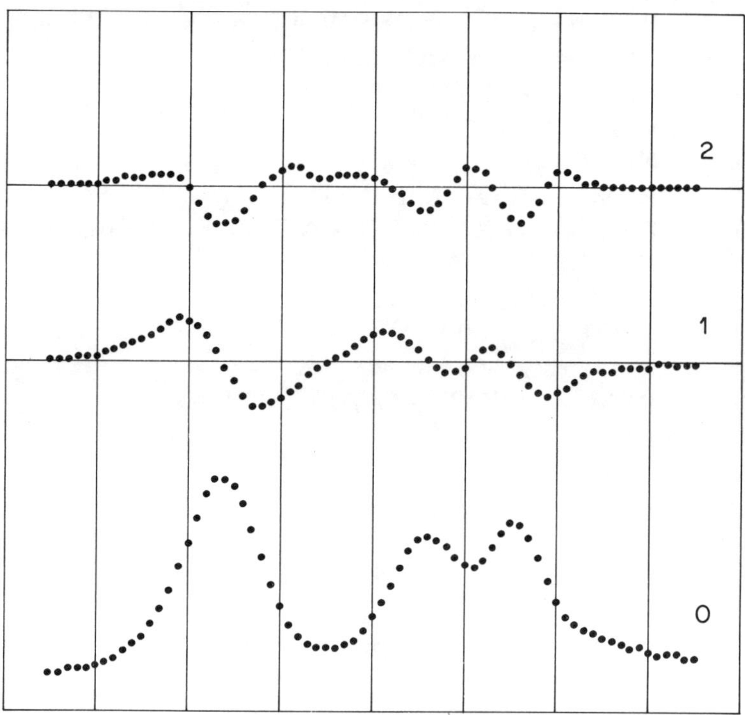

Fig. 5.2c. *(0) Smoothed*
(1) 1st derivative
(2) 2nd derivative

SAQ 5.2c

The simplest and most direct method of producing a first derivative is to plot the differences between successive points (ie plot Y(I%)–Y(I%−1) against I%). Use the data of the example program to do this. Compare the plot obtained by applying this simple method to the raw data with that obtained after smoothing the data.

5.3. PEAK AREAS

5.3.1. Graphical Integration

The height of a chromatographic peak may be taken as an indication of the amount of the substance responsible for the peak but the area under the peak is a better measure of amount. The measurement of area is in effect a determination of the integral of signal amplitude with respect to time over the time that the peak is passing the detecting point. In coulometry, the amount of electricity which has passed in a given time is obtained by integrating current with respect to time and so the area under a current-time plot between two times gives a measure of the electric charge which has passed. In some spectroscopic procedures an integral of absorbance with respect to wavelength or frequency is sometimes required and again this involves measuring the area under a peak.

Many modern analytical instruments have facilities for integrating or measuring areas either during the time data are being collected or in subsequent procedures. When the instrument does not provide this service the analyst can collect the necessary data and carry through the calculation either by hand or, more efficiently, by computer. The essential process may be visualised as the plotting of a graph followed by calculation of the area under the graph. Provided the graphical data are in digital form one of a number of numerical methods may be used to perform the calculation. As we look at a few of the methods that are relatively easy to apply we shall assume that readings of amplitude (Y) are taken at regular intervals (X) and that the amplitudes are stored in an array or in data statements or in memory locations.

5.3.2. Addition of Ordinates

The simplest method of computing an approximate integral is to add up all the amplitudes and multiply the sum by the interval between readings (ie between the X values). This may not be the most accurate method of estimating an area but it has the merit of being readily applied to *real time* measurement. That is, a Y amplitude can be added to the sum immediately it is measured instead of being stored for later computation. Of course, it is necessary to decide where to start and finish the additions and this can cause problems if the base line does not lie along the X axis. At present we shall ignore the problems and write the equation used to calculate the area when the base line always has zero amplitude. If H represents the interval between readings and N values of amplitude, Y(1), Y(2), ... Y(N), are read then the area is given by:

$$\text{area} = H * [Y(1) + Y(2) + Y(3) + \ldots + Y(N)]$$

This equation is given here because its simplicity makes it a convenient starting point for a discussion of more accurate methods. We shall return to this method in Section 5.4.

5.3.3. The Trapezoidal Rule

This is the next simplest of the numerical methods of integration. It is superior to the method of adding ordinates and is a good introduction to other, more accurate, methods.

The essence of the technique is to divide the area to be measured into a number of trapezoids. Fig. 5.3a shows a trapezoid of width H drawn to include points 2 and 3 which lie on a curve. The area of the segment under the curve between these points is approximately equal to the area of the trapezoid ABCE. This area is the sum of two areas:

$$\text{area of rectangle ABDE} = H * Y(2)$$

$$\text{area of triangle BDC} = H * [Y(3) - Y(2)]/2$$

As before, * is used to indicate multiplication and array notation, Y(1), Y(2), ... is employed instead of the more usual y_1, y_2 ...

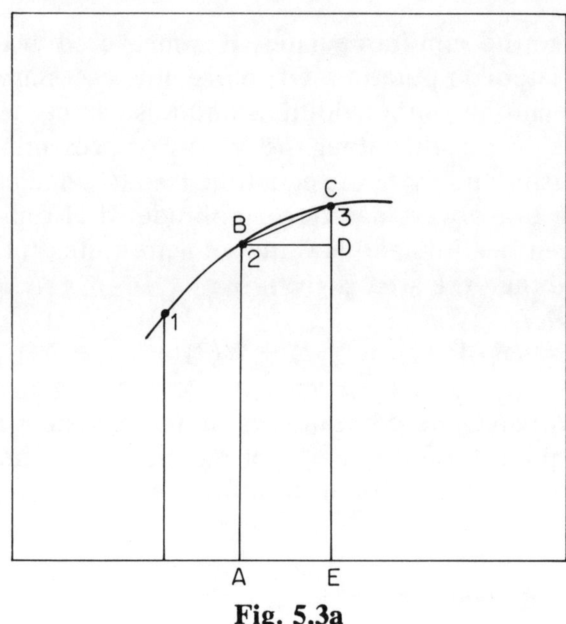

Fig. 5.3a

The area of the trapezoid is the sum of the two contributions:

$$(H/2) * [Y(3) + Y(2)]$$

If the whole of the curve between points P and Q is divided into a number of strips of width H then the area under the curve between these points is approximately equal to the sum of the areas of the strips:

$$(H/2) * \{[Y(2)+Y(1)] + [Y(3)+Y(2)] + [Y(4)+Y(3)] + \ldots\}$$

That is, the area between P and Q approximates to:

$$(H/2) * [Y(1) + 2*Y(2) + 2 * Y(3) + \ldots 2 * Y(N-1) + Y(N)]$$

where N is the number of strips of width H.

Clearly, in this case, the calculated area is smaller than the true area enclosed by the curve, the difference being equal to the sum of the areas between the curve proper and chords like BC. This *truncation* error depends on the strip width and in principle it would appear that the error could be reduced to negligible proportions by making the width very small. The situation is more complicated than this, however, because reduction of the strip width means an increase in the number of small areas to be added. Every addition is accompanied by a *round-off* error and the accumulation of round-off errors acts in opposition to the reduction of strip width. For a full discussion of the errors involved in the application of the trapezoidal and other rules you should read the relevant chapters in Norris (1981) or Dorn and McCracken (1972). In practice the width of a strip is usually decided on the basis of convenience together with a certain amount of trial.

To write a computer program which performs an integration by the trapezoidal rule the essential steps are:

Decide the number of strips and the strip width. We shall use variables N% and H for these. (The number must be an integer).

Ensure that the amplitude signals or Y values are available. These may be in memory locations or in an array or in data statements.

Add the first and last Y values to give a sum (S).

Construct a loop to compute twice the sum of the other Y values and add this to the previous sum to get a new sum.

Evaluate the area as H*S/2.

When the data to be used come from an instrument in digital form the Y values required may be stored in known memory locations from which they can be retrieved as desired. Because of variations in the methods of storage and retrieval used by different microcomputers, however, we shall place the relevant information in data statements. You should be able to adapt the program to read from memory if you so desire.

In the program below the calculation is done as the amplitudes are read from data statements. An alternative method would be to read the data into an array and then perform the calculation. There are 31 points (N%) and the strip width (H) is 10 units.

```
100 REM ** TRAPEZOIDAL RULE
110 REM ** N% & H BOTH IN DATA STATEMENT
120 READ N%,H
130 S=0
140 REM ** START READING Y VALUES
150 READ Y
160 REM ** THIS IS FIRST Y-CALL IT F
170 F=Y
180 REM ** GO INTO LOOP TO SUM
190 FOR I%=2 TO N%-1
200       READ Y
210       S=S+Y
220       NEXT I%
230 REM ** LOOP DOES NOT READ LAST Y
240 REM ** ADD SUM TO FIRST
250 S=S + F/2
260 REM ** READ LAST Y & ADD
270 READ Y
280 S=S + Y/2
290 PRINT "AREA ="; H*S
300 END
310 REM ** DATA STATEMENTS
320 DATA 31,10
330 DATA 0,1,2,5,8,12,18,28,42,65,100
340 DATA 141,172,190,199,199,190,174,151
350 DATA 126,101,76,53,35,21,12,6,3,2,1,0
```

Run this program and see if you get an answer near our 21330 units for the area of the peak. Of course the units of the 'area' will depend on the units of H and Y but normally some calibration procedure using standards leads directly to a relationship between area and amount of material.

SAQ 5.3a

The simple method of adding ordinates may be regarded as an approximate form of the trapezoidal rule which takes the sum of amplitudes and multiplies the sum by the strip width (no multiplying or dividing by 2).

Apply this approximation to the data in the text example but read the data into an array before carrying out the calculation.

5.3.4. Simpson's Rule

Simpson's Rule is probably the most popular of the simple methods of finding an area. It is only slightly more complex than the trapezoidal rule but is rather more accurate. The total area is divided into small intervals or strips as before but, instead of replacing a section of the curve by a straight line between two points as in the trapezoidal method, a parabola is fitted to three points and the area under the parabola is taken as an approximation to the strip area. The final result gives the total area as:

(H/3) * [Y(1) + 4 * Y(2) + 2 * Y(3) + 4 * Y(4) + 2 * Y(5) + ...

... + 2 * Y(N–2) + 4 * Y(N–1) + Y(N)]

Note that the first and last Y values stand alone but between them the multipliers are 4..2..4..2 ... ... 4. There is an even number of strips and an odd total number of Y values.

A program to compute an area by Simpson's Rule can be written in a number of ways. The one below may not be the most elegant but it should direct attention to the nature of the terms involved in the summation.

The data are first read into an array and then the calculation is done on the array. The first data items are the number of points (N% = 31) and the strip width (H = 10 units).

```
100 REM ** SIMPSON'S RULE
110 REM ** N% (ODD) & H FROM DATA
120 DIM Y(50)
130 READ N%,H
140 REM ** READ Y VALUES INTO ARRAY
150 FOR I%=1 TO N%
160       READ Y(I%)
170       NEXT I%
180 REM ** ADD FIRST AND LAST
190 S=Y(1)+Y(N%)
200 REM **ADD TERMS TO BE MULT'D BY 4
210 T=0
220 FOR I%=2 TO N%-1 STEP 2
230       T=T+Y(I%)
240       NEXT I%
250 REM ** MULTIPLY BY 4 & ADD
260 S=S+4*T
270 REM ** ADD TERMS TO BE MULT'D BY 2
280 T=0
290 FOR I%=3 TO N%-2 STEP 2
```

```
300      T=T+Y(I%)
310      NEXT I%
320 REM ** MULTIPLY BY 2 & ADD
330 S=S+2*T
340 REM ** SUM NOW COMPLETE
350 PRINT "AREA ="; H*S/3
360 END
400 REM ** DATA STATEMENTS
410 DATA 31,10
420 DATA 0,1,2,5,8,12,18,28,42,65,100
430 DATA 141,172,190,199,199,190,174,151
440 DATA 126,101,76,53,35,21,12,6,3,2,1,0
```

This program gave an area of 21340 units.

Simpson's Rule is quite easy to apply and is remarkably accurate. It is also not difficult to program and so we suggest that you give it some careful study. You might care to develop modifications of the program above. Could you replace the array by a block of memory? Could you write a program to perform the integration as the amplitudes are received into the microcomputer (ie in *real time*)?

In the two programs used as examples the area found was the area lying between the curve and the X axis. In many cases it is necessary to make allowance for a base line which does not lie conveniently on the X axis. In the simplest case the base line is parallel to the X axis so that each Y value need only be reduced by a constant amount before being used in the computation. In a more complex case the base line might vary between the limits of integration. Ideally of course an experimental base line should be obtained by running a 'blank' but this is not always possible or convenient. If it may be assumed that the base line shows a linear variation with variable X (or I%) it is relatively easy to make the necessary allowance. For example, suppose we want the area below the curve and the straight line between points P and Q of Fig 5.3b. We proceed as before to find the area under the curve between X1 and X2 on the X axis and then subtract the area defined by the points X1, P, Q and X2. That is, we subtract an area

(X2–X1) * (Y(X2) + (Y(X1) – Y(X2))/2).

In application, it is only necessary to supply the values of X1 and X2 (or the appropriate 1% values) for the program to make this correction. In routine work these values might be included in the program but, more generally, the curve can be displayed on a VDU and the values of X1 and X2 input by the analyst.

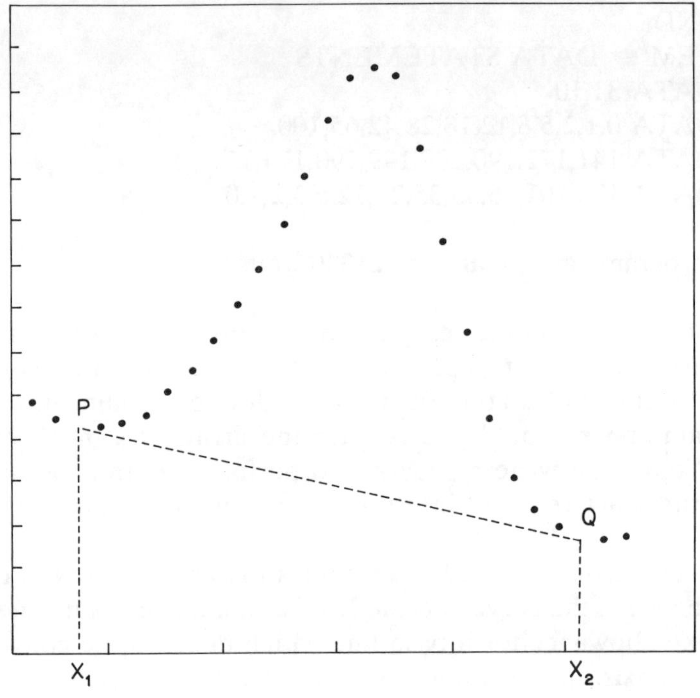

Fig. 5.3b

SAQ 5.3b

The 27 numbers below represent signal amplitudes measured at regular intervals of 5 units. Determine the area of the peak by Simpson's Rule, taking the base line as a straight line between the amplitudes at points 3 and 25.

Data:

58 54 52 52 53 55 60 65 72
80 88 99 110 123 132 135 133 116
95 74 54 40 33 29 26 26 27

5.4. A CHROMATOGRAPHY PROGRAM

5.4.1. Locating Peaks

In a final programming study we look at how peak positions, heights and areas might be measured in real time by a relatively simple program. Modern commercial instruments for gas chromatography and HPLC can measure retention times and calculate peak areas while a chromatogram is being drawn. Though the operator is not expected to be familiar with the hardware and software necessary to achieve the output from such instruments a good analyst should have an awareness of the kind of procedures used and possibly also be able to apply similar procedures to data collected from less sophisticated equipment.

Let us suppose that signal amplitudes are measured at regular intervals of time by means of an ADC and that the base line amplitude is always zero. (eg Fig. 5.4a).

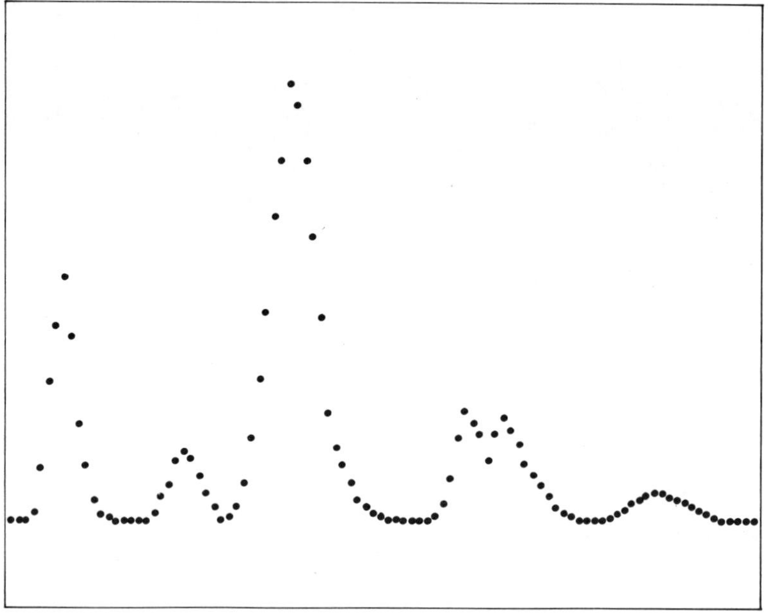

Fig. 5.4a

To locate the positions of the peak maxima and the limits of peaks the program must be able to detect how the amplitudes change:

At the start of a peak the plot of amplitude against time has the shape of (*i*) or (*ii*). Before a peak starts the gradient is either zero or negative and after the start it is positive. We use variable F% to represent how the amplitude has changed from the previous point, +1 indicating an increase, −1 a decrease and 0 indicating no change. The sign of F% is therefore the sign of the gradient immediately before the point being processed. Similarly, variable G% represents the value which F% had when the *previous* point was processed. (Similar variables were used in 5.2 and in SAQ 5.2a.) Using this convention the start of a peak can be detected one point after the start by the criteria:

either $G\% = 0$ and $F\% = 1$

or $G\% = -1$ and $F\% = 1$

The first of these is appropriate when the peak starts from a horizontal base line (*i*), the second when one peak follows another (*ii*). If we do not need to distinguish between the two cases the criteria may be combined to give:

$G\% < 1$ and $F\% = 1$

At the end of a peak the trace has the shape depicted by diagram (*ii*) or (*iii*) and the corresponding criteria are:

either $G\% = -1$ and $F\% = 1$

or $G\% = -1$ and $F\% = 0$

These may be combined to give:

$$G\% < 0 \quad \text{and} \quad F\% > -1.$$

At the maximum of a peak the gradient changes from positive to negative (*iv*), but to allow for the possibility of a peak that is two points wide (F% = 0) the condition is best written:

$$G\% = 1 \quad \text{and} \quad F\% < 1$$

5.4.2. The Program

For simplicity and ease of formatting text output we use integer variables for the signal amplitudes. This is usually possible though real variables become necessary as soon as amplitudes are adjusted by smoothing or by making allowances for a base line. The program uses subroutines for certain operations which are performed after maxima and ends of peaks have been detected but the general structure is well suited to the use of procedures rather than subroutines. Variable F% is found by taking the difference between the current amplitude (A%) and the immediately previous amplitude (L%). The usual BASIC statement for returning the sign as +1, 0 or −1 is:

```
F%=SGN(A%-L%)
```

After an amplitude has been processed the value of F% is transferred to G% and the value of A% to L%. In this way the program remembers how the curve was changing when the previous amplitude was processed. The area under a peak is taken as the sum of the amplitudes between the start and end of each peak. This may not give the best area but the method provides useful information on the positions and heights of peaks and the relative areas under each when great accuracy is not essential. As written below the program carries out the tests and performs the integration as data are read from DATA statements. With slight modification the same program could be used in a real-time application, the data being read directly through an ADC. It will be appreciated, however, that this kind of programming can only be successful if the data are reasonably free from noise. If this is not the case the readings should be collected into an array and smoothed before running the detection and integrating routines. It would then be inappropriate to use integer variables for the smoothed data.

```
100 REM ** LOCATE & INTEGRATE
110 REM ** INITIALISE **********************
120 DIM A%(12):REM AREA OF PEAK J%
130 T%=0:REM TOTAL AREA (SUM OF A%(J%)
140 S%=0:REM AREA AS SUM OF AMPLITUDES
150 J%=0:REM PEAK INDEX NUMBERS
160 G%=0:REM SIGN OF SLOPE LAST TIME (REMEMBERS F%)
170 PRINT "PEAK LOCATION AND INTEGRATION"
180 PRINT
190 PRINT "SEPARATE INTEGRATION OF"
200 INPUT "DOUBLE PEAKS ? (Y/N)"A$
210 REM ** SET B% TO DEFINE END OF PEAK (SEE NOTE)
220 IF A$="Y"OR A$="y"THEN B%=1000 ELSE B%=1
230 CLS: REM CLEAR SCREEN
240 PRINT "PEAK TIME HEIGHT AREA TOTAL %"
250 PRINT
260 READ N%:REM TOTAL NO. OF AMPLITUDES
270 READ A%:REM FIRST (OR ZEROTH) AMPLITUDE
280 L%=A%:REM L% REMEMBERS LAST AMPLITUDE
290 REM ** (PLOT ZEROTH POINT)
300 REM ** (I%=0,A%=ZEROTH DATUM)
310 FOR I%=1 TO N%-1
320       READ A%
330       F%=SGN(A%-L%):REM HOW CHANGING?
340       REM ** TEST FOR START FROM BASE
350       IF G%=0 AND F%=1 THEN S%=0
360       REM ** TEST FOR MAXIMUM
370       IF G%=1 AND F%<1 THEN GOSUB 600
380       REM ** TEST FOR END OF PEAK
390       IF G%=-1 AND F%>-1 THEN GOSUB 700
400       S%=S%+A%:REM ADD TO SUM
410       L%=A%:G%=F%:REM REMEMBER THESE VALUES
420       REM ** (PLOT THE POINT)
430       NEXT I%
440 REM ** CALCULATE & PRINT AREAS AS PERCENT
450 FOR I%=1 TO J%
460       PRINT TAB(30,1+I%)
470       A=INT(1000* A%(I%)/T%+0.5)/10
```

```
 480      IF A>0 THEN PRINT A ELSE PRINT
 490      NEXT I%
 500 END
 600 REM ** PEAK MAXIMUM DETECTED **********
 610 J%=J%+1:REM NEW PEAK
 620 PRINT J%,I%-1,L%;
 630 RETURN
 700 REM ** END OF PEAK DETECTED ***********
 710 REM ** SEE IF END REACHED
 720 IF L%<B% THEN GOSUB 800 ELSE PRINT
 730 RETURN
 800 REM ** PRINT PEAK AREA ****************
 810 T%=T%+S%:REM TOTAL AREA
 820 PRINT S%,T%
 830 A%(J%)=S%:REM STORE AREA OF PEAK J% IN ARRAY
 840 S%=0:REM START SUM FOR NEXT PEAK
 850 RETURN
1000 REM ** DATA STATEMENTS ***************
1010 DATA 100:REM NO. OF POINTS
1020 DATA 0,0,0,0,2,15,40,56,70,53
1030 DATA 28,16,6,2,1,0,0,0,0,0
1040 DATA 2,7,10,17,20,18,13,8,4,0
1050 DATA 1,4,11,24,41,60,88,104,126,120
1060 DATA 104,82,58,31,21,16,11,6,4,2
1070 DATA 1,0,0,0,0,0,0,1,5,12
1080 DATA 24,32,28,25,23,25,30,26,22,17
1090 DATA 13,10,7,4,2,1,0,0,0,0
2000 DATA 1,2,3,5,6,7,8,8,7,6
2010 DATA 5,4,3,2,1,0,0,0,0,0
```

The data correspond to Fig. 5.4a and the following table shows the output obtained when separate integration of double peaks was requested (line 200):

Peak	Time	Height	Area	Total	%
1	8	70	289	289	17.2
2	24	20	99	388	5.9
3	38	126	915	1303	54.5
4	61	32	150	1453	8.9
5	66	30	157	1610	9.4
6	86	8	68	1678	4.1

(Some format control was necessary.)

Fig. 5.4a

When separate integration was not selected the time and height only were printed for peak 4 and an area of 307 or 18.3% was reported for peak 5.

Variable B% is used to decide when a peak is complete. In this case we use one of two values (1 or 1000) and if the lowest point in the valley between adjacent peaks is below the value selected the peaks are taken to be separate. Obviously, the value of B% can be selected to suit the circumstances.

Though the sum S% is set to zero at the start of a peak (line 350) this is not really necessary for the particular set of data used because every amplitude is zero except during a peak and the program sets S% to zero at the end of every peak.

To use the program properly it is necessary to include routines for clearing the screen, drawing scales and plotting the points or the lines between points (line 420). Formatting of the printed results must be controlled to tabulate the results properly. The TAB statements used above assume a 6-character field.

SAQ 5.4a

> Suggest how the location and integration program might be modified (*i*) to yield peak times relative to that of a selected peak, (*ii*) to take account of a base line which is not zero but which may change in a predictable manner.

5.4.3. Further Development

The example of real-time data processing discussed above can readily be adapted to deal with arrays of real numbers which may or may not have been smoothed by one of the procedures discussed in 5.1 and it can form the basis of procedures for processing data when the base line is not so well-behaved and peaks are less well defined. It has already been mentioned that the simple sum of amplitudes is not the best estimate of area but in practice greater errors are probably caused by incorrect prediction of the base line and by difficulties in locating where the peaks leave and re-join the base line. We cannot consider how the program might be modified to take account of all possible variations but some pointers to methods of developing this or a similar program might be useful.

In the example program the areas of overlapping peaks were calculated after effectively dropping a vertical from the valley to the base line. Variable B% represented a threshold level of amplitude in the valley between peaks and a peak was considered to be complete when the amplitude fell below this threshold. A slight modification could make this variable depend on the time (ie on I%) or on the area of the peak being measured. A much superior method of dealing with overlapping peaks is to extrapolate the sides of both peaks to the base line and then calculate the areas of both peaks separately. Such a procedure requires a reasonable number of measurements to be made during each peak and involves some assumption about peak shape.

When a straight line drawn between the start and end of a peak is an adequate base line (eg Fig. 5.3b) there is little difficulty in integrating over the peak and then programming a correction calculated from the start and end coordinates (see SAQ 5.3b). When the base line is not constant but curves in a reasonably predictable way it may be possible to derive an equation for the curve from measurements made before any peaks are detected and then to use the equation to calculate the base amplitude at later times.

Detection of peak limits can be difficult if the base line is itself curved. The start of a peak may have to be detected by noting when the *slope* of the experimental curve increases above some threshold slope and the value of the threshold slope may be fixed or may be relative to the slope of the base line at the particular point. If the base line decreases with time the end of a peak may be taken as the time when the signal becomes equal to the base signal after a maximum has been passed rather than by the method adopted in the example (gradient zero or positive following a negative gradient). Another method of locating the termination of a peak associates the peak end with return of the slope to some threshold value after a maximum while another is based on the assumption that the peak ends after the lapse of a certain time following a maximum. An alternative to all of these 'analytical' methods is to plot the signals on the screen and then to 'tell' the program where to draw a base line either by an INPUT statement or through a procedure which allows the operator to place 'markers' on the screen; when the markers are in the correct positions a key is pressed to initiate an integration procedure.

All of the procedures indicated are quite straightforward in principle though the actual programming may be somewhat demanding, particularly if real-time operation is required. If time is relatively unimportant or if accurate computations are required the data may be collected and stored in an array or in memory locations and then processed to locate peak maxima and calculate areas by means of separate programs.

We have concentrated on programming aspects of data processing but it must not be forgotten that the data are obtained through hardware. The analogue–digital converter is obviously the most important piece of hardware but other devices such as mutiplexers and switches must also be considered. For example, the measurement of a large solvent peak in chromatography may require detection of signals in a voltage range which is too insensitive to allow accurate measurement of subsequent peaks. The problem can be solved by using an ADC with a number of *channels* which can be selected by a program. Each channel amplifies (or attenuates) the signal by a different amount and the program selects a low amplification channel until the first peak is passed. An alternative method is to make no measurements for a definite time after injection of the sample but if this method is adopted it is necessary to ensure that the ADC is not harmed by being connected to a source of high signals even when it is not being used.

If the time required by a BASIC program is too long and real-time procedures are essential then part of the program must be written in machine code or a machine code routine employed to acquire and store the data before it is processed by a program written in a high-level language. The principles involved are similar to those discussed above but the time required to read and store a signal is measured in microseconds rather than centi- or milli-seconds. The sensitivity (number of bits) of the ADC and the conversion time are the usual factors which must be considered when machine code programs become necessary.

Objectives

On completion of Part 5 of Unit you should be able to write BASIC programs which present data collected from analytical instruments in graphical form and which:

- smooth the data by the moving average and least-squares methods (SAQs 5.1a, b, c);
- locate the positions and determine the heights of peak maxima (SAQs 5.2a, b, 5.4a);
- plot derivatives of curves and peaks (SAQ 5.2c);
- determine areas under curves and peaks (SAQs 5.3a, b, 5.4a).

6. Case Study: On-line Measurements in Atomic Absorption Spectroscopy

Overview

Before studying this final part of the Unit you should have mastered the material of previous parts. In particular you should know enough BASIC to follow competently a listing given appropriate supporting documentation. You also need to be familiar with subroutines and should appreciate how they can be used to create a well structured program. On the interfacing side you will need to recall the operation of an 8255 Programmable Peripheral Interface, how data are represented by binary coded decimal, and how such data can be decoded. All of this material is covered in detail in earlier Parts of this Unit.

This case study is devoted to the addition of an on-line data processing capability to an older atomic absorption spectrophotometer which only has a recorder output available for external use. Those readers who are unfamiliar with atomic absorption measurements should consult the ACOL Unit dealing with the topic to obtain the necessary background knowledge, or one of the many texts covering instrumental analytical techniques, such as:

Analytical Chemistry: an introduction, D.A. Skoog and D.M. West, Saunders College Publishing, 1986

Part of the data processing in the program discussed in this case study involves the determination of the parameters in a mathematical representation of the calibration curve of reading against concentration. No attempt has been made to discuss the basis of the statistical techniques of linear regression used since they are amply dealt with elsewhere. The interested reader is therefore asked to consult the *ACOL Unit: Measurement Statistics and Computation* or one of the many standard works on statistical methods such as,

Statistical Theory with Engineering Applications, A. Hald, J. Wiley and Sons, 1952

or,

Computational Chemistry, A.C. Norris, J. Wiley and Sons, 1981

Those who would like a more in-depth treatment of digital methods of peak detection applied to atomic absorption spectroscopy should consult the book,

FORTRAN IV in Chemistry, G. Beech, J. Wiley and Sons, 1975

6.1. THE NATURE OF THE PROBLEM

6.1.1. Assumptions About the Instrument

This section deals with examples of on-line data processing which are standard features on many modern atomic absorption (AA) spectrophotometers. For the purpose of illustration we shall assume we are dealing with an older instrument which provides a DC output signal to drive a recorder. The aim is to upgrade the instrument to provide some of the data processing capabilities of a modern AA spectrophotometer.

We shall also assume that the recorder output signal is proportional to absorbance and ranges from 0 to 100 mV, and that it is free from AC components associated with the modulation of the lamp which provides the source of light for a particular element. If you are going to repeat this case study using an older instrument you should use an oscilloscope to check the nature of the signal that is normally fed to the recorder, for the presence of AC components. Usually AC interference can be eliminated by strapping a capacitor across the two output leads which provide recorder output.

The analogue signal has to be converted to digital form to allow on-line data processing. In this case study we used a digital volt meter (DVM) with binary coded decimal output and a computer with a memory-mapped 8255 PPI interface. A full description of the operation of the interface is given in Section 6.4.

6.1.2. The Specification of the Program

The program should be written to cope with each of the following tasks:

(*a*) The operator will be asked for the element to be determined as one of several options in an opening menu. Information stored on the disc will be examined and the appropriate operating conditions displayed (wavelength, lamp current, slit width, working range, fuel and oxidant). If the information for the element is not on disc then an appropriate message should be displayed.

(*b*) Other options in the opening menu will include:

(*i*) adding details for an element to the disc file;

(*ii*) deleting details for an element;

(*iii*) changing details for an element already on file;

(*iv*) exiting from the program;

(*v*) proceeding with the analysis for the chosen element;

(*vi*) listing elements on file.

(*c*) Having selected the element to be determined, the computer must prompt the operator to set up the instrument according to the specified conditions. Each instrumental condition should be clearly presented on the screen in turn and the operater should be invited to press the space bar, after setting the instrument parameter correctly on the AA spectrophotometer. In this way the operator is led through the setting-up procedure and the computer waits for an acknowledgment at each stage.

(*d*) When setting-up the instrument, readings from the AA spectrophotometer will be accumulated by the computer whilst water is sprayed into the flame. To reduce noise, each measurement stored in the computer should be the mean of at least 30 readings. The average reading should be displayed on the screen. At no time should the screen be left blank. For example, during the accumulation of data to calculate the mean, an appropriate message should be displayed. During the initial setting-up phase, when water is being sprayed, the averaged reading will be the digital equivalent of the mV signal corresponding to an absorbance of zero. During optimisation of the choice of wavelength, the strongest standard solution will be sprayed into the flame, and once again the computer will display an averaged reading. In all operations to accumulate instrumental readings the user must be prompted to spray the appropriate liquid into the flame and then asked to press the space bar by way of acknowledgement. No data readings must be taken until the operator has responded in this way. The readings should be collected over at least a 3-second period, as fast as BASIC will allow (with some computers more than 30 readings may be possible in the time). Before and after measurement of each standard solution or sample, the operator must be asked to spray water, with a prompt as described above, and the baseline monitored again.

(*e*) Once data have been collected for all the analytical measurements, the data pairs (averaged reading, and concentration in parts per million, ppm) should be fitted to a calibration curve of the type:

$$Y = a + bX + cX^2 \tag{6.1}$$

where a, b and c are constants, X is the concentration and Y is the reading. If the calibration plot is curved, that is when c is non-zero, the user should be able to choose to have linearised output displayed. Linearisation is to be achieved by adding a curvature correction to the actual reading so as to give a total which varies linearly with concentration as shown schematically in Fig. 6.1a. The linear plot to which Eq. 6.1 approximates at low values of X is given by:

$$Y = a + bX \qquad (6.2)$$

The method of linearisation to be adopted is to determine the coefficients a, b and c in Eq. (6.1) and then add $-cX^2$ to the reading Y to give the equivalent linear output.

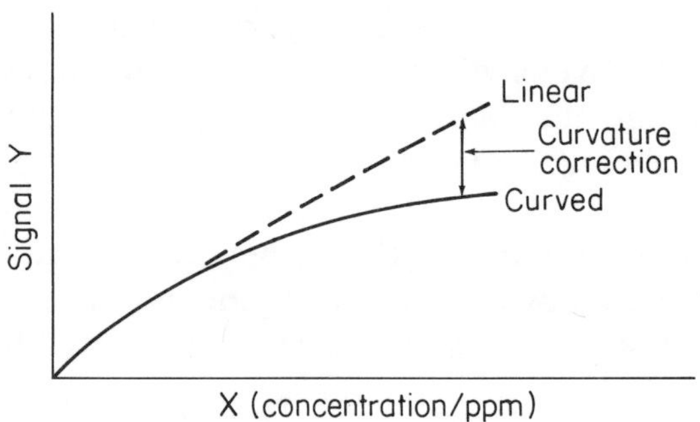

Fig. 6.1a. *Linearisation of a curved calibration plot by addition of a curvature correction to the observed readings*

The parameters which define the equation for the 'best' fit are determined by a least-squares analysis. A full discussion of this method is given in the Unit on Measurement, Statistics and Computation to which the interested reader is referred. In the present context we assume that the errors all lie in the measured values Y and there are none in the concentrations X. The equation for the 'best' fit is obtained by minimising the squared deviations of the experimental values of Y from the corresponding predicted values. Formally the technique to be used is described as multiple linear regression in the two variables X and X^2 to obtain the best fit for the parameters a, b and c. This approach has the advantage that an expression can be obtained which gives an estimate of the uncertainty in the predicted value of Y for a given value of X. As analytical chemists we are faced with the opposite problem. Having established a calibration curve, we wish to find the value of X which corresponds to a given value of Y. We can certainly obtain a value of X corresponding to the value of Y by rearranging the equation (6.1) to give X in terms of Y. Unfortunately we cannot similarly invert the equations for the uncertainty in Y for a given value of X because the equation so obtained would be invalid. So although we can compute a value for X given a reading Y we cannot say anything about the uncertainty in X. A full treatment of this problem is beyond the scope of the present work so we have adopted a rather simpler approach to give an indication of the uncertainty in predicted values of X. As well as computing the value of X, corresponding to Y, we will also compute the values of X corresponding to Y plus or minus 5%. If our measurements fall on a part of the calibration curve which is flattening out, the observed differences between these predicted concentrations will then be much larger than for the linear sections at lower concentrations.

(*f*) Once the concentration of a sample has been computed, the program should display the result along with linearised output if required. The user should then be asked if another sample is to be measured. If so he or she should be prompted to spray solvent, then the sample and then solvent again, whilst the appropriate measurements are accumulated by the computer. The results obtained are then displayed as indicated above. This procedure should be repeated until no more samples are to be processed and then the user should be presented with the options in the opening menu again.

SAQ 6.1a

The program specification requires that the computer accumulates readings to obtain a baseline value before and after measuring each standard and unknown solution. With reference to these measurements, indicate whether or not each of these statements is true (T) or false (F).

(*i*) A drifting baseline is fully compensated for by taking the mean baseline before and after each measurement of sample or standard solution.

(T / F)

(*ii*) The output to the recorder, which is used as an input signal to the computer, should be adjusted so that a zero reading is obtained by the computer when solvent is sprayed into the flame.

(T / F)

(*iii*) The digital readings, corresponding to the analogue voltage in the millivolt range, should be scaled to give a proper absorbance values before using them to compute analytical results.

(T / F)

SAQ 6.1b

The readings, Y, obtained from the AA spectrophotometer, produced the following calibration curve over the range 0 to 5 ppm.

$$Y = 10 * X - 0.2 * X^2$$

(*i*) Calculate the values of X for a reading of Y = 40, by use of the standard solution, given below, for a quadratic equation expressed in the form:

$$AX^2 + B * X + C = 0.$$

$$X = \frac{-B \pm \sqrt{B^2 - 4 * A * C}}{2 * A}$$

(general solution)

Of the two values of X obtained from the solution of the quadratic equation, which one is physically reasonable?

(*ii*) Under what circumstances is it impossible to calculate a value for X from the general solution given in (*i*) above?

SAQ 6.1b

6.2. PROGRAM DESIGN

6.2.1. Initialisation and the Opening Menu

Fig. 6.2a shows a flow sheet of the overall structure of the proposed program. The first block is the initialisation section where arrays are dimensioned and data relating to the operating conditions are read from disc. It is good practice to dimension arrays in this block rather than in the body of the program to increase readability of the program and to avoid errors arising by accidentally dimensioning an array twice. The initialisation block is also used to configure the interface and produce an initial information page giving the program title and details of what equipment should be used with the program.

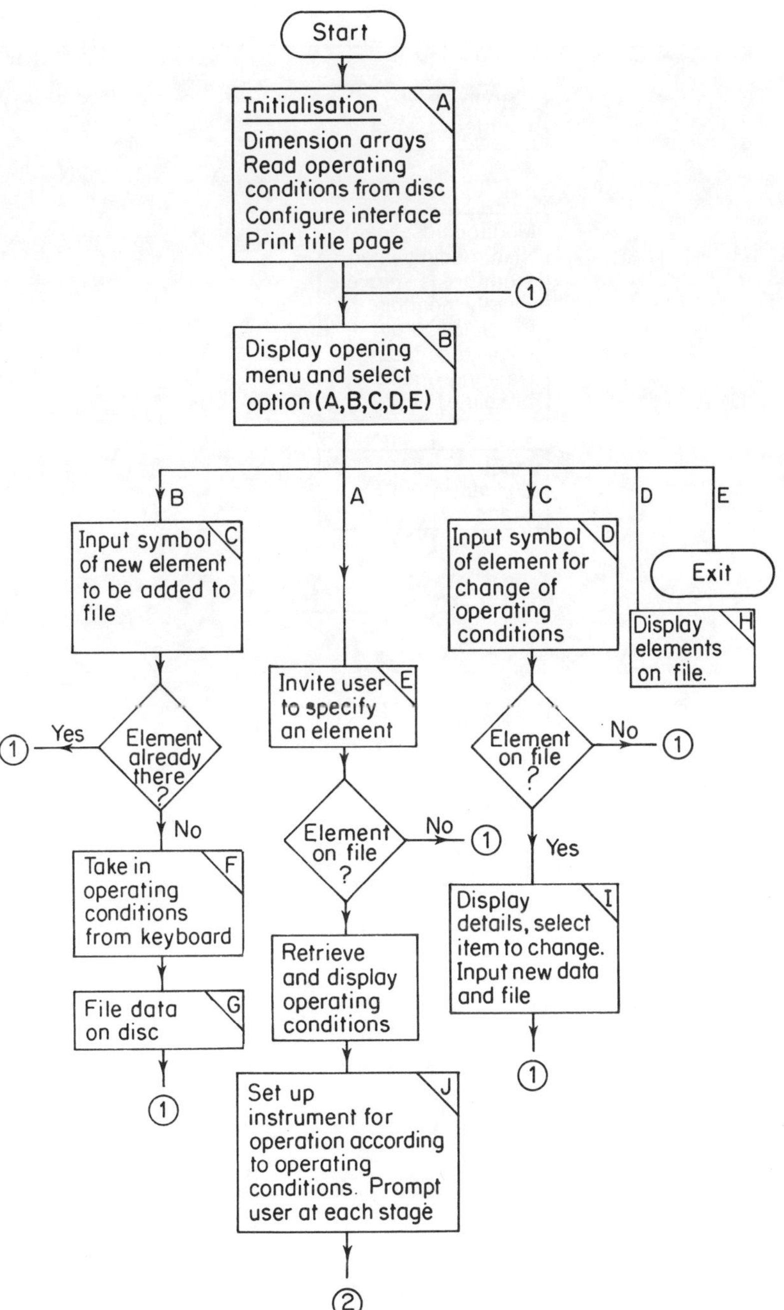

Fig. 6.2a. *Flow chart showing overall structure of the program*

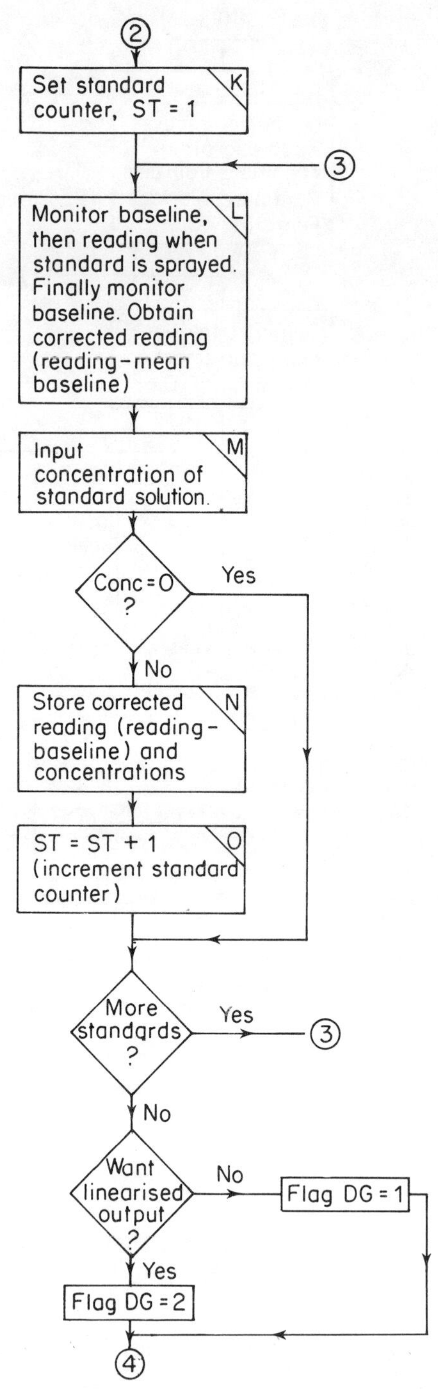
2
Set standard counter, ST = 1
K
3
Monitor baseline, then reading when standard is sprayed. Finally monitor baseline. Obtain corrected reading (reading − mean baseline)
L
Input concentration of standard solution.
M
Conc = O ?
Yes
No
Store corrected reading (reading − baseline) and concentrations
N
ST = ST + 1 (increment standard counter)
O
More standards ?
Yes
3
No
Want linearised output ?
No
Flag DG = 1
Yes
Flag DG = 2
4

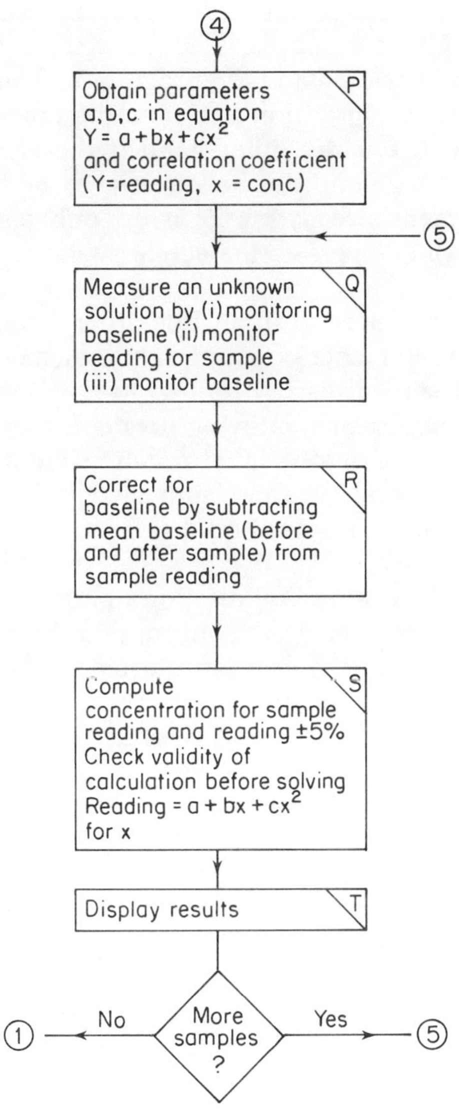

After a brief delay, the opening 'menu' is displayed on the screen. The use of a list of permitted options (the menu) is an accepted method of eliciting one of a range of responses. Unexpected responses can be rejected and the user can be prompted to input one of the allowed responses. The function of each of the responses A, B, C, D and E is discussed below.

Option A produces an invitation to specify the element to be determined. If the element chosen is not recognised as being on file, an error message is printed and the program returns to the opening menu which includes an option to define the operating conditions for a new element. On the other hand if the element is already on file, the operating conditions are retrieved and displayed. The choice of an element already on file is the only entry into the rest of the program which oversees the actual analysis.

Option B allows the user to input operating details for a new element. The computer checks to see if the element is already on file and if so the screen display returns to the opening menu with its options for deleting or modifying details for an element. For a new element the operator keys in information about the wavelength, slitwidth, combustion mixture and lamp current. These data are then added to the information stored on disc.

Option C allows data, stored on disc for a particular element, to be altered. First a check is made to confirm that the element is in fact on file. If it is not then the user is returned to the opening menu. If the element exists on disc, the operating conditions are displayed as a series of items. The user selects which item is to change and the new data are read from the keyboard. The user can change a succession of items in this way for the chosen element. When no further changes are required the data are filed on disc after the user is given a final chance to cancel the changes. If cancellation is selected, no information is written to disc and the opening menu is displayed once more.

Option D displays elements on File and E simply terminates the run. After each series of analytical measurements the opening menu is presented and so option E is the only exit from the program.

6.2.2. Setting up the Instrument

If option A is chosen from the opening menu, the program passes to the section which prompts the user at each stage of setting up the instrument. The required instrumental settings are displayed at each stage using the operating conditions previously read from disc.

The programmed sequence is as follows:

(*a*) Switch on the power to the AA spectrophotometer and computer interface.

(*b*) Switch on the lamp for the element being determined.

(*c*) Set the lamp current to the value specified by the computer.

(*d*) Set the wavelength to the specified value.

(*e*) Correctly set the fuel and oxidant flow rates and then ignite the flame.

(*f*) Spray solvent into the flame.

(*g*) Set the % transmittance to 100.

(*h*) Set the output to absorbance rather than transmittance.

(*i*) Spray solvent again.

(*j*) Now spray the strongest standard solution.

(*k*) Optimise the wavelength by making slight adjustments to increase the absorbance (without obstructing the beam of radiation from the hollow cathode lamp).

(*l*) Next optimise the burner height to increase the absorbance, whilst continuing to spray the strongest standard.

(*m*) Finally spray solvent into the flame.

Successful completion of the above sequence should leave the apparatus ready to collect data and process them under computer control. At each step the operator is prompted to take the appropriate action and the computer waits for the operator to acknowledge completion of the task before moving on to the next step.

6.2.3. Calibration

The next stage is to calibrate the instrumental response to a series of standard solutions of known concentration. For each standard solution the baseline is first monitored whilst the solvent is sprayed into the flame. Then the operator is prompted to spray a standard solution. The computer monitors the signal, then the baseline again, and finally the operator is asked to key in the concentration of the standard. A concentration of zero is used as an abort signal and the instrumental readings obtained for that standard solution are not saved. If the concentration of the standard is non-zero, the reading corrected for baseline absorbance, is stored in an array element Y(ST), for the STth standard. The corresponding concentration is retained in an array element X(ST). The variable ST, which counts the number of standards, is incremented by 1 following the acceptance of data for each standard. The above sequence is repeated until there are no further standards to process, although the dimension statement in the initialisation block in fact limits the number of standards to 10.

6.2.4. Curve Fitting

The curve fitting subroutine uses the NS standards to obtain the best fit to the parameters a, b and c. The readings Y and concentrations X are stored in arrays Y and X. The subroutine which evaluates the coefficients a, b and c uses standard statistical equations in terms of the elements of arrays X() and Y(). The method used is linear multiple regression in two variables. In fact we only have one independent variable X, but, as X^2 is not linearly related to X, it is permissible to treat X and X^2 as independent. Those who would like more detail on regression are recommended to consult a specialist text such as that by Hald (see Overview).

Not only does the program compute the parameters a, b and c to give the 'best fit' of the data, but also the correlation coefficient is obtained. This has a value of 1 if the curve fits perfectly through the points, and a value of 0 if there is no fit at all. For further details the reader is again referred to a specialist text (Hald).

6.2.5. Measurement of Samples

Having carried out the calibration procedure and obtained the equation for the calibration curve, the program then allows the operator to analyse solutions of unknown concentrations of the element in question. As before the operator is prompted through each of the following stages. First solvent is aspirated and the baseline reading so obtained is stored in B0. The sample is then sprayed and the measured signal is assigned to RS. The baseline is checked again by monitoring the signal when solvent is sprayed again. This result is assigned to B1 and the corrected sample reading CR is computed as CR = RS − (B0 + B1)/2. The final stage is to substitute CR for Y in Eq. 6.1 and solve the resulting quadratic equation for the concentration X. To give a feel for the uncertainty of the concentrations so obtained, concentrations are also computed for CR plus or minus 5% as explained earlier. One slight problem is that it is possible that no solution exists for Eq. 6.1 for the reading Y = CR + 5%. This would happen if the reading Y = CR corresponds to the almost flat part of a curved calibration plot, as CR + 5% may well not be on the curve at all. To guard against this difficulty the program checks to see if the computation of the concentration, for a given instrumental reading would involve taking the square root of a negative number. If so the computation is by-passed and an error message is printed.

6.2.6. Linearised Output

If the calibration plot is curved, parameter c in Eq. 6.1 will be non-zero and negative. To linearise the ouput signals we need to enhance the reading Y by an amount equal to $-cX^2$. This is done only if the operator requested linearised output earlier in the program. The linearised reading is printed on the screen along with the computed concentration, correlation coefficient and parameters for the plot following the processing of each sample. Samples are repeatedly processed on the basis of the calibration until the operator indicates that no more samples are to be analysed. In this case the program returns to the main menu.

SAQ 6.2a

The flow chart in Fig. 6.2a has individual blocks identified by letters in the top right-hand corner. In answering the questions below, refer to these block identification letters.

The user of the program starts his run by adding a new element to disc. He then continues by processing 4 calibration standards followed by 2 unknown solutions. By reference to the flow chart in Fig. 6.2a, which one of the following sequences correctly describes the user's route through the system.

(*i*) A,B,E,B,C,F,G,B,E,J,K,L,M,N,O,L,M, N,O,L,M,N,O,L,M,N,O,L,M,N,O,P,Q,R, S,T,B,exit

(*ii*) A,B,C,F,G,B,E,J,K,L,M,N,O,P,Q,R,S,T, Q,R,S,T,B,exit

(*iii*) A,B,C,F,G,B,E,J,K,L,M,N,O,L,M,N, O,L,M,N,O,P,Q,R,S,T,Q,R,S,T,B,exit

SAQ 6.2b

How would you modify the program design given in Fig. 6.2a to prevent the user from trying to produce a calibration curve based on less than 3 standard solutions?

SAQ 6.2c

Suppose the the operator made a mistake and put the same standard solution in twice (with the correct concentration in each case), would this lead to any computational problems for the program as designed?

6.3. STORAGE, RETRIEVAL AND EDITING OF OPERATING CONDITIONS

The program specification requires the facility of storing and retrieving operating conditions for a variety of elements. This amounts to the implementation of a mini database which allows not only storage and retrieval but also the means of deleting and changing existing material and adding new material. To include a full data base system within the program is beyond the scope of this case study, but for our purpose a restricted facility will serve well. To appreciate the problems and how they are solved we shall briefly examine some of the ways information can be stored on disc.

Two sorts of data file can be distinguished in computing. First we have the serial file which is effectively a sequence of data items. To read an item halfway through a sequential file we have to read the data from the beginning until we reach the required item. This can be a lengthy process for big databases and there are more efficient methods that can be used. Another method is to use a so-called random access file. Here information contained within the body of the file can be accessed directly without having to read the data from the beginning.

We are interested in storing a relatively small amount of data. We may have data for say fifty elements on file, and for each element we need to store the following information:

(*a*) the symbol for the element,

(*b*) the slit width and wavelength,

(*c*) the fuel and oxidant,

(*d*) the working range,

(*e*) the lamp current.

These data can all be stored as individual strings of characters and the whole lot can be held in the microcomputer memory at one time. In this special case it is advantageous to store all the data on disc as a sequential file, but to access or manipulate the data we read all of it into a number of one-dimensional arrays. The array EL$ could be used to store the symbols for the elements, so that the *i*th element would be assigned to EL$(I). Similarly we could assign the slitwidth SW$(I), wavelength WV$(I), the fuel oxidant mixture FO$(I), the working range RA$(I), and the lamp current LC$(I). At the head of the data we place the current number of chemical elements (NN say) in the file. The structure of the data file is shown in Fig. 6.3a.

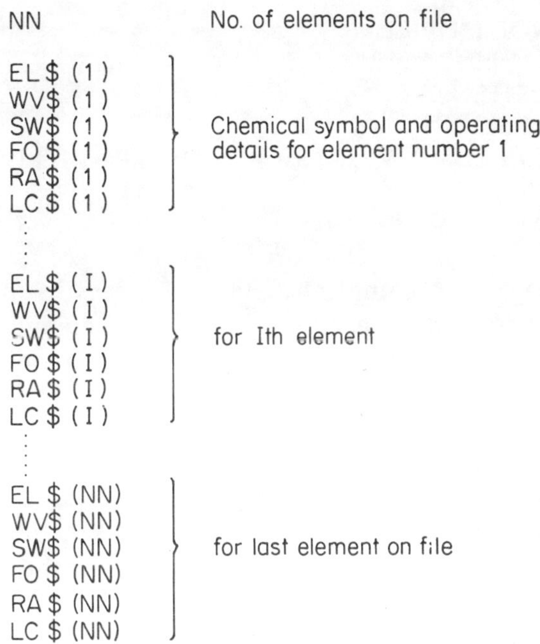

Fig. 6.3a. *Structure of data file for operating conditions*

This structure can be read directly into the computer by the lines of program given below. The precise way a file is opened for reading or writing and the way data are actually read from, or written to, a serial file depends to some extent on the microcomputer used. The following is correct for the BBC microcomputer where line 10 opens a file named DFILE for input and lines 20 and 50 read the data from channel X which has been allocated to DFILE by line 10.

```
10 X=OPENIN"DFILE"
20 INPUT#X,NN:REM read the number of chemical elements on file
30 IF NN=0 THEN 70
40 FOR I = 1 TO NN
50     INPUT#X,EL$(I),WV$(I),SW$(I),FO$(I),RA$(I),LC$(I)
60     NEXT I
70 CLOSE#X
```

Line 70 closes the file for input so that it is released to be written to subsequently by coding such as that given below.

```
10 X=OPENOUT"DFILE"
20 PRINT#X,NN:REM write the number of chemical elements to file
30 IF NN=0 THEN 70
40 FOR I = 1 TO NN
50     PRINT#X,EL$(I),WV$(I),SW$(I),FO$(I),RA$(I),LC$(I)
60     NEXT I
70 CLOSE#X
```

The advantage of reading the data into arrays in the computer is that we have immediate access to any array element. If for example we wish to print the lamp current for the 10th chemical element on the disc we simply use the BASIC statement:

```
100 PRINT LC$(10)
```

Similarly to change the lamp current we could read in the new value from the keyboard as a string of characters using the following statement.

```
200 INPUT LC$(10)
```

If we want to add a new chemical element to the disc, and there are currently NN elements filed, we could read the new data from the keyboard into the appropriate arrays at the NN + 1th array element by the following statement for example.

```
300 I=NN+1
310 INPUT EL$(I),WV$(I),SW$(I),FO$(I),RA$(I),LC$(I)
320 NN=NN+1:REM increment number of chemical elements
```

The operation of insertion and deletion involve more complex operations than those described above. However insertion is not required here since new elements are added onto the end of the data arrays and deletion is a facility which is not permitted to avoid the operator accidentally removing data from the disc. The program does allow the details for an element to be changed but once an element is entered onto the disc the program has no facility for deleting it.

The program is built from a number of subroutines dedicated to a particular task. In Section 6.5, which contains the program listing, the subroutines for reading data from disc into the arrays, for writing data to disc, for adding new data, and for changing existing data, are all referred to by their function and line numbers for ease of reference. An examination of these subroutines will show that they are all built around the principles discussed above.

SAQ 6.3a

Would the following section of program fail if line 30 were omitted?

```
10 X=OPENIN"DFILE"
20 INPUT#X,NN:REM read the number of
                        chemical elements on file
30 IF NN=0 THEN 70
40 FOR I = 1 TO NN
50     INPUT#X, EL$(I),WV$(I),SW$(I),
                        FO$(I),RA$(I),LC$(I)
60     NEXT I
70 CLOSE#X
```

SAQ 6.3a

6.4. DATA CAPTURE

As discussed in Part 3 of this Unit, several approaches are available to convert the analogue signal from the spectrophotometer's recorder output to digital form. Each method requires its own specific equipment and software to drive it. We shall illustrate what can be done using a digital voltmeter with binary coded decimal output. The part of the program devoted to reading the analogue signal is written as a self-contained subroutine which can be replaced with appropriate software to communicate with another type of analogue interface (eg a DVM which uses an IEEE interface).

We shall assume that the DVM provides a BCD readout with 4 digits which are connected to the computer through Ports A and B of an 8255 PPI as shown in Fig. 6.4a.

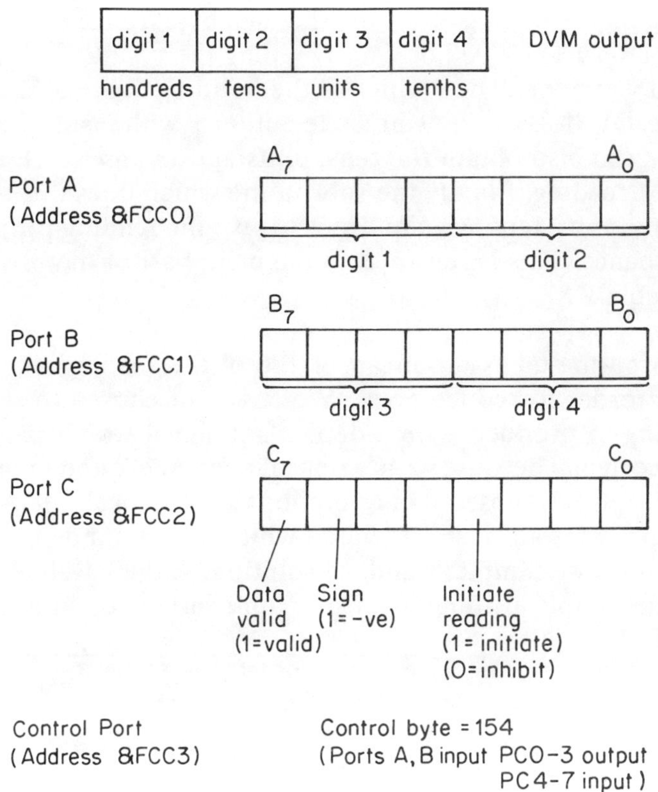

Fig. 6.4a

The subroutine which obtains readings from the DVM works by first outputting a positive pulse to pin C3 of Port C to initiate a reading. The data valid pin, C7, is then monitored. If C7 = 1 then the datum is valid and can be read from Ports A and B. Digits 1 and 2 are obtained together from Port A, with digit 2 occupying the least significant 4 bits and digit 1 fills the other 4 bits. Digits 3 and 4 are read from Port B and correspond to the upper and lower nibbles respectively as shown in Fig. 6.4a.

To decode the individual bytes to yield the values of each digit we need to use the logical AND operator to mask out unwanted bits as discussed at some length in the early parts of this Unit. For example to obtain digit 1 we need to AND the data byte from Port A with the binary pattern 11110000 or 240. Dividing the result by 16 gives the required value for digit 1 which is the hundreds part of the reading from the DVM. By using similar techniques with each of the other digits we can also obtain the tens, units and tenths which make up the DVM reading. Finally the subroutine which does this combines the tenths, units, tens and hundreds to give the reading proportional to absorbance. This is returned to the main part of the program *via* the variable RX.

The subroutine takes the mean of 100 of the readings obtained as above in order to reduce noise. We have not chosen to use digital smoothing to produce a time-dependent signal which is less noisy than the original because we are not so much interested in producing a trace for use in subsequent operations such as peak identification, but simply we want just one value which corresponds to the measurement of the sample, standard solution or the baseline. For this reason the simple arithmetic mean is appropriate to our application.

SAQ 6.4a

Which of the following binary patterns corresponds to DVM reading of +89.4 mV?

	Port A	Port B	Port C
(*i*)	01001001	00001000	10000000
(*ii*)	10000000	01001001	11000000
(*iii*)	00001000	10010100	10000000

6.5. PROGRAM LISTING

Fig. 6.5a contains a complete program listing. The main part of the program starts execution at line 50 and carries on through to line 140. This main block is largely made up from a sequence of subroutine calls to accomplish distinct tasks as detailed below with cross references to identification letters in the design in Fig. 6.2a.

Initialisation (Block A): lines 700–880, this calls the subroutine at 890 to check that the data file called 'DFILE' actually exists. If it does not the error which is produced is trapped and a file named 'DFILE' is created with the only data entry NN = 0. This indicates that no operational details are filed on disc. Having checked the presence of DFILE the subroutine at 2370 is called to load data from disc. The only other functions of this block are to dimension various arrays and configure the interface.

Display opening menu and get response (Block B): This is accomplished through the subroutine at 1060 which scans the keyboard in line 1170 and waits for the user to press a key. The character corresponding to the key press is stored in X$.

SAQ 6.5a

In obtaining a response to the opening menu, would the computer accept lower case rather than capital letters?

Add a new element (Block C): This subroutine at 2610, first checks that the element is not already on file, and, if it is not, the necessary data are read directly into the appropriate array elements and then, by calling another subroutine at 2840, the data as a whole are written to the disc.

Modify data for an element (Block D): This is dealt with by the subroutine at 1220. It first checks to confirm that the element is indeed on file. If it is, the data are displayed on the screen and the user is asked to specify which item (labelled A,B,C,D,E) is to change. Having obtained the new data, the whole data set is written to disc.

Display data on disc (Block H): This is accomplished through the subroutine at 1710 which prints on the screen a table of the elements for which data are already on disc.

Specify the element to be determined (Block E): This option is selected if X = 1 (option A), and is accomplished first through the subroutine at 2480, which scans the array EL$ to find the sequential number of the element in the stored data. If the element is not on file the flagg EFLAG = 1 and this is taken as a cue to print out an error message. If the element is on file then the user is allowed to proceed with the analysis in accordance with line 210 onwards as described below.

The first step in the analytical run is to display the operating conditions using the subroutine at 2270. The next step is to set-up the instrument (Block J) via the subroutine at 1820. It is merely a sequence of print instructions punctuated by scans of the keyboard to check the user has responded.

Having set the standard counter to 1 at line 230 we start the calibration sequence. Here we enter a loop during each cycle of which we process one standard solution *via* the subroutine at 2080. The sequence for each standard is to spray solvent, then standard and then solvent again. During each of these three stages another subroutine is called at 3000 to read the signal from the DVM. If you want to run the program without the DVM connected you should insert the statement,

```
3040 PRINT "INPUT READING": INPUT RX: RETURN
```

in place of the REM statement. Finally the concentration of the standard solution is read and provided it is greater than zero, the reading and concentration are stored in the appropriate elements of the arrays Y and X respectively. This process is repeated until no more standards are required.

Curve fitting (Block P) is done by the subroutine at 3400 by using standard statistical formula involving sums of products and powers of the readings Y and concentrations X. This subroutine returns values for the parameters a,b and c as well as the correlation coefficient which reflects the goodness of fit.

Analysis of unknown solutions (Block Q,R,S) is carried out in lines 430 to 670. Readings are obtained from the DVM during the spraying of solvent before and after the sample and during measurement of the sample. A sample reading corrected for baseline absorbance is calculated and this reading is used to compute a concentration *via* the subroutine at 1520. Similar calculations are also performed for the sample reading plus and minus 5%. Finally, if the user requested it earlier, linearised output is printed on the screen. The whole of this process is repeated until there are no more unknown solutions to analyse.

```
 10 REM **************************
 20 REM ON-LINE DATA PROCESSING
 30 REM APPLIED TO ATOMIC ABSORPTION
 40 REM **************************
 50 GOSUB 700:REM INITIALISATION
 60 REPEAT
 70 GOSUB 1050:REM OPENING MENU
 80 REM RETURNS X=1,2,3,4,5
 90 IF X=1 THEN GOSUB 150
100 IF X=2 THEN GOSUB 2600
110 IF X=3 THEN GOSUB 1220
120 IF X=4 THEN GOSUB 1710
130 UNTIL X=5
140 END
150 REM ****** DISPLAY OPERATING CONDITIONS
160 GOSUB 2480:REM FIND ELEMENT
```

```
170 IF IW>0 THEN 210
180 PRINT TAB(0,22);"Element not on file";
190 Q=INKEY(200)
200 GOTO 680
210 GOSUB 2270:REM DISPLAY DATA
220 GOSUB 1820:REM SET UP INSTRUMENT
230 ST=1: REM INITIALISE STANDARD COUNTER
240 PRINTTAB(0,20);"Hit a key to measure standards": X$=GET$
250 PRINTTAB(0,20);"                              ";
260 GOSUB 2080:REM MEASURE STANDARD
270 PRINTTAB(0,20);"Any more standards (Y/N)?"
280 X$=GET$
290 IF X$<>"Y"AND X$<>"N"THEN 280
300 IF X$="Y"THEN GOTO 260
310 REM ************************
320 REM CURVE FITTING
330 REM ************************
340 CLS
350 NS=ST-1:REM NS STANDARDS
360 PRINTTAB(0,5);"Do you wish to have linearised output";
370 PRINTTAB(0,7);"if calibration plot is curved (Y/N)?";
380 X$=GET$
390 IF X$<>"Y"AND X$<>"N"THEN 380
400 IF X$="Y"THEN DG=2 ELSE DG=1
410 GOSUB 3400:REM GET PARAMETERS
420 REM **************************
430 REM MEASUREMENT OF AN UNKNOWN SOLUTION
440 REM ***************************
450 PRINTTAB(0,20);"Spray solvent"; :REM BASELINE
460 GOSUB 2920
470 GOSUB 3000:REM GET READING
480 B0=RX:REM STORE INITIAL BASELINE
490 PRINTTAB(0,20);"Spray sample";
500 GOSUB 2920
510 GOSUB 3000:REM GET READING
520 RS=RX
530 PRINTTAB(0,20);"Spray solvent";
540 GOSUB 2920
550 GOSUB 3000:REM GET READING
560 B1=RX:REM STORE BASELINE AFTER MEASUREMENT
570 CR=RS-(B1+B0)/2:REM CORRECTS FOR BASELINE
                                        ABSORBANCE
580 GOSUB 1520:REM OBTAINS CX
```

```
590 PRINTTAB(0,10);"Concentration = "; INT(1000*CX(2))/
                                                  1000; " ppm"
600 PRINT TAB(0,12);"(+,-5% in reading: ";INT(1000*CX(1))/
                          1000; "->";INT(1000*CX(3))/1000;")";
610 IF DG=2 THEN PRINTTAB(0,14);"Linearised output: ";
                                          INT(1000*LR)/1000;
620 GOSUB 2920
630 CLS
640 PRINTTAB(0,10);"Any more samples?"
650 X$=GET$
660 0 IF X$<>"Y"AND X$<>"N"THEN 650
670 IF X$="Y"THEN GOTO 420
680 RETURN
690 REM ****************************
700 REM INITIALISATION ROUTINE
710 REM ****************************
720 DIM EL$(50),SW$(50),WV$(50),FO$(50),RA$(50)
730 DIM RS(10),LC$(50),Y(10),X(10),CS(3),CX(3),R(3)
740 GOSUB 890:REM CHECK PRESENCE OF DFILE
750 GOSUB 2370:REM READ OPERATING CONDITIONS
760 ?&FECC3=154
770 REM PORTS A & B INPUT, PC0-PC3 OUTPUT, PC4-PC7 INPUT
780 REM NOW THE SCREEN
790 CLS
800 PRINT TAB(0,5);"On-line Data Processing ";
810 PRINT TAB(0,7);"for AA Spectrophotometry";
820 PRINTTAB(0,11);"This program requires an AA instrument";
830 PRINTTAB(0,13);"to be connected on-line via a DVM."
840 PRINT TAB(0,20);"PLEASE CHECK THE EQUIPMENT IS
                                                  CONNECTED";
850 PRINT TAB(0,22);"and hit a key when you are ready.";
860 X$=GET$
870 CLS
880 RETURN
890 REM ****************************
900 REM CHECKS PRESENCE OF DATA FILE
910 REM ****************************
920 GOTO 1040
930 ON ERROR GOTO 970
940 XX=OPENIN"DFILE"
950 ON ERROR OFF
960 GOTO 1030
970 ON ERROR OFF
```

```
 980 XX=OPENOUT"DFILE"
 990 NN=0
1000 PRINT#XX,NN
1010 CLOSE#XX
1020 GOTO 940
1030 INPUT#XX,NN
1040 RETURN
1050 REM **************************
1060 REM OPENING MENU
1070 REM **************************
1080 CLS
1090 PRINTTAB(0,5);"Opening Menu";
1100 PRINTTAB(0,10);"(A) Specify an element to be determined";
1110 PRINTTAB(0,12);"(B) Define operating conditions";
1120 PRINTTAB(0,14);"(C) Alter existing conditions";
1130 PRINTTAB(0,16);"(D) List elements on file";
1140 PRINT TAB(0,18);"(E) Exit";
1150 PRINT TAB(0,20);"WHAT IS YOUR CHOICE?";
1160 REPEAT
1170 X$=GET$
1180 X=ASC(X$)
1190 UNTIL X>64 AND X<70
1200 X=X-64
1210 RETURN
1220 REM **********************
1230 REM MODIFIES DISC DATA
1240 REM **********************
1250 IF NN>0 THEN 1300
1260 CLS
1270 PRINT TAB(0,10);"No data on file to alter!";
1280 Q=INKEY(200)
1290 GOTO 1510
1300 CLS
1310 GOSUB 2480:REM SELECT AN ELEMENT
1320 IF IW>0 THEN 1350
1330 PRINTTAB(0,22);"Element";WE$;"not on file";
1340 Q=INKEY(200):GOTO 1510
1350 GOSUB 2270 :REM DISPLAY DATA
1360 PRINTTAB(0,22);"Which item is to change(A,B,C,D,E)?";
1370 REPEAT
1380 X$=GET$
1390 Z=ASC(X$)
1400 UNTIL Z>64 AND Z<70
```

```
1410 PRINTTAB(0,22);"                              ";
                              TAB(0,22);"Input the new information";
1420 INPUT X$
1430 ON Z-64 GOTO 1440,1450,1460,1470,1480
1440 WV$(IW)=X$:GOTO 1490
1450 SW$(IW)=X$:GOTO 1498
1460 FO$(IW)=X$:GOTO 1490
1470 RA$(IW)=X$:GOTO 1490
1480 LC$(IW)=X$
1490 REM
1500 GOSUB 2840:REM WRITE NEW DATA
1510 RETURN
1520 REM ******************
1530 REM FINDS CONC OF SAMPLE READING CR
1540 REM AND FOR CS +,-5%
1550 REM *******************
1560 R(1)=CR-0.05*CR
1570 R(2)=CR
1580 R(3)=CR+0.05*CR
1590 QA=C:QB=B:REM B,C ARE PARAM. IN BEST FIT EQ.
1600 FOR J=1 TO 3
1610 QC=A-R(J):REM A IS A PARAM IN BEST FIT EQ.
1620 XX=SQR(QB*QB-4*QA*QC)
1630 IF ABS(QA)<0.00001 THEN CX(J)=-QC/QB:GOTO 1670
1640 CX1=(-QB+XX)/(2*QA)
1650 CX2=(-QB-XX)/(2*QA)
1660 IF CX1<0 THEN CX(J)=CX2 ELSE CX(J)=CX1
1670 NEXT J
1680 LR = CR -C*CX(2)^2:REM LINEARISED OUTPUT
1690 RETURN
1700 REM *************************
1710 REM DISPLAYS ELEMENTS ON FILE
1720 REM *************************
1730 CLS
1740 PRINT "List of elements on file :": PRINT
1750 IF NN=0 THEN 1810
1760 FOR I=1 TO NN STEP 9
1770 PRINT:PRINTEL$(I);" "; EL$(I+1);" "; EL$(I+2);" ";
          EL$(I+3) : " "; EL$(I+4);" "; EL$(I+5);" "; EL$(I+6);" ";
                                          EL$(I+7);" "; EL$(I+8)
1780 NEXT I
1790 PRINT:PRINT:PRINT"Hit a key";
1800 X$=GET$
```

```
1810 RETURN
1820 REM ******************
1830 REM SETS UP INSTRUMENT
1840 REM ******************
1850 PRINTTAB(0,20);"Switch on the power to the AA, DVM";
1860 GOSUB 2920:REM WAIT FOR RESPONSE
1870 PRINTTAB(0,20);"Switch on the lamp for "; WE$;
1880 GOSUB 2920
1890 PRINTTAB(0,20);"Set the lamp current to "; LC$(IW);
1900 GOSUB 2920
1910 PRINTTAB(0,20);"Set the wavelength to "; WV$(IW);
1920 GOSUB 2920
1930 PRINTTAB(0,20);"Set fuel/oxidant flow & ignite flame";
1940 GOSUB 2920
1950 PRINTTAB(0,20);"Spray solvent into the flame";
1960 GOSUB 2920
1970 PRINTTAB(0,20);"Set 0% transmission";
1980 GOSUB 2920
1990 PRINTTAB(0,20);"Switch to absorbance readings";
2000 GOSUB 2920
2010 PRINTTAB(0,20);"Spray the strongest standard";
2020 GOSUB 2920
2030 PRINTTAB(0,20);"Optimise wavelength to give max. abs.";
2040 GOSUB 2920
2050 PRINTTAB(0,20);"Optimise the burner height";
2060 GOSUB 2920
2070 RETURN
2080 REM *********************
2090 REM MEASURES A STANDARD SOLN.
2100 REM *********************
2110 PRINTTAB(0,20);"Spray solvent";
2120 GOSUB 2920
2130 GOSUB 3000:B0=RX:REM GET READING
2140 PRINT TAB(0,20);"Spray standard No. "; ST;
2150 GOSUB 2920
2160 GOSUB 3000: STAN=RX:REM GET READING RX
2170 PRINTTAB(0,20);"Spray solvent": GOSUB 2920
2180 GOSUB 3000:B1=RX
2190 PRINTTAB(0,20);"What was the concentration (ppm)";
2200 PRINTTAB(0,22);"To cancel the measurement put 0 ppm";
2210 INPUT CS
2220 IF CS=0 THEN 2260
2230 Y(ST)=STAN-(B0+B1)/2
```

```
2240 X(ST)=CS
2250 ST=ST+1
2260 CLS:RETURN
2270 REM ********************
2280 REM DISPLAYS WANTED DATA
2290 REM ********************
2300 PRINT TAB(0,2);"Operating conditions for "; EL$(IW);
2310 PRINT TAB(0,6);"(A) Wavelength: "; WV$(IW);
2320 PRINT TAB(0,8);"(B) Slit width: "; SW$(IW);
2330 PRINT TAB(0,10);"(C) Fuel/Oxidant: "; FO$(IW);
2340 PRINT TAB(0,12);"(D) Working range: "; RA$(IW);
2350 PRINTTAB(0,14);"(E) Lamp current "; LC$(IW);
2360 RETURN
2370 REM **************************
2380 REM LOADS DATA
2390 REM **************************
2400 XX=OPENIN"DFILE"
2410 INPUT#XX,NN
2420 IF NN=0 THEN 2470 :REM EXIT IF NO ELEMENTS ON FILE
2430 FOR I=1 TO NN
2440 INPUT#XX,EL$(I),WV$(I),SW$(I),FO$(I),RA$(I),LC$(I)
2450 NEXT I
2460 CLOSE#XX
2470 RETURN
2480 REM **********************
2490 REM CHOOSE AN ELEMENT
2500 REM **********************
2510 CLS
2520 IW=0
2530 PRINT TAB(0,10);"Which element";
2540 INPUT WE$
2550 IF WE$="0"THEN 2590
2560 FOR I=1 TO NN
2570 IF EL$(I)=WE$ THEN IW=I
2580 NEXT I
2590 RETURN
2600 REM ************************
2610 REM ADD NEW DATA
2620 REM ************************
2630 GOSUB 2480
2640 IF IW<1 THEN 2680
2650 PRINTTAB(0,22);"Element already on file";
2660 Q=INKEY(200)
```

```
2670 GOTO 2830
2680 NN=NN+1
2690 CLS
2700 PRINT"Data for "; WE$;
2710 EL$(NN)=WE$
2720 PRINTTAB(0,8);"What is the wavelength";
2730 INPUT WV$(NN)
2740 PRINTTAB(0,10);"What is the slitwidth";
2750 INPUT SW$(NN)
2760 PRINTTAB(0,12);"What is the Fuel/Oxidant";
2770 INPUT FO$(NN)
2780 PRINTTAB(0,14);"What is the working range";
2790 INPUT RA$(NN)
2800 PRINT TAB(0,16);"What is the lamp current";
2810 INPUT LC$(NN)
2820 GOSUB 2840:REM WRITE DATA
2830 RETURN
2840 XX=OPENOUT"DFILE"
2850 PRINT#XX,NN
2860 IF NN=0 THEN 2900
2870 FOR I=1 TO NN
2880 PRINT#XX,EL$(I),WV$(I),SW$(I),FO$(I),RA$(I),LC$(I)
2890 NEXT I
2900 CLOSE#XX
2910 RETURN
2920 REM ************
2930 REM GET KEYBOARD RESPONSE
2940 REM ************
2950 PRINTTAB(0,22);"Press the space bar when you are ready.";
2960 X$=GET$
2970 PRINTTAB(0,22);"                                        ";
2980 PRINTTAB(0,20);"                                        ";
2990 RETURN
3000 REM ******************
3010 REM GETS READING FROM DVM
3020 REM ***********************
3030 CLS
3040 REM
3050 REM *** START MULTIMETER ***
3060 REM *** BY SENDING Hold* HIGH ***
3070 REM *** BIT 6 OF PORT C (&FCC1) ***
3080 ?&FCC2=8
3090 REM *** NOW TAKE 10 * 5 SECOND READINGS ***
```

```
3100 NR=0
3110 SU=0
3120 PRINT TAB(0,5);"MONITORING";
3130 FOR KK=1 TO 100
3140 REM *** SEND Hold* LOW ***
3150 ?&FCC2=0
3160 REM *** NOW WAIT FOR VALID DATA ***
3170 REPEAT UNTIL ?&FCC2 AND 128 = 128
3180 REM *** READ PORT A ***
3190 PORTA=?&FCC0
3200 REM *** READ PORT B ***
3210 PORTB=?&FCC1
3220 REM *** READ PORT C ***
3230 PORTC=?&FCC2
3240 REM *** RESTART MULTIMETER ***
3250 ?&FCC2=8
3260 REM *** DECODE BOTH BCD DIGITS IN EACH PORT ***
3270 D2=PORTA AND 15
3280 D1=INT(PORTA/16)
3290 D4=PORTB AND 15
3300 D3=INT(PORTB/16)
3310 REM *** WEIGHT AND ADD EACH DIGIT FOR
                                        ORIGINAL NUMBER ***
3320 RX=D1*1000+D2*100+D3*10+D4
3330 REM *** CORRECT FOR NEGATIVE NUMBER ***
3340 IF PORTC AND 64 THEN RX = -RX
3350 SU=SU+RX
3360 NEXT KK
3370 RX=SU/100
3380 PRINT TAB(0,10);"READING = "; RX;
3390 RETURN
3400 REM ************************
3410 REM FIT Y = A + BX + CX^2
3420 REM CHOICE OF LINEAR EQ. SETS C=0
3430 REM ************************
3440 REM INITIALISE SUMMING VARIABLES
3450 SX2=0:SY2=0:SZ2=0:SX=0:SZ=0:SXZ=0:SY=0:SZY=0:S XY=0
3460 RECAL=0
3470 FOR I=1 TO NS
3480 SX2=SX2+X(I)*X(I)
3490 SZ2=SZ2+X(I)^4
3500 SX=SX+X(I)
3510 SZ=SZ+X(I)^2
```

```
3520 SXZ=SXZ+X(I)^3
3530 SY=SY+Y(I)
3540 SY2=SY2+Y(I)*Y(I)
3550 SZY=SZY+X(I)*X(I)*Y(I)
3560 SXY=SXY+X(I)*Y(I)
3570 NEXT I
3580 MX=SX/NS:MY=SY/NS:MZ=SX2/NS:REM MEANS
3590 PA=SX2-SX*SX/NS
3600 PB=SZ2-SZ*SZ/NS
3610 PC=SXZ-SX*SZ/NS
3620 PD=SXY-SX*SY/NS
3630 PE=SZY-SZ*SY/NS
3640 XX=PA*PB-PC*PC
3650 IF ABS(XX)>0.001 THEN 3700
3660 PRINT TAB(0,20);"EQUATIONS ILL-CONDITIONED
                                          -RECALIBRATE";
3670 GOSUB 2920
3680 RECAL=1
3690 GOTO 3800
3700 B=(PD*PB-PE*PC)/XX
3710 C=(PA*PE-PC*PD)/XX
3720 A=MY-B*MX-C*MZ
3730 R2=(B*PD+C*PE)/(SY2-SY*SY/NS)
3740 PRINTTAB(0,10);"Calibration Y = A + B*X + C*X*X";
3750 PRINT TAB(0,12);"with,"; TAB(0,14);"A = "; INT(A*10000)/10000
3760 PRINT"B = "; INT(B*10000)/10000
3770 PRINT"C = "; INT(C*10000)/10000
3780 PRINT:PRINT"Correlation coefficient= "; INT(R2*10000)/10000;
3790 GOSUB 2920
3800 RETURN
```

Fig. 6.5a. *Program listing*

Summary

Section 6.1 specified the problem to be solved and provide a detailed specification for the program. This should have illustrated to you the importance of thinking about the problem in detail before you start to design a program. You should be clear about the nature of the data, how they are to be processed and stored, and finally how the results should be presented.

Section 6.2 concentrated on program design and included a flow chart for the program to be developed. The process of producing a program design from a program specification is to some extent intuitive and the ability to do this improves with practice. In this case we used a number of simple techniques to build a program. First we had sequences of identifiable blocks of program, as indicated in the flow chart. Second we used selection, where the sequence of blocks to be executed was determined by choice introduced *via* a menu. Finally we used iteration where repeated similar processing was needed for a series of analytical standard solutions, or for processing a number of unknowns. If you try to design your programs with these concepts (sequence, selection and iteration) in mind you may find it easier to translate a specified problem into a design.

Sections 6.3 and 6.4 discussed specific aspects of the program, namely information storage and retrieval, and data capture using the 8255 PPI. The material of 6.4 should have been fairly straightforward, but you should have obtained the impression from 6.3 that there is a great more to learn about information retrieval, particularly with respect to large databases, and specialised texts on computer science should be consulted for further information.

Finally Section 6.5 contained the program listing which you should have been able to relate to the design given in Section 6.2.

Objectives

On completing this part you should be able to:

- read and understand a program specification involving on-line data processing in analytical chemistry; (SAQs 6.1a, 6.1b)

- relate a program design, presented as a flow chart with supporting information, to the original specification and describe typical routes through the program; (SAQs 6.2a, 6.2b, 6.2c, 6.5a)

- describe how a general-purpose digital voltmeter can be used to capture data from an analytical instrument. (SAQ 6.4a)

Self Assessment Questions and Responses

SAQ 1.1a

What are the decimal numbers which correspond to each of the following bytes?

(*i*) 01101101

(*ii*) 10000000

(*iii*) 11111111

Response

(*i*) The answer is 109. For any other answer check your working against the following in which the value of each bit is multiplied by its weight to determine its contribution to the number.

Bit					
Bit d0 = 1	which contributes	1 ×	1 =	1
Bit d1 = 0	,, ,,	0 ×	2 =	0
Bit d2 = 1	,, ,,	1 ×	4 =	4
Bit d3 = 1	,, ,,	1 ×	8 =	8
Bit d4 = 0	,, ,,	0 ×	16 =	0
Bit d5 = 1	,, ,,	1 ×	32 =	32
Bit d6 = 1	,, ,,	1 ×	64 =	64
Bit d7 = 0	,, ,,	0 ×	128 =	0
	TOTAL			109

(*ii*) The answer is 128 as only bit d7 is set to 1 and its weight is 128 as indicated in the answer to (*i*) above.

(*iii*) Here all the bits are set to 1 so we obtain the highest number that can be represented in terms of 8 bits, which is 255. If you did not obtain this answer just add up the weights of all the bits following the method used in (*i*) above. You should obtain 1 + 2 + 4 + 8 + 16 + 32 + 64 + 128 which equals 255.

****

SAQ 1.1b

Write the 8-bit binary numbers which correspond to each of the following decimal numbers.

(*i*) 38
(*ii*) 240
(*iii*) 15

Response

(*i*) The answer is 00100110. If you did not obtain this answer check the steps in your working against the following:

The given number is 38. Starting with the *most significant bit* (d7) we compare its weight (128) with 38. The weight of bit d7 is too large to contribute so the value of d7 must be 0. Similarly d6, which has a weight of 64, is too large so d6 = 0. The next bit is d5 with a weight of 32 which is less than 38 and so d5 contributes and has a value of 1. Before undertaking further tests we must first subtract the contribution of d5 (32) from 38 which leaves a residue of 6. We now compare the weights of the remaining bytes in turn against the residue 6 to complete the sequence shown below.

d7(weight 128):	as $128>38$	d7 = 0	residue 38
d6(,, 64):	as $64>38$	d6 = 0	,, 38
d5(,, 32):	as $32<38$	d5 = 1	,, 6
d4(,, 16):	as $16>6$	d4 = 0	,, 6
d3(,, 8):	as $8>6$	d3 = 0	,, 6
d2(,, 4):	as $4<6$	d2 = 1	,, 2
d1(,, 2):	as $2=2$	d1 = 1	,, 0
d0(,, 1):	as $1>0$	d0 = 0	,, 0

Therefore the binary equivalent of 38 is 00100110.

(*ii*) The binary equivalent of 240 is 11110000 which can be seen by summing the weights of the bits with a value of 1 (128 + 64 + 32 + 16 = 240). If you did not obtain this check your working again using the type of sequence given in (*i*) above.

(*iii*) The 8 bit binary equivalent of 15 is 00001111. You can check this by working back from the binary. If you obtained any other answer you are advised to follow through the corresponding steps of the sequence in (*i*) above.

****

SAQ 1.1c

Write each of the following binary numbers in hexadecimal.

(*i*) 01110011

(*ii*) 11111111

(*iii*) 11100001

Response

(*i*) The answer is &73. If you obtained some other value work through the following steps:

(1) Divide the given byte (01110011) into two groups of 4 bits: 0111 0011.

(2) Considering each nibble as separate quantities, convert each to a hexadecimal digit: 0111 = &7 and 0011 = &3.

(3) Write the two hexadecimal digits out as a single hexadecimal number: &73 = 01110011.

As a check the decimal equivalent of 01110011 is 115 and the decimal equivalent of &73 is 7 x 16 + 3 = 115.

(*ii*) The answer is &FF which is the highest value which can be held in 8 bits. You can easily check the answer by realising that each of the two nibbles is binary 1111 which corresponds to the hexadecimal digit &F.

(*iii*) The answer this time is &E1. The &E corresponds to the most significant nibble (1110) and the other digit is obtained from the lowest 4-bits (0001).

****

SAQ 1.1d

Which of the following nibbles represent valid binary-coded decimal (BCD) digits?

	Nibble	Valid
(*i*)	1000	Y/N
(*ii*)	1001	Y/N
(*iii*)	1010	Y/N
(*iv*)	1111	Y/N

Response

The answers are given below with the corresponding decimal digit in brackets:

	Nibble	Valid (Y/N)	Decimal digit and comment
(*i*)	1000	Y	8, needs single decimal digit
(*ii*)	1001	Y	9, ,, ,, ,, ,,
(*iii*)	1010	N	10, needs two decimal digits
(*iv*)	1111	N	15, ,, ,, ,, ,,

If you found difficulty with any of the above you probably have not appreciated that BCD 4 bit patterns represent only single decimal digits. Since these range from 0–9 the 4-bit numbers range from 0000 to 1001 only.

****

SAQ 1.1e

Match each of the memory addresses on the left with one of the alternatives (*a*)–(*g*) on the right.

(*i*) &F000	(*a*) 61440
	(*b*) &0400
(*ii*) 1023	(*c*) &00FF
	(*d*) &03FF
(*iii*) &00F0	(*e*) 65535
	(*f*) 240
(*iv*) Binary 1111111111111111	(*g*) 15

Response

Left Column		Right Column	Comment
(*i*) &F000	⟶	(*a*) 61440	The weight of the most significant hexadecimal digit is 4096, and $15 \times 4096 = 61440$
(*ii*) 1023	⟶	(*d*) &03FF	1024 is &0400 and we want one less which is &03FF.
(*iii*) &00F0	⟶	(*f*) 240	The weight of the digit F in this hexadecimal number is 16. The decimal equivalent of F is $15 \times 16 = 240$.

(*iv*) 1111111111111111 ⟶ (*e*) 65535 This is the highest number which can be contained in 16 bits. In hexadecimal it is represented as &FFFF or $15 \times 4096 + 15 \times 256 + 15 \times 16 + 15$

****

SAQ 1.2a

> By choosing the most appropriate phrase from the list below, complete the following sentence.
>
> The microprocessor is .... part of a microcomputer system.
>
> (*i*) the most important
>
> (*ii*) the most expensive
>
> (*iii*) a desirable optional

Response

The correct answer is (*i*) so that the sentence reads:

'The microprocessor is the most important part of a microcomputer system'.

If you chose (*ii*) you were wrong because microprocessor chips are now quite cheap since the volume of production has been very large. In fact many items of equipment needed to complete a viable microcomputer system tend to be more expensive than the microprocessor itself.

Option (*iii*) is incorrect because you cannot have a microcomputer system without a microprocessor.

****

SAQ 1.2b

By ringing either T or F, indicate whether each of the following statements is true or false.

Microprocessors can undertake the following operations directly:

(*i*) multiplication or division (T / F)

(*ii*) addition or subtraction (T / F)

(*iii*) comparison of data bytes (T / F)

Response

Answer (*i*) is false as microprocessors are limited in their arithmetic capability to addition and subtraction. Clearly microcomputers can do multiplication and division, as well as quite complex mathematical functions, but only by using small programs provided by the manufacturer of the microcomputer system.

Answer (*ii*) is correct.

Answer (*iii*) is also correct. This is a valuable facility which allows the microprocessor to branch to different parts of the program depending on the result of a comparison between bytes of data

****

SAQ 1.2c

Circle T or F to indicate whether each of the following is true or false.

(*i*) RAM can be used for the long-term storage of programs.

(T / F)

(*ii*) ROM is used for the storage of programs written by analytical chemists because the program is always available for use.

(T/ F)

(*iii*) EPROM can be programmed using an ordinary microcomputer in the same way as RAM.

(T / F)

(*iv*) The contents of RAM can be made non-volatile by the provision of supplementary power, in the form of a battery, which becomes active when the mains power is switched off.

(T / F)

(*v*) Prototype programs should never be committed to ROM.

(T / F)

(*vi*) All microcomputer systems use some programs contained in non-volatile memory but not all may require programs in RAM.

(T / F)

Response

(*i*) This is false because as soon as the power supply is removed the contents of RAM (volatile-memory) are lost. See the answer to (*iv*) for the use of batteries to preserve the contents of RAM.

(*ii*) This is false because the user of the microcomputer cannot store his program in ROM. This type of memory is filled with information by the manufacturer and cannot be changed by the user. You cannot therefore write programs to ROM to store them permanently.

(*iii*) This is also false. The electronic signals in a microcomputer which are used to change the contents of RAM are not able to alter the contents of EPROM locations. This can only be done by firstly erasing the contents by irradiation of the chip with ultra-violet light and then reprogramming with a PROM programmer at considerably higher voltage levels than used by a microcomputer.

(*iv*) This is true. Some RAM packs have batteries which maintain the power when the mains supply to the computer is switched off.

(*v*) This is true. Only final versions of program should be committed to ROM. The reason is that ROM is used for microprocessor-based products with high volume sales. It is expensive to produce the first ROM chip due to the nature of the process. However, once the first one has been produced it is relatively easy to produce copies so that the unit cost falls rapidly with the volume of production. If you put a program on ROM which is not fully tested you could have thousands of faulty ROMs to replace!

(*vi*) This is true. When power is applied to a microcomputer system a program is run automatically which, for example, will display a message and/or scan the keyboard or keypad to take in the user's instructions. Most microcomputer systems have a certain amount of RAM for user programs or data, but it is possible to imagine a microcomputer whose actions have been completely pre-programmed in ROM and which requires no data. A sequence controller, eg for a washing machine, could be programmed like this to carry out a sequence of actions with appropriate time delays between the various steps. The actions and time delays could in principle be fixed in advance and the control program committed to ROM. In this case there would be no need for any RAM. On the other hand it is difficult to imagine any microcomputer system without some program stored in ROM.

****

SAQ 1.2d

Circle T or F to indicate whether the statements below are true or false.

(*i*) During execution of the program the machine code instructions to be accessed by the microprocessor must be placed on either magnetic disc or tape.

(T / F)

(*ii*) All microcomputer systems must have facilities for either disc or tape storage.

(T / F)

(*iii*) Tape storage is in general inferior to disc storage ONLY because of its lower reliability with respect to retrieval of information.

(T / F)

(*iv*) Magnetic materials can adversely affect the integrity of information stored on tape or disc.

(T / F)

Response

(*i*) This is false because the microprocessor has no means of directly accessing information stored on tape or disc. The program must first be placed in main memory (RAM) by loading it from peripheral memory (tape or disc). The microprocessor is only able to fetch and execute instructions which are contained in main memory.

(*ii*) This is also false because the program may be wholly contained in non-volatile memory (eg ROM or EPROM). The program would then not be lost when the power is removed from the system and there would be no need for disc or tape storage.

(*iii*) This is false because audio-tape is inferior to disc storage for reasons other than unreliability. Programs take much longer to load from tape than disc and also they take longer to save. Tape is a linear medium which can mean a lot of tape winding to locate the spot where a particular program or data file is stored. In contrast rapid access is possible to any part of a disc which saves a great deal of time particularly when information is being filed and retrieved during the execution of a program.

(*iv*) This is true and is an important point to remember. A magnet, magnetic strip, a transformer or even a telephone can corrupt the information on a disc. Problems can also arise if you leave a disc clamped in a disc drive when the mains power is switched on or off. It is possible for spurious signals at the disc write-head to corrupt the disc. This may not happen often but it is a wise precaution to open the disc drive doors before switching the power on or off. Floppy discs can also be damaged by bending them, writing on them with anything other than a felt-tipped pen, by touching the magnetic surface or even by leaving them out of their protective sleeve to gather dust.

****

SAQ 1.2e

Assuming that you are using a microcomputer which has a keypad equipped with the digits 0–9, the letters A, B, C, D, E and F, and other specialist keys appropriate to the application in question, indicate by circling Y or N whether you could (Y) or could not (N) accomplish the following:

(*i*) Input a machine code program in hexadecimal.

(Y / N)

(*ii*) Specify numeric data such as the number of a sample.

(Y / N)

(*iii*) Program the microcomputer in BASIC.

(Y / N)

(*iv*) Specify the operating parameters of an instrument such as an infra-red spectrophotometer.

(Y / N)

Response

(*i*) Yes. The keypad would suffice for this as all the instructions used by the microprocessor, often called machine code instructions, can be expressed in hexadecimal which needs the digits 0–9 and the letters A to F.

(*ii*) Yes, this is possible too. Indeed many microprocessor-based analytical instruments provide a simple keypad for such a purpose.

(*iii*) No. This would be impossible as you would need a full QWERTY keyboard to type in the program.

(*iv*) Yes, provided that a menu appeared on the display screen which relates the numbers on the keys to the functions needed to operate the spectrophotometer. An alternative approach is to arrange the keypad so that a sequence of keys is placed along the bottom of the screen. The significance of each key can then be indicated by a label which can be printed on the screen directly above the key. The purpose of each key can then be changed depending on the type of operation (such as setting the scan rate or resolution). This is often called a 'soft keypad' and is becoming more common with the more sophisticated laboratory instruments.

****

SAQ 1.2f

A small microcomputer allows the programmer to select one of a number of modes of screen display with different resolutions. The highest resolution (640 × 256) has 640 pixels in a row and the screen is filled by 256 such rows. A medium resolution mode is available with 320 × 256 pixels. Both modes use main memory to store data on each pixel. For a monochrome display, which you can assume is in use in this example, each pixel is represented by 1 bit.

(*i*) How much memory has to be reserved when using

(*a*) high resolution,
(*b*) medium resolution?

(Express your answer in Kbytes (1K = 1024 bytes).

(*ii*) If the program takes 13K of memory and the computer has a total of 32K of memory available, could the high resolution graphics mode be used?

Response

(*i*) As each pixel is represented by 1 bit, the total number of bits is simply the number of pixels in a row times the number of rows. To get the number of bytes we simply divide by 8, and then divide by 1024 to get the number of kilobytes. The arithmetic is then:

(*a*) (640 × 256)/(8 × 1024) = 20K

(*b*) (320 × 256)/(8 × 1024) = 10K

(*ii*) If the program takes 13K of memory, then we have only 32 − 13 = 19K for the screen display. Since the high resolution mode takes 20K, we would have to be content with medium resolution graphics. This would be adequate for many purposes but not for the representation of infra-red spectra for example.

****

SAQ 1.2g

(*i*) Is an interface necessary if we wish to read data into a computer from an instrument which provides digital signals (0 and 5 volts)?

(*ii*) Is an interface necessary if we wish to switch on or off, under computer control, an electric motor which runs on 5 volts?

Response

(*i*) Yes an interface is necessary. Although the voltage levels are compatible with the computer system, for correct operation of the microprocessor when reading data, the electronic signals from an external device are only allowed to reach the microprocessor data lines when an instruction reading the data port is being executed. Therefore we need some way of allowing the input signals to reach the data lines at precisely the right instant, and then we must electronically disconnect the data lines from the instrument. This would be the job of the interface in this case.

(*ii*) Again yes! This time we have a different problem in which the output signal produced by the microcomputer lasts for only a few microseconds, and although it maybe of the order of 5 volts it would supply very little drive current. The interface in this case has to catch the transient output signal and also increase the current to drive the motor.

****

SAQ 1.3a

(*i*) What is the start address of the program in Fig. 1.3a?

(*ii*) By choosing T(true) or F(false) comment on the validity of the the following statements.

(1) A machine code program designed to run on one type of 8 bit microprocessor can be executed by any other 8 bit microprocessor.

(T / F)

(2) Machine code programs execute at the maximum possible speed for a given computer system.

(T / F)

SAQ 1.3a (cont.)

(3) Machine code programs are difficult to edit.

(T / F)

(4) Machine code programs are difficult to read and understand.

(T / F)

Response

(*i*) The start-address is the address in memory of the first executable instruction in the program. In this case it is &19B7.

(*ii*) (1) This is false as there are considerable difference in the instruction codes from one microprocessor to another.

(2) This is true. The microprocessor executes instructions very rapidly in strict synchronisation with timing pulses which are generated typically at a rate of a few million per second. It is therefore quite possible that the countdown from 255 to 0 would be completed in a few milliseconds.

(3) This is certainly true. If we need to change the place of a label in an assembler program, no alteration is imposed on any other part of the program. On the other hand machine code programs use memory addresses in place of labels. A change in the address (machine code label) of an instruction, referred to by a second instruction, would require a modification of the second instruction too. This type of problem makes editing machine code programs both tedious and difficult.

(4) This is also true. You could only read machine code programs fluently if you took the trouble to memorise the entire set of instruction codes and their function.

****

SAQ 1.3b

The 6502 assembler program below counts down from decimal 100 to 0, by steps of 1, repeatedly.

```
.START  LDA  ?100
        TAX
.LOOP   DEX
        BNE  LOOP
        JMP  START
```

(*i*) With reference to the above program, match the terms in the list on the left, below, with the correct explanatory comment in the list on the right.

(1) .START	(*a*) Branch to the instruction labelled LOOP
(2) TAX	(*b*) Jump to the instruction labelled START
(3) DEX	(*c*) Transfer the contents of A to X
(4) JMP START	(*d*) Load register A
	(*e*) Increment the contents of A by 1
	(*f*) Decrement the contents of X by 1
	(*g*) A label used to identify a point in the program

(*ii*) By selecting T(true) or F(false), indicate which of the following statements are true or false.

(1) Assembler code, like machine code, can be directly executed by the microprocessor.

(T / F)

⟶

SAQ 1.3b (cont.)

(2) An assembler program for one type of microprocessor can be assembled and run on another type of microprocessor. (T / F)

(3) Once an assembler program has been converted to machine code it will run as fast as one which was written directly in machine code. (T / F)

Response

(*i*) The correct matches are as follows:

(1) .START (*g*) A label used to identify a point in the program.

(2) TAX (*c*) Transfer contents of A to X.

(3) DEX (*f*) Decrement the contents of X by 1.

(4) JMP START (*b*) Jump to the instruction labelled START.

(*ii*) (1) This is false because an assembler program must first be converted to machine code before it can be executed.

(2) This is also false because the instruction codes differ from one type of microprocessor to another as do the mnemonic codes.

(3) This is quite correct which is why there is no point in programming in machine code directly.

****

SAQ 1.3c

The above FORTRAN and BASIC programs both decrement the value of X from 255 down to 0 by steps of 1. Which program will execute the fastest, FORTRAN or BASIC?

Response

The FORTRAN program will execute much faster than the BASIC program. The reason is that, once the FORTRAN program is compiled it in fact becomes a machine code program which executes at the maximum possible speed. The BASIC program would execute comparatively slowly because each line of the program has to be interpreted before execution. The extra convenience in interactive program development is paid for at the expense of the speed of execution. It is interesting to note that BASIC compilers are becoming increasingly available so that we can use the interpreter for development work and compile the finished program to get the best of both worlds.

****

SAQ 1.3d

Examine the following pair of simultaneous equations and decide if the calculation of the unknowns A and B, from the two measured quantities M1 and M2, would be a well formulated problem. Assume that M1 = 10.1 and M2 = 60.0 in a typical measurement.

$$M1 = 5 \times A + 7 \times B \tag{1.1}$$

$$M2 = 31 \times A + 41 \times B \tag{1.2}$$

⟶

SAQ 1.3d (cont.)

(Hint: For M1 = 10.1 and M2 = 60.0 obtain an expression for A by multiplying Eq. (1.1) by 31/5 and subtracting Eq. (1.2) from the result. This will give an equation which can be solved for B. Having obtained B, substitute its value back in Eq. (1.1) to obtain an expression which can be solved for A. To test for instability, assume that on remeasurement, M1 became 10.2 and M2 = 60.4, and recalculate the results for A and B).

Response

The simultaneous equations are solved as follows:

Multiply Eq. (1.1) by 31/5 to give equation (1.3),

$$62.62 = 31 \times A + 43.4 \times B \tag{1.3}$$

Subtract Eq. (1.2) from Eq. (1.3) to obtain:

$$2.62 = 0 + 2.4 \times B \tag{1.4}$$

This gives B = 1.092, which can be substituted in Eq. (1.1) to obtain the following expression involving only the unknown A.

$$10.1 = 5 \times A + 7 \times 1.092 \tag{1.5}$$

This gives A = 0.491.

For the two measurements M1 = 10.1 and M2 = 60.0 the unknown quantities A and B are 0.491 and 1.092 respectively. If we repeat the above calculations with M = 10.2 and M2 = 60.4 we have the result that A = 0.406 and B = 1.167. Therefore with changes in the measured quantities of less than 1% we have analytical results which vary widely (about 18% for A and 8% for B). The reason is the equations are 'ill-conditioned' or not well formulated. In general, to solve a problem with two unknowns we need two independent equations. The problem here is that the two equations are very similar and we have effectively only one equation and two unknowns.

****

SAQ 1.3e

Which one of the following equations would you use to fit the curve given below assuming the values of the constants M, C, A and B are non-zero in the equations which include them?

(*i*) $R = M.X + C$

(*ii*) $R = M.X$

(*iii*) $R = A.X^2 + B.X + C$

(*iv*) $R = A.X^2 + B.X$

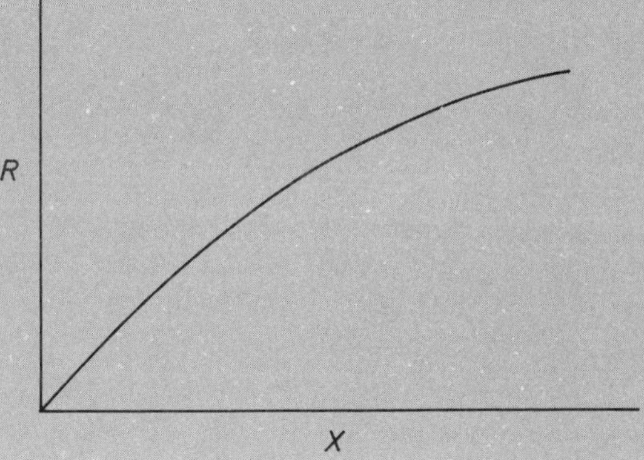

Response

(*i*) and (*ii*) No, these equations apply to a straight line plot. The second one passes through the origin and the first one does not if C is non-zero. Both are invalid because the given plot is curved.

(*iii*) Close, but not quite right. Remember that the question specified C to be a non-zero constant which means that when X = 0, R is not. In other words the plot according to this expression would not go through the origin whereas the given plot does.

(*iv*) Yes that is correct. This is the equation for a curve which passes through the origin. Notice that equation (*iii*) is identical if the intercept C is 0.

****

SAQ 1.3f

The output from a colorimeter which is part of an auto-analyser system is monitored by a microcomputer. A sample is taken every 0.5 second, the run lasts for 20 minutes and each data value read takes 6 bytes of memory. There is 10K of memory available in the computer to store the data. Can the program be designed to accumulate all the data in memory and then transfer the whole lot as a batch to disc, or will each data item have to be transferred to disc as it is obtained?

Response

Sampling every 0.5 second for 20 minutes means $20 \times 60 \times 2 = 2400$ readings will be taken. Since each reading needs 6 bytes of memory, the total memory requirement is 14,400 bytes. If there is only 10K (10 × 1024 = 10,240 bytes) of memory available then we cannot write a successful program to store all the data in memory before transferring the whole lot to disc later. The solution is to write the data to disc a reading at a time, providing the transfer takes less than 0.5 seconds.

****

SAQ 1.3g

For each of the following tasks, which programming language would you use (MACHINE CODE, ASSEMBLER, BASIC, FORTRAN)?

(*i*) The solution of a set of 10 simultaneous equations to determine the concentration of 10 unknowns from 10 measurements of the absorbance of a mixture at different wavelengths. The computation has to be repeated every second to provide rapid continuous analysis of a reagent stream.

(*ii*) The monitoring of the number of times a microswitch closes in a control application involving the dispensing of liquids.

(*iii*) The measurement of pH every 5 seconds in an effluent stream from a chemical plant.

Response

(*i*) There is quite a lot of arithmetic processing to be done in this problem. If this has to be done every half a second then BASIC is definitely out because of its slow execution. The best choice in this case is to use FORTRAN for which ready written programs are certainly available to solve the equations. (See for example A C Norris, *Computational Chemistry*, Wiley, Chichester, 1981).

(*ii*) This could be done in BASIC quite easily provided that the switch was not changing state too rapidly. For dispensing liquids this is not likely to be the case as the microswitch would be used to indicate the end of traversal of a shaft in filling or emptying a syringe. If BASIC was too slow then the job could be done quite easily in ASSEMBLER.

(*iii*) Ion selective electrodes do not in general have very rapid responses to change and so even in BASIC the program would have to wait for the electrode to stabilise. BASIC would be good choice, although it could be done in FORTRAN. The logarithmic transformations required to relate the measured voltage to concentration makes ASSEMBLER language unattractive.

****

SAQ 2.1a

In the tables below the entry in column X, row A is 2.0 and that in column Y, row B is 4.0. The product of these is 8.0. That is, AX times BY gives 8.0. What other combinations produce the same result, viz 8.0?

	X	Y
A	2.0	0.4E-1
B	2.0E1	4.0
C	200E-2	4.0E3
D	0.2E3	40.0E-1
E	0.002E0	4.0E-1

Response

The following products give the result 8.0:

AX by BY   CX by BY   BX by EY   CX by DY

AX by DY   DX by AY   EX by CY

You should test yourself and your computer by making it accept and print some of these numbers. A suitable procedure is to type in lines like:

```
LET P = 0.2E3

LET Q = 0.4E-1

PRINT P*Q
```

****

SAQ 2.1b

What type of variable might be employed for the following:

(*i*) the date of a month;

(*ii*) the name of a day;

(*iii*) the weight of a sample.

Response

(*i*) A date must always be a whole number and therefore either an integer or a real variable would be suitable.

(*ii*) A name must be held by a string variable.

(*iii*) A weight is unlikely to be a whole number and should therefore be represented by a real variable.

****

SAQ 2.1c

Some of the following statements are unacceptable and some will lead to errors. Comment on each statement.

(*i*) LET X% = 12

(*ii*) LET X + Y% = 12

(*iii*) LET residue = 0.456

(*iv*) LET PPT = 1.347g

(*v*) LET B% = 3.456

(*vi*) LET 3rd% = 4.67

(*vii*) LET sample% = 3.92

(*viii*) LET Name$ = "Tom Brown"

(*ix*) LET PRINTER$ = "Caxton"

(*x*) LET FIVE = 6

Response

(*i*) LET X% = 12 is all right since 'X%' is an integer variable.

(*ii*) LET X+Y% = 12 is not correct because it includes the plus sign, 'XY%' or 'X_Y%' would probably be acceptable as integer variables.

(*iii*) LET residue = 0.456 is all right, 'residue' being the name of a real variable.

(*iv*) LET PPT = 1.347g will not do. 'PPT' is all right as the name of a real variable but the 'g' after the number will cause an error.

(*v*) LET B% = 3.456 would be accepted by the computer but the value held by integer variable 'B%' would be 3, the number being truncated to the lower integer.

(*vi*) LET 3rd% = 4.67 has two errors. The computer would not accept this assignment because a variable name must not start with a number. Also, the % sign indicates an integer variable which is unsuitable for a value which includes a fractional part.

(*vii*) LET sample% = 3.92 would be accepted but the integer variable 'sample%' would have the value 3.

(*viii*) LET Name$ = "Tom Brown" is all right. Note that this string includes 3 spaces.

(*ix*) LET PRINTER$ = "Caxton" would not be accepted because the name starts with the reserved word, PRINT. The corresponding name in lower case, *viz* "printer$", would probably be acceptable.

(*x*) LET FIVE = 6 is all right as far as the computer is concerned since FIVE can be the name of a real variable. In the interests of clarity and sanity, however, an assignment like this is inadvisable.

****

SAQ 2.2a

Problems (*i*) to (*iii*) below should be done by hand (or mentally) and the results checked by computer.

(*i*)

Z1 = 100/(X + Y)   Z2 = 100/X + Y

If X = 5 and Y = 15, what are the values of Z1 and Z2?

(*ii*)

Z3 = 100/X/Y   Z4 = 100/X + Y
Z5 = 100/(X + Y)

If X = 5 and Y = 2, what are the values of Z3, Z4 and Z5?

(*iii*)

Z6 = 100/X*Y   Z7 = 100/(X*Y)

If X = 2 and Y = 10, what are the values of Z6 and Z7?

(*iv*)

Write computer-type expressions corresponding to:

$Z8 = Ax^2 + Bx + C$

$Z9 = Ax^{(a + b)} + Bx^2 + 1/C$

Response

Each expression is evaluated in parts. Pay particular attention to the order in which each step is done.

(*i*) Z1 = 100/(5 + 15) = 100/20 = 5

Z2 = 100/5 + 15 = 20 + 15 = 35

(*ii*) Z3 = 100/5/2 = 20/2 = 10

Z4 = 100/5 + 2 = 20 + 2 = 22

Z5 = 100/(5 + 2) = 100/7 = 14.3

(*iii*) Z6 = 100/2*10 = 50*10 = 500

Z7 = 100/(2*10) = 100/20 = 5

(*iv*) Z8 = A*x^2 + B*x + C

Z9 = A*x^(a+b) + B*x^2 + 1/C

Examine Z6 and Z7 closely to see the effect of divide and multiply without brackets.

The lower-case X has been used in (*iv*). Some computers require all variable names to be in upper-case. You will soon discover what your machine allows.

When expressions are evaluated the order of priority is usually a matter of mathematical common sense, *viz*:

Expressions within parenthesis (brackets).

Exponentiation, ie Raise to a power (^).

Multiplication and Division (* and /).

Addition and Subtraction (+ and −).

Watch the division sign (/). Some computers have the sign (\) called the *backslash*. It does *not* mean divide.

****

SAQ 2.2b

> In the standardisation of acid solutions a known weight of anhydrous sodium carbonate is titrated with an acid. One mole of monobasic acid neutralises 53.00 g of Na_2CO_3.
>
> Write and run a program to calculate the molarity of a hydrochloric acid solution from the weight of sodium carbonate (W g) and the volume of hydrochloric acid (V cm^3) required for neutralisation.
>
> Reminder:
>
> molarity = number of moles per litre of solution.

Response

The essential theory is that the number of moles acid is W/53.00, the volume of acid is V/1000 dm^3 and the molarity is the number of moles per dm^3. The molarity of the acid is therefore 1000W/(53.00V) or, using computer notation, the molarity is represented by variable C in:

C = 1000*W/53.00/V

Since 1000/53.00 always has the same value we may define a constant K by

K = 1000/53.00

We have now done the chemistry and can concentrate on the computation. It is essential that problems be tackled in this order: a computer does not know any chemistry. The program to be written must:

accept values of weight and volume;

calculate concentration;

print out concentration.

We can see clearly how to do each of these and so we are in a position to write a program.

There are several possible ways of handling the data. The following is a very short program without prompts:

```
 99 REM ** SHORT MOLARITY
100 LET K = 1000/53.00
110 INPUT W
120 INPUT V
130 LET C = K*W/V
140 PRINT C
150 END
```

The inclusion of some prompts improves the program:

```
 99 REM ** MOLARITY
100 LET K = 1000/53.00
110 PRINT "WEIGHT (GRAMMES)?";
120 INPUT W
130 PRINT "VOLUME (CUBIC CENTIMETRES.)?";
140 INPUT V
150 LET C = K*W/V
160 PRINT "CONCENTRATION = ";C
170 END
```

You should check the program with known data. A weight of 0.1325 g and a titration of 25.00 cm^{-3} will give a 0.1 mol dm^{-3} solution. Try several weights and titrations but do not worry if the output gives a large number of figures. We shall tidy this up later.

In these two programs a variable K was defined to combine values that are constant. The programs could be made more general by allowing K to be defined from the keyboard. Why not work out what value K must have if the standard substance is borax or if the titrant is sulphuric acid?

If you have applications requiring it, an obvious extension is to more complex calculations involving oxidising and reducing agents. You could ask for the molar mass and the number of electrons involved in the reaction before working out the value of K.

****

SAQ 2.2c

Write a program which accepts numbers from the keyboard and prints out the reciprocal of the number correct to 3 decimal places.

Response

We use the INT function with an expression as argument. For 3 decimal places we multiply and divide by 1000. In general, for n decimal places we multiply and divide by 10 to the power n.

```
 99 REM ** 1/X TO 3 PLACES
100 INPUT X
110 Y = 1/X
119 REM ** NOW ADJUST
120 Z = INT (1000*Y+0.5)/1000
130 PRINT Z
140 END
```

Even with formatting like this some computers may not make it obvious that there has been any control. Thus, if an answer is 0.6 exactly this may be printed 0.6 instead of 0.600 as we might expect and desire. The computer manual must be consulted to sort out such problems.

In the above program three lines, 110 to 130, are used where one would do. We could have used a single statement:

PRINT INT(1000/X+0.5)/1000.

****

SAQ 2.2d

Write programs to calculate:

(*i*) absorbance from percent transmittance;

(*ii*) percent transmittance from absorbance;

(*iii*) pH from hydrogen ion concentration;

(*iv*) hydrogen ion concentration from pH.

Reminder:

absorbance = $\log_{10}(100/\%\text{transmittance})$

pH = $-\log_{10}[H^+]$

Response

(*i*) In this program the value of absorbance first found is adjusted by means of the INT function to give 2 figures after the decimal point.

```
 99 REM ** PERCENT T to A
100 PRINT "ENTER % TRANSMITTANCE"
110 INPUT T
119 REM ** ASSUME LOG( ) IS TO BASE 10
120 A = LOG(100/T)
130 A = INT(100*A+0.5)/100
140 PRINT "ABSORBANCE IS ";A
150 END
```

Notice how the same variable name, *viz* A, can be re-used after adjusting the number of decimal places: the INT function obtains a value which is 'placed in the box labelled A'.

If several conversions are to be made it would be useful to change line 150 to:

```
150 GOTO 100
```

To stop the program, however, you would have to use the ESCAPE or BREAK key. Better methods of controlling a program are discussed in the next Section.

(*ii*) In this program a prompt-string is used at the INPUT line and the transmittance is given to one decimal place.

```
 99 REM ** A to PERCENT T
100 INPUT "ENTER ABSORBANCE", A
110 T = 100*10^(-A)
120 T = INT(10*T+0.5)/10
130 PRINT "TRANSMITTANCE IS ";T;"%"
140 END
```

At line 110 the base 10 is raised to the power −A and the result multiplied by 100. Then INT is used at line 120.

An alternative procedure would be to multiply A by −2.303 and then take the exponential.

```
110 T = 100*EXP(-2.303*A)
```

The factor 2.303 converts the common log to the natural log.

(*iii*) We revert to 2 decimal places though whether or not this is justified will depend on the precision with which the hydrogen ion concentration is known.

```
 99 REM ** pH CALCULATION
100 INPUT "ENTER HYDROGEN ION CONCN",H
110 P = -LOG(H)
120 P = INT(100*P+0.5)/100
130 PRINT "pH = ";P
140 END
```

(*iv*) This program does not attempt to fix the decimal point.

```
 99 REM ** pH to H ion
100 INPUT "ENTER pH",P
110 H = 10^(-P)
120 PRINT "H+ CONCENTRATION IS ";H
130 END
```

****

SAQ 2.2e

We wish to write a program which asks a child his/her name and age and then prints the message

"(NAME) WILL BE (Y) NEXT YEAR"

with the correct name and number.

(*i*) Write the program on the assumption that the child will always enter the age as a number.

(*ii*) Write the program to allow the age to be entered as a number followed by a word like 'years'. For this case assume that the VAL function can be used to return a number which appears before any non-numeric character.

Response

Some parts of the programs listed below may not be suitable for your machine because of the way it controls spacing and prompts for input. Thus, if your machine does not use the question mark as a prompt a '?' should be placed at the end of the string at line 100.

(*i*) Since the age is expected to be a whole number an integer variable, X%, is advisable to hold the number of years in the first program.

```
 99 REM ** NAME & AGE
100 PRINT "WHAT IS YOUR NAME ";
110 INPUT N$
120 PRINT
130 PRINT "HELLO ";N$;" HOW OLD ARE YOU ";
140 INPUT X%
150 PRINT
160 PRINT N$;" WILL BE ";X%+1;" NEXT YEAR"
170 END
```

The PRINT statements at lines 120 and 150 help the layout on the screen.

When running programs like this one of the chief problems is to ensure that input data are of the right type. The program above would probably fail through a child entering '5 years' or 'six and a half'. Provided no arithmetic had to be done a string variable, X$, could be used instead of X% at line 130 but then it would only be possible to repeat exactly what was entered:

```
140 PRINT N$;" IS ";X$
```

The use of the VAL function may get round this difficulty.

(*ii*) The following program will only be successful if the VAL function accepts the numerical part of a string like '12 years old' as a numeric variable.

```
 99 REM ** NAME & AGE
100 PRINT "WHAT IS YOUR NAME ";
110 INPUT N$
120 PRINT
130 PRINT "HELLO ";N$;" HOW OLD ARE YOU ";
140 INPUT X$
142 X% = VAL(X$)
150 PRINT
160 PRINT N$;" WILL BE ";X%+1;" NEXT YEAR"
170 END
```

****

SAQ 2.3a

Write programs to display on the screen:

(*i*) the squares of the numbers 1 to 10 one per line;

(*ii*) the squares of the numbers 1 to 10 in a row with a space between each number shown;

(*iii*) the *even* numbers from 0 to 10 and their squares in two columns (take 0 as an even number.)

(*iv*) as (*iii*) but starting with 10 and finishing with 0.

Response

(*i*) Since the step size in the FOR-NEXT loop is +1 it may be omitted at line 10 below.

```
99 REM ** TABLE OF SQUARES
10 FOR I = 1 TO 10 [STEP 1]
20 LET X = I^2
30 PRINT X
40 NEXT I
50 END
```

It is usually possible to replace lines 20 and 30 by a single line:

```
30 PRINT I^2.
```

(*ii*) If your computer uses the semicolon to make next printing follow on directly from the previous then line 30 should end with ;" "; (that is, semicolon, space, semicolon).

(*iii*) To select even numbers only line 10 becomes:

```
10 FOR I = 0 TO 10 STEP 2
```

To get column or zone format replace line 30 by:

```
30 PRINT I,X
```

(*iv*) To go down in steps of two from 10 replace line 10 by:

```
10 FOR I = 10 TO 0 STEP -2
```

Note that the only step that may be omitted is STEP 1. Any other step must be stated.

****

SAQ 2.3b

(*i*) Alter the adding program of Section 2.3.3 so that the program stops and prints the sum when a negative number is entered. The negative number should not be included in the summation.

(*ii*) Alter the program so that the number of items to be added is entered at the keyboard before the loop commences. The program must add this number of items and finish by printing not only the value of the sum, S, but also the value of the control variable, I.

Is the value of I the same as the number of items added?

If I is different from the number of items added why is this?

Response

(*i*) The following is one possibility:

```
 99 REM ** ADDING PROGRAM
100 LET S = 0
110 FOR I = 1 TO 100
120     INPUT X
130     IF X<0 GOTO 160
140     LET S = S+X
150     NEXT I
160 PRINT S
170 END
```

At line 110 "STEP 1" is understood.

This program limits the number of items to 100 but it could easily be modified to make it more general.

The program is made to 'jump out' of the loop by entering a negative value for X. The negative value is not included in the sum.

In general 'jumping out' of a loop is not recommended. It is acceptable with a short program in which we exit the loop simply to stop the program but with longer programs the language may object if several loops are not completed. This is discussed more fully in 2.3.4.

(*ii*) The necessary changes are:

```
105 INPUT N
110 FOR I = 1 TO N
130 (OMIT)
160 PRINT S,I
```

To cancel line 130 you simply type 130 and press RETURN.

The final value of I will indicate how the loop operates.

After the first pass the value of I is increased by the step size (1 in this case); then I is tested to see if its value exceeds the end value which is N in the program above. If the program continues until N items are entered the final value of I will be one more than N. When I becomes N + 1 the test at line 110 shows that I is greater than the end value and so execution of the loop terminates and the program continues at line 160.

****

SAQ 2.3c

Write a program which:

(*i*) accepts a maximum of 20 numbers;

(*ii*) computes the sum of the numbers;

(*iii*) computes the sum of the squares of the numbers;

(*iv*) stops when a negative number is entered;

(*v*) calculates the total number of items;

(*vi*) prints out the sum, the sum of squares, the total number of items and the mean of the numbers entered.

Run the program using the following numbers:

89 82 67 72 86 80 86 91 80 80

Response

In the program below variable S represents the sum of the numbers, variable P the sum of the squares:

```
 99 REM ** MEAN PROGRAM
100 S = 0
110 P = 0
119 REM ** MAKE LOOP BIG ENOUGH
120 FOR I = 1 TO 20
130     PRINT "ENTER NUMBER (-1 TO STOP)";
140     INPUT X
150     IF X<0 GOTO 190
160     S = S+X
170     P = P+X^2
180     NEXT I
189 REM ** I IS 1 TOO MANY
190 N = I-1
199 REM ** DIVIDE TO GET MEAN M
```

```
200 M = S/N
209 REM ** ROUND OFF M
210 M = INT(10*M+0.5)/10
220 PRINT "SUM OF NUMBERS = ";S
230 PRINT "SUM OF SQUARES = ";P
240 PRINT "TOTAL ENTRIES = ";N
250 PRINT "MEAN VALUE = ";M
260 END
```

Note that the number of items is one less than the final value of I. The final number input was the negative number to stop the program and we do not count this.

Using the numbers given our answers were:

S = 813 P = 66591 N = 10 MEAN = 81.3

The mean is the only number which has been rounded though it was not really necessary in this particular calculation because the sum was divided by 10; had there been 11 samples, however, rounding would have been very desirable.

****

SAQ 2.4a

Write a program which includes DATA statements and which:

(*i*) reads the number of items from the data list;

(*ii*) calculates and prints out the mean of several numbers;

(*iii*) restores the data pointer;

(*iv*) calculates and prints out the standard deviation of the numbers. ⟶

SAQ 2.4a (cont.)

Test the program using the following numbers which were obtained by 11 analysts for the concentration (g dm^{-3}) of copper in a solution:

49.89	49.82	49.67	49.72	49.86	49.80
49.86	49.96	49.80	49.80	49.83	

Response

In the program below the variable S is used first to sum the data and then to sum the squares of the deviations from the mean. Re-use of the same variable is all right and indeed is advisable if a microcomputer has a low memory but you must be sure that the variable is set to the proper value when it is used for the second time.

```
 98 REM ** STANDARD DEVIATION
 99 REM ** SUM IS S
100 S = 0
110 READ N
120 FOR I = 1 TO N
130     READ X
140     S = S + X
150     NEXT I
170 M = S/N
180 PRINT "NO. OF ITEMS = ";N
190 PRINT "MEAN VALUE = ";M
198 REM ** POINTER BACK
200 RESTORE 410
209 REM ** S NOW SUM OF SQUARED DEVS.
210 S = 0
220 FOR I = 1 TO N
230     READ X
240     S = S + (X-M)^2
250     NEXT I
260 V = S/(N-1)
270 PRINT "VARIANCE = ";V
```

```
280 PRINT "STANDARD DEVIATION = ";SQR(V)
300 END
399 REM ** DATA
400 DATA 11
410 DATA 49.89,49.82,49.67,49.72
420 DATA 49.86,49.80,49.86,49.96
430 DATA 49.80,49.80,49.83
```

Results with this program are:

MEAN = 49.82 VARIANCE = 0.00615

STANDARD DEVIATION = 0.0784

(These results have been tidied up to get rid of extra figures. Probably the program should include some rounding but this has been omitted to concentrate on other points).

You will notice that N is not read after the RESTORE statement because the pointer is restored only to line 410.

This program might be worth developing. You could add a section of program to restore the data pointer again and then count the number of items which are more than 2 and 3 standard deviations from the mean. Such calculations can be very useful to check for systematic errors. Thus, following a large number of measurements of the same property it is expected that about 95% of the results will be within two standard deviations of the mean result and that about 68% will be within one standard deviation. If a series of results departs markedly from these percentages some kind of strange error or blunder should be suspected, eg there was an unusual variation of the base-line in a spectroscopic measurement or the wrong pipette was used in a volumetric analysis. Checks of this nature are particularly important in the processing of data obtained from on-line instruments.

An alternative, and possibly better, method of programming the same kind of calculation uses an array. This is discussed in Section 2.4.3.

****

SAQ 2.4b

In a book on quantitative analysis there are tables of data on common acids. Typical entries are:

	% w/w	kg/litre	mol/litre
Hydrochloric acid	35	1.18	11.3
Sulphuric acid	96	1.84	8.0

Write a program which employs TAB statements to produce a table like this from data in DATA statements.

Response

The program below allows for expansion by placing the names of the acids in DATA statements and reading the names into a string variable. The first data item gives the number of entries in the table. This number would be updated every time data for another acid are added.

```
 98 REM ** ACID DATA
 99 REM ** START WITH HEADINGS
100 PRINT TAB(18);"% W/W";
110 PRINT TAB(25);"KG/LITRE";
120 PRINT TAB(35);"MOL/LITRE"
130 PRINT
139 REM ** HOW MANY?
140 READ N
150 FOR I = 1 TO N
160     READ N$,A$,B$,C$
170     PRINT N$;
180     PRINT TAB(20);A$;
190     PRINT TAB(27);B$;TAB(36);C$
200     NEXT I
210 END
299 REM ** DATA
```

```
300 DATA 2
310 DATA "HYDROCHLORIC ACID"
320 DATA 35,1.18,11.3
330 DATA "SULPHURIC ACID"
340 DATA 96,1.84,18.0
```

Note that the print lists at lines 100, 110, 170 and 180 finish with the punctuation mark which suppresses an 'end of line' message but those at lines 120 and 190 do not have this mark.

****

SAQ 2.4c

Write a program to accept up to 10 real numbers into an array. After testing the program with simple numbers (eg 2,4,6, . . .) alter the latter part of the program to make it print out:

(*i*) the numbers and their cubes in column format;

(*ii*) the numbers, and their reciprocals correct to 3 decimal places

Response

(*i*) To print out numbers and cubes in column format.

```
 99 REM ** TABLE OF CUBES
100 DIM X(10)
110 PRINT "HOW MANY VALUES? (UP TO 10) ",N
120 INPUT N
130 FOR I = 1 TO N
140     PRINT "ENTER ITEM ";I
150     INPUT X(I)
160     NEXT I
```

```
169 REM ** ALL IN. CALCULATE & PRINT
170 FOR I = 1 TO N
180     PRINT X(I), X(I)^3
190     NEXT I
200 END
```

(*ii*) To print numbers and reciprocals correct to 3 decimal places the only change really necessary is at line 180. This is changed to:

```
180 PRINT X(I), INT(1000/X(I) + 0.5)/1000
```

If the second item in the print list look rather complex you can work it out in stages. Starting with the reciprocal, 1/X(I), multiply by 1000, add 0.5, truncate to the integer and finally divide by 1000.

****

SAQ 2.4d

In a gas chromatography experiment the solvent peak was recorded at a retention time of 0.85 minutes and successive peaks were recorded at retention times of 3.96, 4.33, 5.20, 6.72, 7.58, 9.28 and 10.75 minutes. Write a program which will:

(*i*) read the experimental data from DATA statements, placing the peak times in an array;

(*ii*) compute retention times relative to peak number 4 (6.72 minutes); that is, for each peak compute the ratio

$$\frac{\text{retention time} - 0.85}{\text{peak 4 time} - 0.85}$$

(*iii*) display the retention times and the relative retention times in a table.

Response

The DATA statements are kept well away from the main program. As these statements will be altered every time the program runs the line numbers should be readily remembered.

```
 99 REM ** RETENTION TIMES
100 DIM T(12)
109 REM ** NO. OF PEAKS, WHICH REF.
110 READ N,P
119 REM ** READ SOLVENT TIME
120 READ S
129 REM ** TIMES INTO ARRAY
130 FOR I = 1 TO N
140    READ T(I)
150    NEXT I
159 REM ** CALCULATE & PRINT
160 FOR I = 1 TO N
170    R = (T(I)-S)/(T(P)-S)
180    R = INT(100*R + 0.5)/100
190    PRINT TAB(4);I;TAB(6);T(I);TAB(12);R
200    NEXT I
210 END
499 REM ** DATA LIST
500 DATA 7,4,0.85
510 DATA 3.96,4.33,5.20,6.72,7.58,9.28,10.75
```

The first READ statement (line 110) takes in the number of peaks (excluding the solvent) and the number of the peak which is to be used as reference. The second READ takes in the solvent time.

The peak retention times are read into an array in a loop from lines 130 to 150. The calculations and printing are done in the loop from line 160 to line 200.

The following output was obtained on running this program:

```
1   3.96  0.53
2   4.33  0.59
3   5.2   0.74
4   6.72  1
5   7.58  1.15
6   9.28  1.44
7  10.75  1.69
```

Some additional formatting of output is required to line up the decimal points in the second column. Your computer probably has some means of achieving this but as the method is likely to be specific to the particular machine you will have to consult your manual for details.

It might be a good idea to look again at SAQ 2.4a (on calculating a standard deviation) and perhaps re-write the program using an array. Restoring the data pointer will not then be necessary. With this kind of calculation the use of an array facilitates the examination of individual data items. For example, one might want to count those items which are more than two standard deviations from the mean. Such calculations are easily programmed using arrays.

****

SAQ 2.5a

Write a program which prints out the reciprocal and the common logarithm of any number entered at the keyboard and which warns without crashing when the task set is impossible. All output should be correct to two places of decimals.

Response

The program below assumes that LOG( ) gives or returns the common logarithm. We guard against input of zero for the reciprocal and against zero and negative values for the log. There is, however, no check for a number which is too large or too small for the computer.

```
 99 REM ** 1/X & LOG(X)
100 INPUT "ENTER NUMBER X ";X
110 PRINT
120 IF X>0 THEN PRINT "LOG(X)=";INT(1000*
    LOG(X)+0.5)/1000
130 IF X< = 0 THEN PRINT "LOG IMPOSSIBLE"
140 PRINT
150 IF X<>0 THEN PRINT "1/X = ";INT(1000/X +
    0.5)/1000
160 IF X = 0 THEN PRINT "RECIPROCAL INFINITE"
170 PRINT
180 GOTO 100
```

This program makes no pretence at elegance. Apart from anything else it can only be stopped by pressing the ESCAPE or BREAK key or by making some input error which the language cannot tolerate.

In the answer to question 2.5c a somewhat better program is suggested.

****

SAQ 2.5b

Write a program which accepts words or numbers greater than −1000 from the keyboard and which:

(*i*) prints the message 'POSITIVE' if the first character of the input string is one of the 'numeric' characters 0 to 9; ⟶

SAQ 2.5b (cont.)

(*ii*) prints 'NEGATIVE' if the first character is the minus sign (−);

(*iii*) prints 'FRACTION' and the character of ASCII code 7 if the first character is the period (.);

(*iv*) stops accepting input when the character 'Q' is entered;

(*v*) reports the total number of entries made, the sum of all the numerical entries and the highest negative numerical entry.

Response

It is necessary to note certain ASCII codes:

Character	Code
0	48
9	57
−	45
.	46
/	47

The program below is one possibility:

```
 98 REM ** REM NUMBER TESTING
 99 REM ** SET TOTAL & SUM TO ZERO
100 T = 0
110 S = 0
119 REM ** H = HIGHEST −VE AT START
120 H = −1000
129 REM ** N = CURRENT HIGHEST −VE
```

```
130 N = H
140 PRINT "ENTER A NUMBER >-1000 OR A WORD."
150 PRINT "TO STOP THE PROGRAM ENTER Q"
160 PRINT
170 INPUT "ENTER (Q TO STOP)"X$
180 IF X$ = "Q" THEN 280
190 C = ASC(X$)
200 IF C>47 AND C<58 THEN PRINT "POSITIVE"
210 IF C = 45 THEN PRINT "NEGATIVE"
220 IF C = 46 THEN PRINT "FRACTION";CHR$(7)
229 REM ** USE VAL ONLY IF NUMERIC
230 IF C>44 AND C<58 AND C<>47 THEN S = S +
    VAL(X$)
239 REM ** CHANGE H IF NECESSARY
240 IF C = 45 THEN N =VAL(X$)
250 IF N>H THEN H = N
260 T = T + 1
270 GOTO 160
280 PRINT
290 PRINT "TOTAL ENTRIES ";T
300 PRINT
310 PRINT "SUM OF NUMBERS ";S
320 PRINT
330 IF H<>-1000 THEN PRINT "HIGHEST
    NEGATIVE";H
340 IF H = -1000 THEN PRINT "NO NEGATIVES
    ENTERED"
350 END
```

This program could be made much tidier and more understandable by employing the IF ... THEN//ELSE structure discussed in 2.5.3.

****

SAQ 2.5c

Provided your computer allows the structure answer SAQ 2.5a again but using IF ... THEN ... ELSE.

Response

It is highly unlikely that your program is exactly the same as the one below as there are many possible ways of using IF ... THEN ... ELSE. The construction is used in lines 160 and 190 to prevent errors. Also, in line 130 the input is tested to see if it is time to stop. The input is taken in as a sequence of characters and assigned to string variable, X$. If the string is not the letter Q it is assumed to be a number and is converted to a value of numeric variable X using VAL. This is one way of testing input for a 'stopper' or 'terminator'.

```
 98 REM ** BETTER 1/X & LOG(X)
 99 REM ** ASSIGN 2 STRINGS
100 A$ = "IMPOSSIBLE"
110 B$ = "INFINITY"
120 INPUT "ENTER NUMBER X "X$
130 IF X$ = "Q" THEN END ELSE X = VAL(X$)
140 PRINT
150 PRINT "LOG(";X;") = ";
160 IF X>0 THEN PRINT; INT(1000*LOG(X)+0.5)/1000
    ELSE PRINT A$
170 PRINT
180 PRINT "1/";X;"= ";
190 IF X<>0 THEN PRINT;INT(1000/X+0.5)/1000 ELSE
    PRINT B$
200 PRINT
210 INPUT "NEXT NUMBER OR Q TO STOP "X$
220 PRINT
230 GOTO 130
```

Note how strings A$ and B$ are given 'values' and then used as required. This technique is particularly valuable when the same string of characters is to be printed several times in a program and space is at a premium. This is not the case in the present program but the use of A$ and B$ keeps lines 160 and 190 reasonable short.

Strictly, one should test a string before using VAL to ensure that conversion to a numeric will be successful. Whether or not this kind of test is worth doing depends on the consequences of an impossible or incorrect conversion.

****

SAQ 2.5d

The potential of an ion-selective electrode may depend on the activities of two ions:

E = E° + 58*LOG(A1 + K*A2) (mV at 292K)

A1 is the activity which the electrode is designed to measure, A2 the activity of an interfering ion. K is the selectivity ratio or selectivity constant.

Write a program to create a table showing, for any value of K, how the second term on the right depends on A1 and A2. Suitable ranges might be:

A1: 0.001 to 0.005 mol dm^{-3} in 5 steps
A2: 0 to 0.01 mol dm^{-3} in 5 steps

Output should be given to the nearest millivolt.

....

An example of output for K = 10 and A2 = 0.002 mol dm^{-3} is given below:

```
INTERFERING ION AT 2 MMOL

ION 1 AT 1 MMOL    TERM = -97
ION 1 AT 2 MMOL    TERM = -96
ION 1 AT 3 MMOL    TERM = -95
ION 1 AT 4 MMOL    TERM = -94
ION 1 AT 5 MMOL    TERM = -93
```

Response

While the program below answers the question there are many other methods of formatting the output.

```
 99 REM ** ION-SELECTIVE
100 INPUT "ENTER SELECTIVITY CONST. "K
110 FOR A2 = 0 TO 10 STEP 2
120     PRINT "INTERFERING ION AT ";A2;" MMOL"
130     PRINT
140     FOR A1 = 1 TO 5
150         E = 58*LOG(A1/1000 + K*A2/1000)
160         E = INT(E + 0.5)
170         PRINT "ION 1 AT ";A1;" MMOL","TERM
            =dq;E
180         NEXT A1
190     PRINT
200     A$ = GET$
210     NEXT A2
220 END
```

The GET$ at line 200 is a convenient method of controlling output on the VDU screen. A key must be pressed to allow the next A2 loop to be completed and displayed.

****

SAQ 2.5e

To assess a new analytical method four samples of a product were taken and each sample was divided into two parts. Four different analysts determined the purity using the old method for one part and the new method for the other. The results were listed as they were reported by the analysts, always in the order:

method, sample, % purity.

The results so listed were:

2 2 98.8	1 4 98.4
1 3 98.1	2 4 98.1
1 2 98.1	2 3 98.9
1 1 98.6	2 1 98.6

⟶

SAQ 2.5e (cont.)

Write a program to read these results into a two-dimensional array and present them in something like this table:

SAMPLE	1	2	3	4
METHOD 1	98.6	98.1	98.1	98.4
METHOD 2	98.6	98.8	98.9	98.1

Response

We know that there are eight sets of data to be read. The first two items of each set give the method and the sample (I and J). The third item is the result. When reading the data we must ensure that I and J are known *before* the result is assigned to an array element.

In the following program it has been assumed that TAB can take an argument which includes variables. That is, the tabulation can depend on I as in line 160.

```
 99 REM ** 2-ARRAY
100 DIM R(2,4)
109 REM ** READ IN 8 SETS OF DATA
110 FOR N = 1 TO 8
120    READ I,J,R(I,J)
130    NEXT N
139 REM ** ALL READ IN
140 PRINT "SAMPLE";
150 FOR I = 1 TO 4
160    PRINT TAB(6 + I*6); I;
170    NEXT I
180 PRINT
190 PRINT
200 FOR I = 1 TO 2
210    PRINT "METHOD "; I;
220    FOR J = 1 TO 4
230       PRINT TAB(6 + J*6); R(I,J);
```

```
240         NEXT J
250     PRINT
260     NEXT I
270 END
399 REM ** DATA LIST
400 DATA 1,2,98.1,2,3,98.9,1,1,98.6
410 DATA 2,1,98.6,2,2,98.8,1,4,98.4
420 DATA 1,3,98.1,2,4,98.1
```

On running this program the table reproduced in the question was obtained.

****

SAQ 2.6a

Write a program to display a graph like the example shown below. The figure produced should have a wavelength scale from 400 to 700 printed at 20 unit intervals on or at the base line and each line should represent 5 wavelength units. You may assume that DATA statements contain the correct number of data items.

```
400:    *
   :          *
   :             *
   :          *
420:       *
   :
   :
```

Response

The following is one possible answer:

```
 99 REM ** STAR GRAPH
100 FOR I = 400 TO 700 STEP 5
110     READ A
120     L = 40*A
129     REM ** IS I A MULTIPLE OF 20?
130     X = I/20-INT(I/20)
139     REM ** DECIDE BASE & PRINT
140     IF X>0 THEN PRINT "   :";
150     IF X = 0 THEN PRINT;I;":";
159     REM ** MAKE SPACES
160     FOR J = 1 TO L-1
170         PRINT " ";
180         NEXT J
189     REM ** PLOT POINT
190     PRINT "*"
200     NEXT I
210 END
299 REM ** DATA
300 DATA ......
310 DATA ......
```

The chief problem with this kind of programming is how to decide when a scale number is to be printed at the start of a line and when spaces are to be left. In the program above we calculate X as the difference between I/20 and INT(I/20). If this difference is zero then I must be a multiple of 20 and the value of I is printed. When X is not zero a string of three spaces is printed. A neater way of obtaining X makes use of the MOD operator. If line 130 is replaced by:

```
130 X = I MOD 20
```

The value of X returned is the remainder after dividing I by 20. The MOD operator will be discussed more fully in Section 2.9.

Various other methods are possible for this kind of program. For example lines 140 and 150 together could be combined in an IF ... THEN ... ELSE statement. Line 130 could also be included as in:

```
130 IF (I MOD 20)>0 THEN PRINT "   :"; ELSE
    PRINT I;":";
```

This statement would replace lines 130 to 150.

If you want to use this technique of graphing with text characters you might care to develop a program which will show the figure in the more conventional way, the scale 400 - 700 running horizontally. One method of doing this is to place all the L values in the elements of an array, L(1), L(2), etc, where the indices 1,2 ... indicate wavelengths. Lines are then printed using spaces or stars in sequence. But be warned: the programming is not trivial.

****

SAQ 2.6b

> Write a program to read plotting coordinates from DATA statements and plot points on the screen. Confirm the program by plotting several points.

Response

The program below is not complete but it should enable you to check yours. Liberal use is made of REM statements to show what should be happening next.

```
100 REM ** A PLOT PROCEDURE
109 REM ** SCALE FACTOR A FOR X
120 A = 24
129 REM ** SCALE FACTOR B FOR Y
130 B = 80
139 REM ** READ LOOP - MAKE IT BIG
140 FOR I = 1 TO 1000
150     READ V,P
```

```
159     REM ** CHECK FOR LAST DATA
160     IF V<0 GOTO 240
169     REM ** CONVERT TO PLOTTING COORDS
170     X = A*V
180     Y = B*P
189     REM ** PLOT THE POINT
199     REM ** DO IT YOUR WAY, EG PLOT X,Y
230     NEXT I
240 END
310 DATA x1,y1,x2,y2, ......... -1,-1
```

It is a good idea to keep a program like this handy for insertion in other programs as required.

While this test program should work all right so long as the data items are within the correct range it is usually sensible to guard against X or Y values that are too large. This can be done easily by conditional statements following line 180, eg:

```
182 IF X<0 OR Y<0 GOTO 230
184 IF X>1200 OR Y>900 GOTO 230
```

These statements assume that 1200 and 900 are the largest allowed values of X and Y.

****

SAQ 2.6c

Write and run a program to display on the screen the neutralisation curve for 100 cm^3 of a weak monobasic acid being titrated with sodium hydroxide of the same molar concentration. Calculate the pH at increments of 2 cm^3 titrant from 10 to 90 cm^3, and then in increments of 0.5 cm^3 to 120 cm^3. Make the usual approximations.

Response

The essential chemistry has been discussed in the introduction to the question. Four kinds of calculation are necessary:

(*i*) Calculate pH at start.

(*ii*) Calculate pH for a number of points before end point.

This is the major part of the curve. The scale along the X axis will depend on your computer; if we want to make 0.5 cm^3 increments and there are 4 addressable points per pixel in the X direction we might take 4 points per 0.5 cm^3. This means 800 points to the end point.

There are two parts in this stage, *viz* from 10 to 90 cm^3 in 2 cm^3 additions and from 90.5 to 99.5 cm^3 in 0.5 cm^3.

(*iii*) Calculate pH at the end point.

(*iv*) Calculate several pH values after the end point. To take the titration to 120 cm^3 a further 160 points are needed, giving a total of 960 points in the X direction.

On the basis of these figures the titrant volume or X scale requires to go from 0 to 120 cm^3 at 8 points per cm^3. We might mark off every 20 cm^3 or every 160 points.

The pH or Y scale must extend from about pH 2 to pH 10 or higher. We might use 80 points per pH unit and mark off every pH unit.

A possible program is given below without details for drawing the axes or plotting the points:

```
 99 REM ** WEAK ACID - STRONG BASE
100 PRINT
110 INPUT "pKa OF ACID ", A
120 PRINT
```

```
130 INPUT "CONCENTRATION OF ACID ",C
139 REM ** SET GRAPH ORIGIN AT X0,Y0
140 X0 = 200
150 Y0 = 200
159 REM ** CLEAR SCREEN. DRAW AND
160 REM ** LABEL X AND Y AXES
169 REM ** PH FACTOR: Y% = 80 PER PH
170 Y% = 80
179 REM ** TITR FACTOR: X% = 8 PER CM(3)
180 X% = 8
189 REM ** VOLUME TO END POINT
190 E = 100
199 REM ** STAGE 1
200 P = A/2-LOG(C)/2
210 X = X0
220 Y = Y0 + (P-2)*Y%
230 REM ** PLOT X,Y - YOUR WAY
239 REM ** STAGE 2 PART 1
240 FOR T = 10 TO 90 STEP 2
250     P = A + LOG(T/(E-T))
260     X = X0 + T*X%
270     Y = Y0 + (P-2)*Y%
280     REM ** PLOT X,Y - YOUR WAY
290     NEXT T
299 REM ** STAGE 2 PART 2
300 FOR T = 90.5 TO 99.5 STEP 0.5
310     P = A + LOG(T/(E-T))
320     X = X0 + T*X%
330     Y = Y0 + (P-2)*Y%
340     REM ** PLOT X,Y - YOUR WAY
350     NEXT T
359 REM ** STAGE 3
360 P = 7 + A/2 + LOG(C/2)/2
370 X = X0 + T*X%
380 Y = Y0 +(P-2)*Y%
390 REM ** PLOT X,Y - YOUR WAY
399 REM ** STAGE 4
400 FOR T = 100.5 TO 120 STEP 0.5
410     P = 14 + LOG(C*(T-E)/(E + T))
420     X = XO + T*X%
```

```
430    Y = Y0 + (P-2)*Y%
440    REM ** PLOT X,Y - YOUR WAY
450    NEXT T
460 END
```

You will have noticed that this program employs the same plotting routine five times. It would be more economical in space and in programming time to have one plotting procedure which can be called as necessary. In Section 2.7 we discuss how this is done by means of a subroutine or a procedure.

In a departure from normal custom we give details below for setting up the screen and plotting points on the basis of BBC syntax. This will obviously be of most interest if you use a BBC machine but users of other microcomputers will probably be able to follow the steps with the aid of the REM statements and the notes at the end. Since a *procedure* is employed you might want to return here after Section 2.7.

Changes and Extension for BBC Mode 1

The program as written above must be altered as follows:

Change all REM lines containing PLOT X,Y to

```
(line no.) PLOT 69,X,Y.
```

Add lines 90 and 162:

```
 90 MODE 1
162 PROCsetup
```

Add the following after END:

```
500 DEF PROCsetup
509 REM ** CLEAR, RED, SET FIELD
510 CLS:GCOL 0,1:@% = 3
519 REM ** OUTLINE GRAPH AREA
520 MOVE X0,Y0
```

```
530 PLOT 1,960,0:PLOT 1,0,800
540 PLOT 1,-960,0:PLOT 1,0,-800
549 REM ** DRAW GRID
550 FOR X = X0 + 80 TO X0 + 880 STEP 80
560 MOVE X,Y0:PLOT 25,0,800:NEXT X
570 FOR Y = Y0 + 80 TO Y0 + 720 STEP 80
580 MOVE X0,Y:PLOT 25,960,0:NEXT Y
589 REM ** WHITE, LINK CURSORS
590 GCOL 0,3:VDU5
599 REM ** MARK X AXIS
600 FOR X% = 0 TO 120 STEP 20
610 MOVE X0 + 8*X%-10,Y0-20
620 PRINT X%:NEXT X%
630 MOVE X0 + 300,Y0-60
640 PRINT"TITRATION"
649 REM ** MARK Y AXIS
650 FOR Y% = 2 TO 12
660 MOVE X0 - 120,Y0 + 80*(Y%-2)
670 PRINT Y%:NEXT Y%
680 MOVE X0 - 130,Y0 + 350:PRINT "pH"
689 REM ** SEPARATE CURSORS
690 VDU4
700 ENDPROC
```

Notes on the graphics procedure:

CLS clears the screen.

@% = 3 sets the field size to 3 characters.

GCOL 0,1 sets the graphics colour to red, GCOL 0,3 to white.

MOVE X,Y means move to X,Y *absolute* without drawing.

PLOT 1,X,Y means draw a line *relative.*

PLOT 25,X,Y means draw a dotted line *relative*.

PLOT 69,X,Y means plot a point at X,Y.

Normally text and graphics are controlled by separate cursors. Following VDU5 the cursors are linked so that text is printed at the graphics cursor; VDU4 returns cursor operation to normal.

****

SAQ 2.7a

Construct a flow diagram for a program which repeatedly accepts a value of an electromotive force from the keyboard and prints out a concentration. The program must stop when a negative value is entered. (Calculation details are not expected.)

Response

The test for end of data must be made before a calculation is attempted, though in many cases it may be assumed that the first data item is not the 'stopper'. The following is onc possibility:

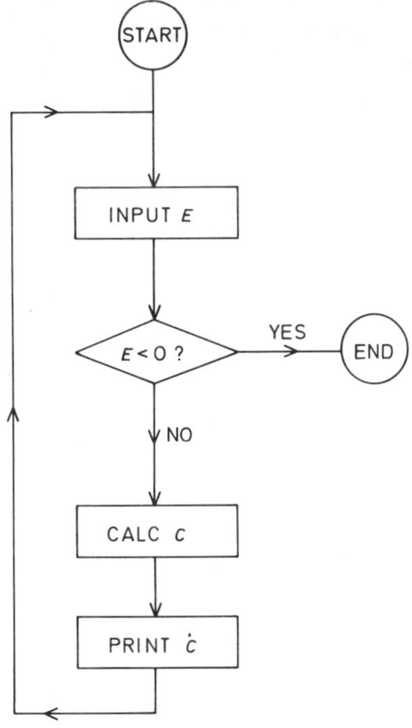

The route from the PRINT operation to the INPUT is shown by an arrow in the diagram but it is not labelled. As the diagram stands a GOTO statement could be used to direct the program back to the INPUT but as the path shown forms the return portion of a loop it could be part of a FOR-NEXT or a REPEAT loop.

Some authors suggest that the route from a diamond should be horizontal if the test is true (yes) and vertical if false (no). This is a fine point. It is more important that the routes are labelled as in the example.

****

SAQ 2.7b

Construct a flow diagram for a program which successively reads three data items from data statements, performs two different calculations and plots a point after each calculation. The 'end of data' is indicated by negative values of three data items. Assume that a subroutine is available for plotting points.

Response

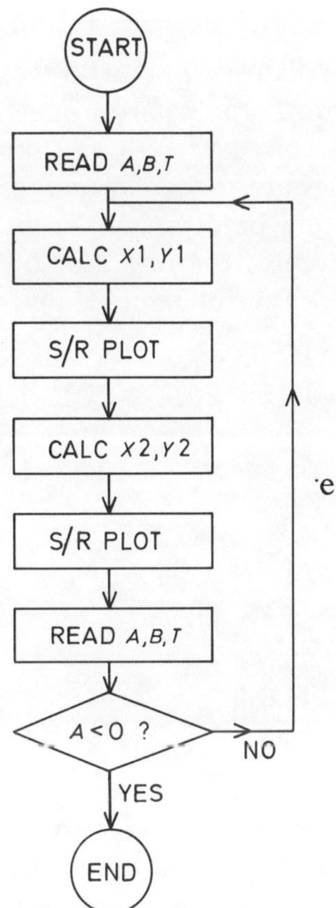

Notice that a READ statement occurs once before the loop and then at the end of each loop cycle. In this way the test is performed and the loop terminated immediately all valid data have been read. The technique was discussed in 2.3.4.

****

SAQ 2.8a

> Three instruments send temperature readings in the range 0 to 100 degrees to a computer which later transfers the readings to an *append* file " temp" on a disc. ⟶

SAQ 2.8a (cont.)

The first entry on the file is the time the file is started. Once started, each instrument reads temperature every three minutes and the three readings are spaced at one minute intervals. Immediately before each series of three readings the name of the operator on duty is written to the file. The following shows typical entries:

1630
JOE
60
55
56
MIKE
63
54
55
TOM
...
...

Construct a flow diagram for a program which uses the filed data to produce a graphical display of the temperature variations over a period of one hour. You may assume that points can be plotted in any of three colours or that three characters are available for use as plotted points.

Response

It would be too much to expect you to produce exactly the same flow diagram as the one below. But have you taken account of all the different parts in some legal way? In particular did you get the correct sequence for reading string, number, number, number, after reading the start time? Did you open and close the files properly?

The program of this diagram reads, calculates and plots repeatedly. An alternative might be to read the data into an array or arrays and then do the plotting.

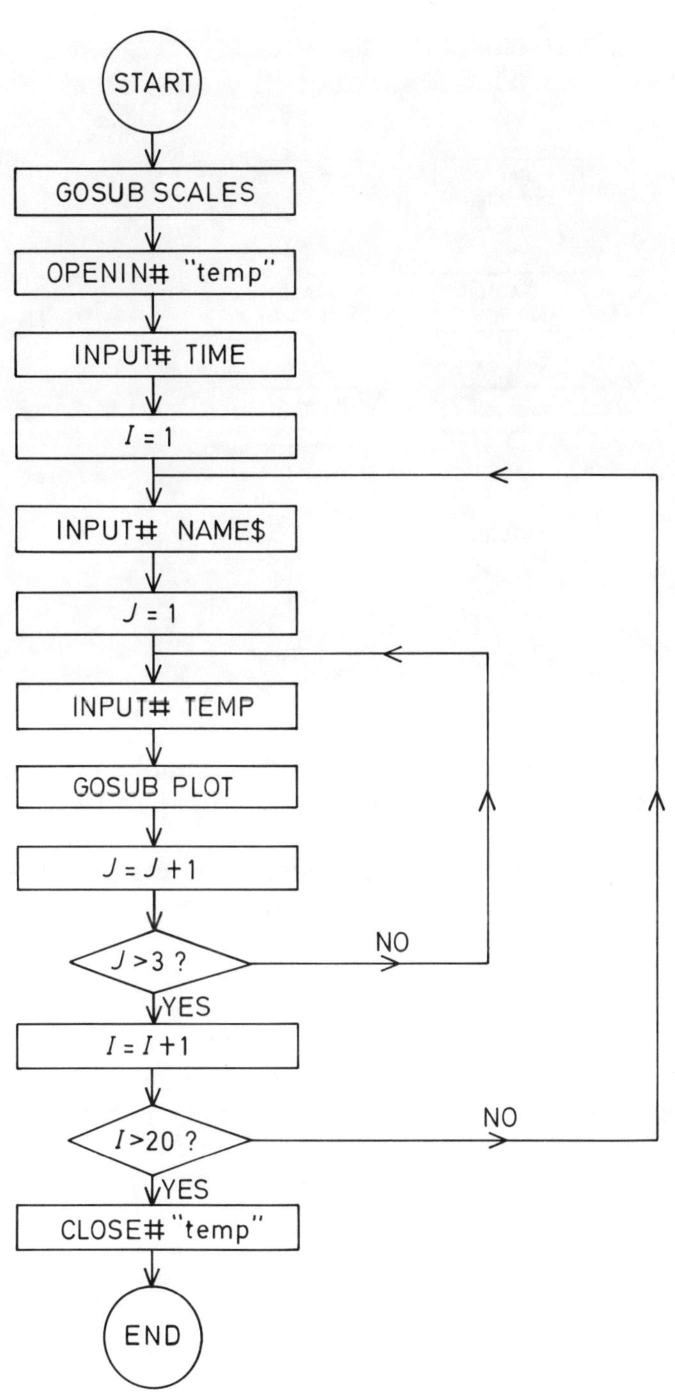
START
GOSUB SCALES
OPENIN# "temp"
INPUT# TIME
I = 1
INPUT# NAME$
J = 1
INPUT# TEMP
GOSUB PLOT
J = J + 1
J > 3 ?
NO
YES
I = I + 1
I > 20 ?
NO
YES
CLOSE# "temp"
END

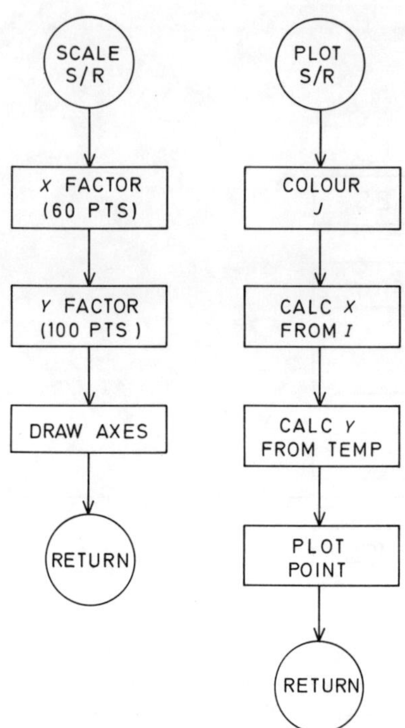

The variable I keeps count of the 3-minute sampling cycles. In the program I goes from 1 to 20 but you could use 0 to 19.

Variable J looks after the three different sources and determines what colour or character is to be used for plotting.

Since a temperature is noted every minute the X scale needs 60 intervals. Dividing the total number of pixels in the X direction by 60 gives the number of pixels per minute and multiplying this by the number of addressable points per pixel gives the number of addressable points per minute.. If this number is K and the scale starts at 50 points in from the screen edge the conversion formula would be X = 50 + K*I. A conversion for Y is worked out similarly.

****

SAQ 2.9a

(*i*) Convert these binary numbers to decimal:

00011000 10101010 01010101

(*ii*) Represent the following numbers in binary notation:

13 35 68 129 212

299 4,000 60,000

(*iii*) What are the maximum numbers that can be stored in (*a*) 2 bytes and (*b*) 3 bytes?

Response

(*i*) Working from the left the decimals values of the 8 bits of a byte are:

128 64 32 16 8 4 2 1

The conversions are done by adding the values (or weights) of the set bits:

$$00011000 = 0 + 0 + 0 + 16 + 8 + 0 + 0 + 0 = 24$$

$$10101010 = 128 + 0 + 32 + 0 + 8 + 0 + 2 + 0 = 170$$

$$01010101 = 0 + 64 + 0 + 16 + 0 + 4 + 0 + 1 = 85$$

(*ii*) In one method the bit values are subtracted, starting with the highest. In another we divide successively by 2 and take the remainders. Using the second method for the first number:

$13/2 = 6 + 1 \quad 6/2 = 3 + 0 \quad 3/2 = 1 + 1 \quad 1/2 = 0 + 1$

Taking the remainders in reverse order decimal 13 is binary 1101.

Let us use the subtraction method for decimal 212:

$212 - 128 = 84$

$84 - 64 = 20$

(miss 32)

$20 - 16 = 4$

(miss 8)

$4 - 4 = 0$

(miss 2)

(miss 1)

Putting a 1 for every subtraction and a 0 for every miss:

decimal 212 = 11010100

The other numbers less than 256 can be converted similarly:

35 = 00100011 68 = 01000100

The last two numbers need two bytes. Let us first divide 4000 by 256:

4000/256 = 15 with remainder 160

Decimal 15 is binary 00001111 while decimal 160 is binary 10100000. The high and low byte required for 4000 are therefore:

00001111 10100000

In the same way the two bytes for decimal 60000 contain 234 and 96:

11101010 01100000

(*iii*) The highest number that can be held by 8 bits is 255. Every unit of a second byte means 256. Hence the maximum decimal number for two bytes is:

256*255 + 255 = 65 535 (*a*)

If we add 1 to get 65536 we need three bytes, the third holding 1 and the other two holding zero. Hence, every unit of the third byte means 65536 and so the three-byte maximum is:

65536*255 + 256*255 + 255 = 16 777 215 (*b*)

****

SAQ 2.9b

An instrument is interfaced to a microcomputer by means of a 5-volt, 12-bit ADC. What resolution is nominally possible and how many bytes are required to store each reading?

Response

The maximum number for 12 bits is decimal 4095. The nominal resolution is therefore 5/4095 volt or 1.22 millivolt. One would not usually rely on the maximum resolution when using an ADC.

The output from a 12-bit ADC needs 2 bytes of memory for storage. The number read from the ADC would be divided by 256, the whole number dividend placed in the high byte, the remainder in the low byte.

****

SAQ 2.9c

> You have a program to accept input from a 10-bit ADC and you need to make 100 readings. You want to store the data immediately below the screen memory which starts at &3000. Where would you place HIMEM?

Response

A 10-bit ADC requires two bytes for each reading and therefore 200 bytes are needed. In hex. this is &C8. Let's play safe and move HIMEM down by &D0 (ie 208 decimal). The new HIMEM should be at &2F30.

****

SAQ 2.9d

> (*i*) Write a program to place the numbers 8 and 10 in two successive memory locations.
>
> (*ii*) Alter the program to make it read the locations after the numbers have been placed in them.

Response

(*i*) Here we use POKE. You will use the version appropriate to your machine.

```
 99 REM ** POKING
100 X = 8
110 Y = 10
119 REM ** USE LOCATION M
```

```
120 M = mmmm
130 POKE M,X
140 POKE M+1,Y
200 END
```

A shorter program would discard lines 100 and 110 and replace X and Y by their actual values in lines 130 and 140.

(*ii*) To read the locations the following lines are added (we use PEEK here):

```
149 REM ** NOW READ
150 X = PEEK M
160 PRINT X
170 Y = PEEK M+1
180 PRINT Y
```

Once you become experienced in reading and writing to memory several steps can be done at once. To read and print out the contents of locations M and M + 1 one could use the statement:

```
PRINT PEEK M, PEEK M+1
```

Perhaps the M and M + 1 should be in brackets. Again this depends on the computer.

****

SAQ 2.9e

(*i*) Write and run a program which will accept any number from the keyboard, divide it by 3, and print the result like this example:

14/3 = 4 + REMAINDER 2 ⟶

SAQ 2.9e (cont.)

(*ii*) Write and run a program to convert a decimal number less than 256 into a binary number and print out the result *in correct order*.

Response

(*i*) This is a straightforward use of DIV and MOD:

```
 99 REM ** DIV & MOD
100 PRINT "ENTER INTEGER"
110 INPUT N
120 D = N DIV 3
130 R = N MOD 3
140 PRINT; N;"/3 = "; D;" + REMAINDER "; R
150 END
```

(*ii*) The essential program is the one used as an example for obtaining the binary representation of a decimal number. However, the remainders are assigned to the elements of an array and then printed out in correct order.

```
 99 REM ** DECIMAL TO BINARY
100 DIM B(8)
110 PRINT "ENTER DECIMAL INTEGER"
120 INPUT N
129 REM ** N MOD 2 INTO ARRAY
130 FOR I = 7 TO 0 STEP -1
140     D = N DIV 2
150     B(I) = N MOD 2
160     N = D
170     NEXT I
179 REM ** PRINT IN REVERSE
180 PRINT "BINARY IS ";
190 FOR I = 0 TO 7
```

```
200       PRINT;B(I);
210       NEXT I
220 PRINT
230 END
```

At lines 130 and 190 we have used array elements 0 to 7. If your machine does not allow the zero element, B(0), you would work with elements 1 to 8.

The program could be made to repeat indefinitely by including a line:

```
212 GOTO 110
```

This is useful if you want to convert several numbers. The repetition could be stopped by a conditional statement or by pressing the escape key.

A similar program for decimal to octal conversion may be written using N MOD 8 but a decimal to hexadecimal program requires some modification to take account of the hex numbers A–E. This could be an interesting exercise if you want to test yourself.

****

SAQ 2.9f

(*i*) Write a program to place into two successive memory locations the low byte and the high byte of an integer entered at the keyboard. The low byte should be placed in the lower of the locations. Check that the program works by placing data and then reading the locations.

(*ii*) Outline how you would store the data from 100 readings from a 10-bit ADC in memory locations immediately below screen memory.

Response

(*i*) When a number greater than 255 is to be held in memory it is necessary to do what this question asks. We use DIV and MOD with 256 as divisor.

The first thing that must be done is to identify the memory locations which are to be used. Make sure that the locations can be 'poked' without upsetting the operating system. In the program below variable M represents the first of the two locations.

```
 99 REM ** INTO MEMORY
100 PRINT "ENTER LOWER MEMORY LOCATION "
110 INPUT M
120 PRINT "WHAT IS TO GO INTO MEMORY"
130 INPUT X
140 D = X DIV 256
150 R = X MOD 256
160 POKE M,R
170 POKE M+1,D
179 REM ** ALL IN. NOW TEST
180 P = PEEK M
190 Q = PEEK M+1
200 PRINT "MEMORY ";M;" HOLDS ";P
210 PRINT "MEMORY ";M+1;" HOLDS ";Q
220 PRINT "ORIGINAL NUMBER WAS ";P + 256*Q
230 END
```

The essential parts of the program are:

Decide the address and input the lower of the two memory locations involved (line 110).

Input the datum to be placed and break it up into a low and a high byte by means of the DIV and MOD operators (lines 130–150).

Place the two bytes of data into the two address bytes remembering that the low byte goes into the lower location (lines 160,170).

To test for success we have deliberately used two different variables, P and Q. The contents of the two address bytes are read into these (lines 180,190).

The final printing at lines 200–220 should display the contents of the two address bytes and also the regenerated number.

(*ii*) In Section 2.9.4 we saw how to reserve memory for data by moving HIMEM. This must be done in the present case.

Since a 10-bit ADC produces a number which requires 2 bytes HIMEM must be moved down by at least 200. A nice round hex figure is &D0.

A FOR-NEXT loop is convenient to control the taking of readings and placing the results in the locations reserved. The step size must be 2 to allow 2 bytes per reading. Without going into details the program above could be modified by the inclusion of statements like:

```
H = (new HIMEM = old HIMEM – &D0)

FOR M = H TO H + 199 STEP 2

(read ADC)

X = (datum from ADC)

D = X DIV 256

R = X MOD 256

POKE M,R

POKE M+1,D

NEXT M
```

****

SAQ 2.9g

Find the results returned by the following:

Z1 = 24 AND 194

Z2 = 3 AND 15

Z3 = 54 AND 240

Response

It is always safer to write out the bytes in full binary and do the comparison as shown below:

```
00011000   00000011   00110110
11000010   00001111   11110000
--------   --------   -------
00000000   00000011   00110000
--------   --------   -------
 Z1 = 0     Z2 = 3    Z3 = 48   (AND)
```

****

SAQ 2.9h

Find the results returned by the following:

Z4 = 24 OR 194

Z5 = 3 OR 15

Z6 = 48 OR 31

Response

As before, it is best to write out the numbers in binary and then do the OR bit by bit:

```
00011000   00000011   00110000
11000010   00001111   00011111
--------   --------   -------
11011010   00001111   00111111
--------   --------   -------
Z4= 218    Z5 = 15    Z6 = 63    (OR)
```

We can use these results to show the differences between OR and EOR.

Z4 would be the same using EOR but Z5 and Z6 would be different. Where both bits are set EOR requires the result bit to be zero:

```
00001111   00110000
00000011   00011111
--------   -------
00001100   00101111
--------   -------
Z5 = 12    Z6 = 47    (EOR)
```

****

SAQ 2.9i

Outline procedures to perform the following operations on memory location mmmm:

(*i*) set bits 0, 3, 5 and 7 to 1 and all other bits to 0;

(*ii*) clear bits 0 and 5;

(*iii*) set bits 4 and 6 without altering other bits;

(*iv*) test bits 0 and 2.

Response

(*i*) To set bits 0, 3, 5 and 7 the binary number required is 10101001. Hence decimal 169 must be placed in mmmm.

(*ii*) To clear bits 0 and 5 the binary with these bits clear is ANDed with the contents of mmmm. This number is 11011110 or decimal 222:

Read contents of mmmm into M

LET X = M AND 222

Place X in mmmm

(*iii*) Setting bits requires an OR. For bits 4 and 6 the binary is 01010000.

We therefore OR with 80:

Read contents of mmmm into M

LET X = M OR 80

Place X in mmmm.

(*iv*) Bit 0 is tested by ANDing the contents of mmmm with 1 while bit 2 is tested by ANDing with 4. In both cases a result greater than zero indicates that the bit is set, a result of zero that it is clear.

It is possible to test both bits together by ANDing the contents of mmmm with binary 00000101 which has bits 0 and 2 both set. This procedure is not recommended generally though it may be of use in special cases. The essential steps are:

Read contents of mmmm into M

LET X = M AND 5

Following these operations the value of variable X indicates the states of bits 0 and 2:

If X = 0 both bits are clear.

If X = 1 bit 0 is set and bit 2 is clear.

If X = 4 bit 2 is set and bit 0 is clear.

If X = 5 both bits are set.

It is important to note that the results obtained by the testing procedures described are not influenced by the state of any other bit of the byte.

****

SAQ 3.1a

A microcomputer is connected to a pH meter which has BCD output with computer-compatible digital signals of 0 and 5 volts. The digits for the tenths and units are connected to an interface which is memory mapped with an address of &FE00. Answer each of the following questions about the interface and the signals involved.

(*i*) Match the name of the signals given in list 1 with the appropriate bus in list 2.

List 1	List 2
(A) Signals corresponding to the pH value from the pH meter.	(X) Address bus
(B) Signals corresponding to &FE00 from the microcomputer.	(Y) Data bus

⟶

SAQ 3.1a (cont.)

(C) A microcomputer signal which indicates that data are to flow from the pH meter to the computer. — (Z) Control bus

(D) A microcomputer signal which indicates when the microprocessor is ready to receive data.

(*ii*) By circling Y (for yes) or N (for no) indicate which of the following are the functions of the interface in this example.

(1) To let data flow on to the data bus when the microprocessor has placed the correct address on the address bus and then is ready to accept data.

(Y / N)

(2) To allow data from the pH meter to reach the data bus whenever the address &FE00 appears on the address bus.

(Y / N)

(3) To protect the microcomputer system from the voltage levels output by the meter.

(Y / N)

Response

(*i*)

(A) Item A corresponds to the data and would be transmitted via the data bus, which is item Y of list 2.

(B) &FE00 is a 16-bit number which is typical of the address of a memory-mapped interface. All addresses are transmitted via the address bus so the correct answer is (X) of list 2.

(C) This is a typical of a control signal which in this case specifies whether a byte is to be written to the interface or read from it. The answer is therefore (Z) of list 2.

(D) This is another example of a control signal. This time it is to do with the timing of signals. Again the answer is (Z).

(*ii*)

(1) Yes this is correct. If data are allowed on to the data bus when it is not expected by the microprocessor, the computer will almost certainly 'hang-up' and any program currently being executed will fail. The only way to recover would be to switch off the power and remove the offending interface.

(2) This statement is false because it does not go far enough. Not only does the address on the address bus have to be correct but signals on the control bus must correctly indicate the direction of data flow and the precise timing of data transfer.

(3) This is false because the BCD signals output by the pH meter would use the same voltage levels as those used by the computer.

****

SAQ 3.1b

All of the following questions refer to the interface given in Fig. 3.1b which allows the computer to read pH values in BCD.

(*i*) Bit d1 of the port with address &FE01 indicates that the BCD datum is valid (d1 = 1 valid, d1 = 0 invalid). ⟶

SAQ 3.1b (cont.)

Suppose the byte X, read from &FE01 has a value 32 and we wish to check that the datum is valid.

(1) What number would you use to mask off all but bit d1 of the byte X ?

(2) What is the result of ANDing the answer from (1) with the byte X = 32?

(3) Is the datum valid?

(*ii*) Bit d0 of the byte X, read from the port with address &FE01, indicates the number of 'tens' in the measured pH. If X = 32, is the pH above or below ten?

(*iii*) The byte Y, read from the port of address &FE00, contains the units and tenths of pH. If the value obtained for Y is 89 what are the values of:

(1) the tenths;

(2) the units of pH?

What is the total pH, assuming no tens are involved?

Response

(*i*)

(1) To mask out all bits apart from bit d1 you should use the byte :

0 0 0 0 0 0 1 0 = 2

To monitor the status of the ith bit we apply a masking byte which has bit i set to 1 and all other bits set to 0. The masking byte therefore has a value 2ˆi. In this case i is 1, so that the byte has a value 2.

(2) To apply the masking byte we use the logical AND operation, as explained in Part 2 of this unit. If X = 32 then the result of ANDing it with 2 is 0 as shown below.

	d7	d6	d5	d4	d3	d2	d1	d0
X = 32 =	0	0	1	0	0	0	0	0
2 =	0	0	0	0	0	0	1	0
32 AND 2 =	0	0	0	0	0	0	0	0

The detailed working above is based on the bit-by-bit operation of the AND operator. The result is 1 only if both bits being ANDed are one, otherwise the result is 0.

Note that had bit d1 been 1 the result of X AND 2 would have been 2, so that the AND operation has essentially concentrated attention on bit d1 irrespective of the value of any other bits. We can therefore easily check the validity of the data from the pH meter. If the result of the AND operation is 0 the datum is invalid but a result of 2 shows that it is valid.

(3) In this case the BCD datum would be invalid since the use of the AND operator indicated that bit d1 was 0.

(*ii*) If X, read from the port with address &FE01, equals 32 the bit zero equals 0. This means that the pH will be less than 10. The status of bit zero could be checked by ANDing the byte with 1. The result would be 0 if the pH were less than 10 and 1 if it were greater than or equal to 10.

(*iii*) The binary pattern corresponding to 89 is 01011001.

(1) The tenths are contained in the least significant 4 bits and they can be isolated by ANDing the byte with 00001111 = 15 as follows.

15 = 00001111
89 = 01011001
15 AND 89 = 00001001 = 9

(2) The units are contained in the 4 most significant bits of Y and can be extracted by dividing Y by 16. The integer part of the result is then equal to the BCD digit for the units:

$$89/16 = 5.56$$

units contribution to pH = 5

You may recall from Part 2 that the computation of the integer part of the result can be programmed using the DIV operator (units = 89 DIV 16).

The pH is made up of 5 units and 9 tenths which can be combined by the expression:

$$pH = 5 + 9/10$$

To generalise this somewhat if the units were stored in A and the tenths in B, the pH would be:

$$pH = A + B/10$$

****

SAQ 3.1c

Recalling the flow-charting method discussed in Part 2, devise a flow chart for a program to read and print pH values repeatedly from a pH meter with BCD output from the interface connections as indicated in Fig. 3.1b. Assume that the pH meter is in 'free-run' mode so that a continuous stream of data is produced. Note that you must check the data valid signal before reading the pH.

Response

If you followed the stepwise design given in Section 3.1.2, your flow chart should resemble that given below. This program runs in a continuous loop and can only be stopped by pressing escape or break, depending on the computer. We could insert an extra few lines of program, after printing the pH, to ask the user if another reading is required. If not the program could be made to terminate by reaching an END instruction.

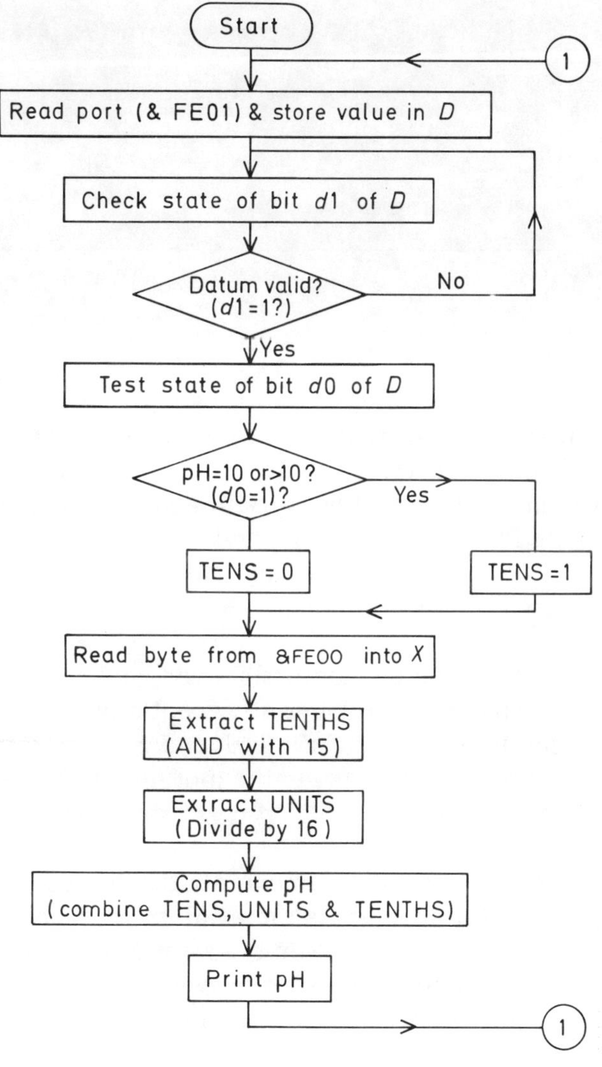

SAQ 3.1d

It is proposed to operate some equipment under microcomputer control by sending output to an appropriate interface. For each of the following items indicate the main functions of the interface.

(*i*) A peristaltic pump operating at 240 volts A.C.

(*ii*) A gas valve operated by a 12 volt D.C. supply.

(*iii*) An indicator lamp operating on 5 volts.

(*iv*) Another computer.

Response

(*i*) The main problem here is the incompatible voltage. A computer controlled relay would be required to switch on and off the mains supply. Specially designed solid state relays are available for mains switching and have the advantage of completely isolating the high voltage side from the low voltage circuitry used by the computer.

(*ii*) Here again we have incompatible voltage levels. The interface would probably be designed to accept a latched output signal from the computer and switch on a transistor which would connect the 12 volt supply to the motor with the required drive capability.

(*iii*) Although the computer can supply the correct voltage in this case, the main problem is the available current drive to light the lamp. The interface would accept a latched output signal from the computer and use it to switch on a transistor to connect a 5-volt supply with plenty of current drive to the lamp.

(*iv*) Here the voltage levels and current drive are not problems. This time the main task of the interface is to control the timing of data flow between the two computers so that data are only passed from the data bus of one machine to that of the other when a valid datum transfer is possible.

****

SAQ 3.1e

(*i*) Complete the DATA statements in the program listed in Fig. 3.1f for the sequence specified in Fig. 3.1e.

(*ii*) The program in Fig. 3.1f uses a time wasting loop to produce the time delay (lines 330, 340).

(1) Is there any other way that a time delay could be produced for use in sequence control?

(2) Are there any disadvantages in using such a time wasting loop?

(*iii*) What modifications would be necessary for the program in Fig. 3.1f if,

(1) the address of the memory-mapped output port were changed to &FF00?

(2) the interface were not memory-mapped but had a port number of 205? (Some versions of BASIC which support output to a numbered port use the instruction OUT n, m where n is the port number and m is the byte to be output).

Response

(*i*) The sequence defined in Figs. 3.1e and 3.1f has 6 steps and so 6 is the first item of data. Therefore we insert 6 into the first DATA statement at line 400. For each of the 6 steps we have to specify the byte to be output and the time for which that output needs to be maintained. The output bytes and the corresponding time periods form pairs of data and are given in the rightmost pair of columns in Fig 3.1e. These data pairs would occupy succeeding DATA statements in the program as shown below:

```
400 DATA 6
410 DATA 2,3
420 DATA 128,10
430 DATA 1,2
440 DATA 128,5
450 DATA 132,10
460 DATA 128,10
```

Alternatively, the data need not be spread over so many lines:

```
400 DATA 6
410 DATA 2,3,128,10,1,2,128,5,132,10,128,10
```

(*ii*)

(1) Yes, other methods are available to produce time delays. Most computers have an internal clock which can be accessed from a program. This 'real-time' clock can take a number of forms from one computer to another.

A simple version increments a counter each time a specific time-period, such as 10 milliseconds, elapses. In this case to produce a 10 second time delay we must firstly set the counter to 0 and then wait for it to reach 1000 counts (1000 × 10 ms = 10 s).

More sophisticated real-time clocks provide counters for seconds, minutes, days, weeks, months and even years, as well as shorter intervals of time.

(2) Yes, there are some disadvantages. Most computers use so-called interrupt routines. These are programs which are executed in preference to your BASIC program to remedy some situation which needs urgent attention. Interrupts may be used in controlling data transfers to and from disc, or even updating the real time clock. The point is that every time an interrupt occurs the execution of your program is suspended. If your program was executing a time wasting loop, then the time delay will be lengthened by the interruption. As a general rule it is better to use a real-time clock for time delay than time-wasting FOR-NEXT loops.

Another problem which can occur with some computers is that the time required to update the contents of a variable may depend on the number of other variables used in the program. So, if a loop with upper limit of 900 gives a time delay of 1 second in one program, it may not give exactly a second in another program.

(*iii*)

(1) The only change would be to line 170 which becomes:

```
170 POKE &FF00,B(I)
```

(2) Again only a small alteration would be necessary. Line 170 would become:

```
170 OUT 205,B(I)
```

where 205 is the port number and B(I) is the byte to be output.

****

SAQ 3.1f

An output port is connected as follows to a series of external devices.

```
d7 d6 d5 d4 d3 d2 d1 d0
1  0  1  1  0  1  1  0
|  |  |  device 4  |  device 0
|  |  device 5  |  device 1
|  device 6  |  device 2
device 7   device 3
```

The current state of the output at the port corresponds to the byte 10110110 or 182 which is stored in a program as variable X. We wish to switch on device 3 without affecting any other device. What byte pattern would you output? In obtaining your answer use the logical OR to assign a value to the variable Z for subsequent output to the port.

Response

The byte to be output would be :

d7	d6	d5	d4	d3	d2	d1	d0	
1	0	1	1	1	1	1	0	= 194

This can be checked by against the original byte with the single change that bit d3 is assigned to 1 rather than 0. To obtain this result using the logical OR operator we proceed as follows.

(*a*) Assign Y the binary value 00001000 in which only bit d3 is set to 1.

(*b*) Apply the OR operator to Y and X and store the result in Z.

(X represents the current state of the output).

```
        Y =  00001000
        X =  10110110
             --------
Z = X OR Y   10111110
```

Thus Z = 194 which could be output to the port to change bit d3 from 0 to 1 without affecting any other bits.

****

SAQ 3.1g

An output port is connected to an external device as indicated below and the bit pattern shown represents the current state of the output.

```
d7  d6  d5  d4  d3  d2  d1  d0
1   0   1   1   0   1   0   1
    not         |   |   |   device 0
 connected      |   |   device 1
                |   device 2
                device 3
```

(*i*) Use the NOT and AND operators to deduce the byte which should be output to reset bit d2 to 0 without affecting other bits.

(*ii*) Use logical operators to deduce the byte which should be output to switch off device 0 and switch on device 3 (assume that 1 = ON and 0 = OFF).

Response

Let us assume that X contains the byte which reflects the current state of the output to the port:

$$X = 10110101$$

(*i*) To reset bit d2 to 0 we firstly define the byte Y with d2 = 1.

$$Y = 00000100$$

We then apply the NOT operator and put the result in W.

$$W = 11111011$$

Finally we AND this result with X.

$$\begin{aligned} W &= 11111011 \\ X &= 10110101 \\ &\text{- - - - - - - -} \\ Z = W \text{ AND } X &= 10110001 \end{aligned}$$

So the byte to be output, Z, equals 177.

(*ii*) Here we need to set 1 bit, d3, to 1 and another, d0, is to be reset to 0.

Again let X contain the current state of the output. To set d3 to 1 we modify X by ORing it with 00001000 = 8 and store the result in Y say:

$$Y = 8 \text{ OR } X$$

We need to modify Y further to reset bit d0 to 0 by ANDing it with 11111110 (as 11111110 = NOT 00000001). Placing the result in Z, the sequence of operations is as follows.

$$\begin{aligned} Y &= 8 \text{ OR } X \\ W &= \text{NOT } 1 \\ Z &= Y \text{ AND } W \end{aligned}$$

This could be condensed to the following statement.

$$Z = (8 \text{ OR } X) \text{ AND } (\text{NOT } 1)$$

The detailed working equivalent to the above statement is given below in case you should doubt its validity.

```
            8 =  00001000
            X =  10110101
                 --------
   Y = 8 OR X =  10111101

            1 =  00000001
   W = NOT 1  =  11111110
            Y =  10111101
                 --------
 Z = Y AND W  =  10111100
```

If we compare the byte pattern for Z with the original pattern X, we see that the sequence of operations has changed bit d0 from 1 to 0 and bit d3 from 0 to 1 as required.

This may appear rather long winded but the power of the method should become apparent when the problem is generalised. Suppose we wish to set bit N to 1 and reset bit M. This could be done by outputting the byte Z computed as follows:

Z = (2^N OR X) AND (NOT 2^M)

where X contains the byte which reflects the current output at the port.

****

SAQ 3.1h

An output port is connected to some external devices as shown below.

```
d7 d6  d5 d4  d3 d2 d1  d0
D  D   1  D   0  D  D   D          D = 0 or 1
|  |   |  device 4| |   device 0   (unknown
|  |   device 5|  | device 1       to program-
|  device 6    |  device 2         mer)
device 7    device 3                      →
```

SAQ 3.1h (cont.)

Although the programmer does not know which devices will be on or off at a particular point in the program he can assume that the last output byte is stored in a variable X.

Write the BASIC statements which will accomplish each of the following tasks (assume that 1 = ON, 0 = OFF, the port has a port number of 202 and the version of BASIC used supports the OUT instruction).

(*i*) Switch on all devices.

(*ii*) Switch off all devices.

(*iii*) Switch on device 3 without affecting other devices.

(*iv*) Switch off device 5 without altering anything else.

Response

(*i*) This is a trivial case since everything has to be switched on together. The instruction OUT 202,255 will achieve this, where the port number is 202.

(*ii*) This is similar to (*i*) above except that everything is switched off together. The BASIC instruction is therefore OUT 202,0.

(*iii*) The required byte pattern can be computed as Z = 8 OR X. The byte pattern to switch on device 3 is 00001000 = 8. The OR operator merges this byte with the original output pattern stored in X.

(*iv*) The answer is Z = X AND (NOT 2^5). Here 2^5 = 00100000 and so (NOT 2^5) = 11011111. When this is ANDed with the original byte X, the result Z has the correct byte pattern to switch off device 5 when it is output to the port.

****

SAQ 3.2a

A schematic pin diagram of the 6522 VIA is shown below. Pins PA0-PA7 refer to Port A, and Pins PB0-PB7 correspond to Port B. If the voltage levels given for each pin apply, choose the option which correctly describes the contents of registers DRA and DRB.

Voltage level (volts)	Pin	Pin numbers on chip	
		1	40
5	PA0	2	39
0	PA1	3	38
5	PA2	4	37
5	PA3	5	36
0	PA4	6	35
0	PA5	7	34
0	PA6	8	33
5	PA7	9	32
0	PB0	10	31
0	PB1	11	30
0	PB2	12	29
0	PB3	13	28
5	PB4	14	27
5	PB5	15	26
5	PB6	16	25
5	PB7	17	24
		18	23
		19	22
		20	21

Pins 25–32: Computer data bus

Option	DRA	DRB
A	11110000	10001101
B	00001111	10110001
C	10001101	11110000
D	50550000	00005555
E	50005505	55550000

Response

The ≃5 Volt level registered at an input pin becomes a digital 1 at the interface, and 0 V becomes digital 0.

The binary data reading for port A is therefore

PA7	PA6	PA5	PA4	PA3	PA2	PA1	PA0
1	0	0	0	1	1	0	1

and for port B we have

PA7	PA6	PA5	PA4	PA3	PA2	PA1	PA0
1	1	1	1	0	0	0	0

The correct answer is therefore option C. All the options which include '5' in the sequence of digits are incorrect because the 5 Volts input appears as a digital value of 1 as part of the input byte.

****

SAQ 3.2b

An instrument provides 16 bits of data to be read as two separate bytes, and requires two control signals from the computer. One signal tells the instrument to hold or 'freeze' the existing datum in order that the computer has time to read one byte and then the other, without allowing the datum to change. We shall call this the 'RUN/HOLD' signal (1 = HOLD, 0 = RUN). The second control signal tells the instrument which byte is to be read (the most significant, or least significant one of the 16 bit number). We shall assume that 1 means the most significant, and 0 means the least significant byte. The connections with the 6522 VIA are as indicated below.

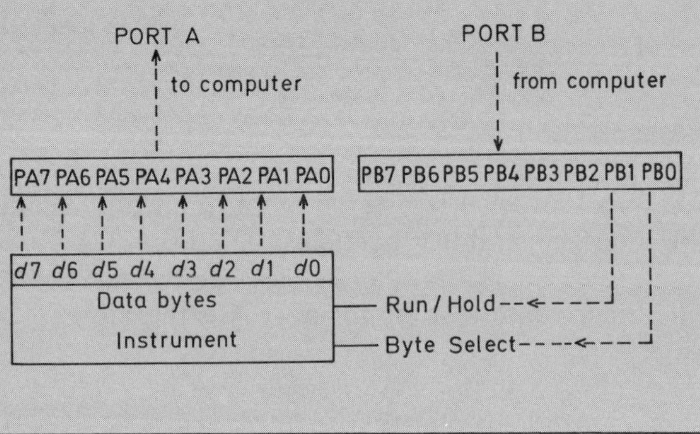

SAQ 3.2b (cont.)

(*i*) What bytes would you write to DDRA and DDRB to configure the 6522 VIA for this application?

(*ii*) Assuming the internal registers of the 6522 VIA have the addresses given below, write the sequence of BASIC statements which would read the least significant byte into X and the most significant one into Y. The sequence of statements should then combine the bytes to give a number stored in Z, print out the value of Z, allow the instrument reading to change and repeat the whole process so that a continual stream of readings is obtained.

Assume that the version of BASIC used supports PEEK and POKE (X=PEEK(ADDRESS) and POKE X, ADDRESS where X is the byte read or written to or from location ADDRESS).

ADDRESS	REGISTER
&FE60	DRB
&FE61	DRA
&FE62	DDRB
&FE63	DDRA

Response

(*i*) The application requires port A to be configured as input (reading data from the instrument) and the least significant two bits of port B as output (controlling the instrument).

Thus send the byte 00000000 = 0 to DDRA to configure port A as all input, and send the byte 00000011 = 3, to DDRB to make bits PB0 and PB1 as output.

(*ii*) The run and hold control is connected to PB1 and the byte select control is at PB0. Here is a sequence of statements to read data from the instrument.

```
80 POKE &FE63,0 : POKE &FE62, 3: REM
   CONFIGURES CHIP
90 REM FREEZE READING AND SELECT LEAST
   SIGNIFICANT BYTE
100 POKE &FE60, 2: REM PB0 = 0 FOR LEAST
   SIGNIFICANT BYTE
110 REM READ 1st DATA BYTE
120 X = PEEK (&FE61)
130 REM AND NOW THE SECOND BYTE MUST BE
   SELECTED
140 POKE &FE60, 3
150 REM OBTAIN 2nd BYTE
160 Y = PEEK (&FE61)
170 COMBINE THE BYTES
180 Z = Y*256 + X : PRINT Z
190 REM UNFREEZE THE READING
200 POKE &FE60, 0
210 GO TO 90
```

****

SAQ 3.2c

Which one of the following control words for the 8255 PPI configures the interface for Port A as input, Port B as output and Port C with PC0-PC3 as input but PC4-PC7 as output?

	D7	D6	D5	D4	D3	D2	D1	D0
(*i*)	0	0	0	1	1	0	0	1
(*ii*)	1	1	1	1	0	1	0	1
(*iii*)	1	0	0	1	0	0	0	1
(*iv*)	0	1	0	0	1	0	0	1

Response

Referring to Fig. 3.2a, the control word is option (*iii*) built up as follows:

D7 must be 1 to change mode, D7 = 1
D6, D5 for mode 0, D6 = 0, D5 = 0
D4 for port A input, D4 = 1
D3 for PC4-PC7 as output, D3 = 0
D2 again for mode select D2 = 0
D1 for port B as output, D1 = 0
D0 for PC0-PC3 as input, D0 = 1

****

SAQ 3.2d

Given the port numbers below for an 8255 PPI, write the sequence of BASIC statements which will allow a stream of bytes of information to be read from an external device, assuming the following connections between the instrument and the 8255 PPI.

Port numbers
Port A = 200; Port B = 201; Port C = 202; Control Port = 203

Connections

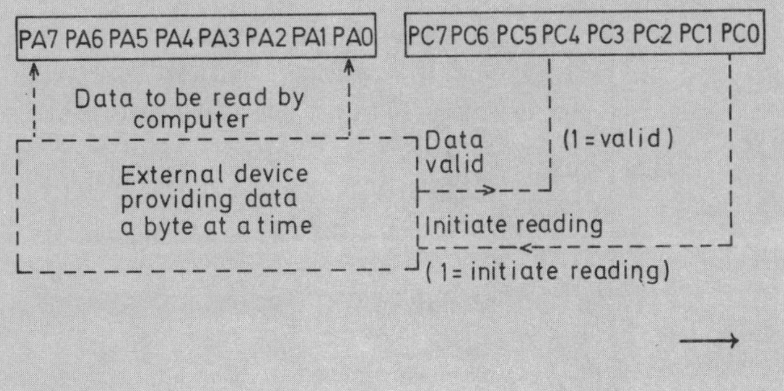

→

SAQ 3.2d (cont.)

Notice that Port A is used for data input and Port C for control purposes. The upper 4 bits of Port C need to be configured for input (to test the validity of data before reading it), and the lower 4 bits for output (to indicate that a new reading from the device is required).

Response

```
10 OUT 203, 152 : REM CONFIGURE CHIP FOR: PA INPUT,
   PC4-PC7 INPUT, PC0-PC3 OUTPUT.
20 OUT 202, 1 : REM OBTAIN A READING
30 X% = INP (202) : REM READ DATA VALID SIGNAL
40 IF (X% AND 16) = 0 THEN 30 ELSE 70
50 REM ABOVE MASKS OFF ALL BUT BIT PC4 AND
   CHECKS TO
60 REM SEE IF IT IS SET TO 1
70 Y% = INP (200) : REM TAKE READING INTO COMPUTER
80 PRINT Y% : GOTO 20
```

****

SAQ 3.3a

(*i*) An analyst wishes to use a computer to sample an analogue signal every second. Each reading is to be stored in the computer's main memory and takes up 5 bytes. If 10K bytes (1K = 1024) are available for data storage, what is the maximum time for which data can be collected?

What could be done to extend the time limit? ⟶

SAQ 3.3a (cont.)

(*ii*) A signal from an instrument is subject to interference from the mains supply operating at 50 Hz. It has been decided to eliminate this interference once the signal has been stored in the computer. Given that 50 Hz is the highest frequency component in the monitored signal, what is the minimum sampling frequency which is acceptable?

Would the sampling be possible using a BASIC program run by an interpreter?

If sampling is required over a 5 minute period, how much storage space would be required if each reading needed 5 bytes of memory

Response

(*i*) 10K bytes of memory = 10,240 bytes. Each reading takes 5 bytes, therefore 2,048 readings are possible.

As a sample is taken each second the maximum time for data collection would be 2,048 seconds or 34 minutes and 8 seconds.

If this time was insufficient the program would have to be modified so as to transfer data straight onto the disk. This can cause problems because disk transfers tend to be done as batches of numbers and although most of the readings would be correctly collected, one may be missed during a disk transfer. Another problem is that we could no longer rely on an absolutely regular sampling period because the sampling process would be interrupted by a disk write. The only way round this would be to read the computer's internal clock at the time of each sample and store both the time and the reading on disk.

(*ii*) According to the Nyquist sampling theorem, the sampling frequency should be at least twice the highest frequency component in the signal. (In practice we would use a higher rate of sampling to be on the safe side!) In this case we would need to sample at a rate of at least 100 Hz or one reading every 1/100 of a second.

When using a BASIC program under the control of an interpreter, the execution of the program is relatively slow. Although for most machines we should be able to achieve sampling rates of the order of about 10 Hz, 100 Hz is too fast. The data would have to be collected using a machine code program, possibly generated by writing the program in a high level language and compiling it to produce the required code.

Over a 5 minute period, sampling at 100 Hz we would collect 5 x 60 × 100 = 30,000 readings. If each takes 5 bytes we would need 150,000 bytes to store the data, which would be beyond many of the smaller microcomputers.

****

SAQ 3.3b

An 8-bit digital to analogue converter (DAC) is accessed via a memory-mapped interface with address &FE01. Write a program which will produce an analogue output to resemble as closely as possible the following waveform.

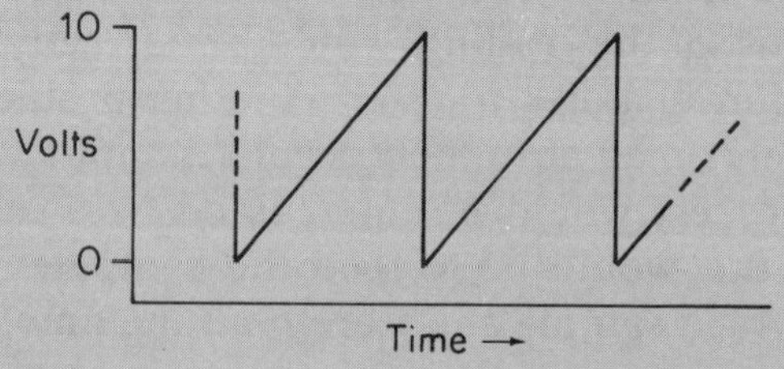

→

SAQ 3.3b (cont.)

Assume that the DAC has an output which ranges from 0 to 10 V and the BASIC used supports POKE.

In what way would the analogue signal generated differ from the target waveform above?

Modify the program so that the user has control of the rate of climb, and therefore the frequency of the waveform.

Response

The waveform consists of a range in which the voltage is raised from 0 to 10 V. The 0 volt level corresponds to digital 0, and in this case 10 volts corresponds to 255. We therefore need to output a digital value which climbs from 0 to 255, which represents the digital ramp, and then reset the digital value to 0 to repeat the cycle. A simple program to do this is given below:

```
 10 REM GENERATES A SAWTOOTH WAVEFORM
 20 REM ADC AT ADDRESS &FE01
 30 REM 8-BIT CONVERTER (0 - 10V)
 40 REM ****
 50 FOR I = 0 TO 255
 60     POKE &FE01, I
 70     REM COULD PUT TIME DELAY HERE
 80     NEXT I
 90 REM RAMP FINISHED
100 GOTO 50 : REM FOR NEXT CYCLE
110 END
```

(For BBC BASIC line 60 becomes ?&FE01 = I).

The above program would produce output of the form given below where the ramp is seen to be a series of discrete steps due to the quantisation of the original digital signal.

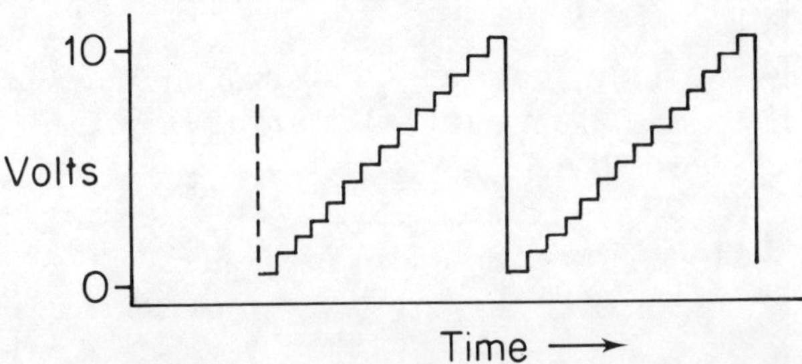

The program could be modified by replacing line 70 by a call to a time delay routine which would determine the time delay in between increments in the FOR – NEXT loop. The simplest approach would be to use a time-wasting loop by amending the above program as follows:

Insert lines 43 and 46 as:

```
43 PRINT "Parameter For Time Delay"
46 INPUT PT.
```

Replace line 70 by:

```
70 GOSUB 1000.
```

Insert the following lines:

```
1000  REM TIME DELAY ROUTING
1010  REM NEEDS PARAMETER PT
1020  REM PT = 100 PRODUCES
1030  REM 0.2 SEC DELAY
1040  FOR K = 1 TO PT
1050        NEXT K
1060  RETURN
```

The parameter PT determines the upper limit of the time wasting loop and hence the time delay.

The correspondence between this parameter and real time would have to be determined empirically and will differ from one computer to another.

****

SAQ 3.3c

A computer is interfaced to a spectrophotometer via an 8-bit analogue to digital converter (ADC). The output from the spectrophotometer is linear in absorbance and 0 absorbance corresponds to 0 read from the ADC. A value of 255 corresponds to an absorbance of 1.2. If an absorbance of 0.6 is obtained, what is the quantisation error? Could an analyst expect to obtain readings in the computer which were within 0.5% of their true values?

Response

Digital 255 corresponds to an absorbance of 1.2 and, as the analogue voltage in this case is proportional to absorbance, an absorbance value of 0.6 would yield a digital value of 127 (or 128). In general the digital valuc may be up to 1 unit out compared with the theoretical value, because of the quantisation in steps of 1 by the ADC. Our reading obtained could therefore be 1 part in 127 out or, 0.79%.

Errors may arise from other sources in this measurement but it is already clear that the analyst could not expect to achieve an accuracy better than 0.5% since the quantisation error is already greater than this. The answer would be to use a 10 bit ADC in which case an absorbance of 0.6 would correspond to a digital value of 512 with a quantisation error of 1 part in 512 or 0.20%.

****

SAQ 3.3d

An 8 bit analogue to digital converter (ADC) is interfaced to a computer via an 8255 PPI.

The 8 bit datum is read through port A and port C is used to control the ADC. Bit PCO is used to output a signal to initiate the start of conversion ($\overline{SC}$) and PC4 as an input to indicate the end-of-conversion ($\overline{EOC}$). Both SC and EOC are active low so that if EOC has a logic value of 0 the conversion is complete. Similarly conversion is started by setting SC low.

(*i*) Produce a step-wise design for a program which will read 1000 numbers from the ADC with one sample taken every 2 seconds. Assume that the computer you are using has a real-time counter which is incremental every 10 ms and the value of the counter is stored in TIME. The 8255 PPI is memory mapped with addresses:

Port A	:	&FE70
Port B	:	&FE7l
Port C	:	&FE72
Control Port	:	&FE73

The control byte needed to configure the 8255 for this application is:

10011000 = 152

(*ii*) After checking your answer to (*i*), write a program to implement your design using either PEEK and POKE or the '?' equivalent of BBC BASIC.

Response

(*i*) Your step-wise design should be along the lines:

(1) Configure the 8255 VIA.
(2) Initialise a counter to zero.
(3) Increment the counter by 1.
(4) Obtain and store a reading from the ADC, and note the time.
(5) Wait for 2 seconds to elapse since the last reading.
(6) If 1000 readings have been obtained, terminate the run or else repeat from (3) above.

(*ii*) Here is an example program which follows the above design:

```
  10 REM COLLECTS 1000 READINGS FROM ADC
  20 REM INTERFACED VIA 8255 PPI
  30 REM DATA PORT A; PC0 SC; PC4 EOC
  40 REM ADDRESSES PORTS A, B, C, CONTROL
  50 REM &FE70, 71, 72, 73
  60 REM ****
  70 DIM RD (1000)
  80 REM CONFIGURE CHIP
  90 POKE &FE73 = 152
 100 S = 0 : REM INITIALISE COUNTER
 110 S = S + 1 : REM START LOOP
 120 GOSUB 1000 : REM SET A READING AND TIME
 130 RD(S) = X : REM STORE READING FROM ADC
 140 IF S = 1000 THEN 150 ELSE 110
 150 END
```

The subroutine which communicates with the ADC and checks the time is as follows:

```
1000 REM OBTAINS A READING FROM ADC
1010 REM AND WAITS 2 SECONDS
1020 POKE &FE72, 0 : REM SETS PC0 TO 0 AND
     STARTS CONV.
1030 X = PEEK (&FE72) : REM READS PORT C
1040 X = X AND 16 : REM MASKS OFF ALL BUT PC4
```

```
1050 IF X = 16 THEN 1030 ELSE 1060
1060 REM ABOVE CAUSES REPEATED SCANS OF
1070 REM PORT C UNTIL PC4 GOES LOW
1080 GOSUB 2000 : REM WAIT 2 SECONDS
1090 RETURN

2000 REM WAIT 2 SECONDS BY REFERENCE
2010 REM TO REAL TIME CLOCK
2020 TIME = 0 : REM RESETS CLOCK TO 0
2030 IF TIME TOO < 2 THEN 2030 : REM WAITS HERE
FOR
     2 SECONDS
2040 REM TIME INCREMENTS BY 1 EVERY 10 MILLI
     SECONDS
2050 RETURN
```

****

SAQ 3.3e

Write a stepwise design for a program which will scan 500 times the first three channels of a 16-channel multiplexed ADC of the type shown in Fig. 3.3h and store the values obtained in arrays. A 2 second delay is required between successive readings of the ADC. Assume the following connections to a 6522 VIA interface through which communication with the ADC is achieved by the computer.

6522 VIA addresses:

Part A	DDRA	: &FE63
	DRA	: &FE61
Part B	DDRB	: &FE62
	DRB	: &FE60

⟶

SAQ 3.3e (cont.)

Connections (see Fig. 3.3h):

$PA_0 \ldots PA_3$	connected to $d_0 \ldots d_3$ respectively (to select the channel number).
$PB_0 \ldots PB_7$	connected to $b_0 \ldots b_7$ (converted 8 bit data)
PA_5	start conversion
PA_7	end conversion

Response

1. Configure the 6522 VIA for port B all input, PA0..PA4 and PA5 as output and PA7 as input. (Thus output 0 to address &FE62 to set all of port B to input, and output 00111111 = 63 to address &FE63 to set PA0..PA5 as output and PA7 as input).

2. Initialise a loop counter (ADCNT) to zero (counts the number of scans of ADC channels) and output 32 to port A to set start conversion line high (inactive).

3. Increment loop counter (ADCNT).

4. Loop over each channel (i = 0, 1, 2) to be sampled.

5. Output channel selection byte (output the value of i to port A, but remember to maintain PA5 high by adding 32 to the byte you output as 32 corresponds to a 1 at PA5.

6. Initiate conversion by sending a negative going pulse on PA5. Do this by sending the value of i, then i + 32 to port A.

7. Monitor the status of PA7 (by reading a byte from port A, masking off all but PA7 by ANDing with 128) and wait until end of conversion is observed (by PA7 going low).

8. Store the value obtained in the next vacant element of an array, ie A (ADCNT, i) for example for scan ADCNT and channel i.

9. Terminate loop over channels.

10. If 500 scans are up the exit else repeat from step 3.

****

SAQ 3.3f

By circling T or F, indicate whether or not each of the following statements is true or false.

(*i*) The quantisation error in a 16 bit ADC is likely to be about twice that of an 8 bit ADC because there are twice as many bits in which an error could arise.

T / F

(*ii*) One must always sample analogue signals at least twice as fast as the highest frequency component, even if the analytical information is contained in a lower frequency signal, and the highest frequency component corresponds to a small amount of noise.

T / F

(*iii*) Because BASIC is slow, in many applications using a fast ADC, it is sufficient to initiate a conversion by setting the $\overline{SC}$ line momentarily low and then reading the data without checking the $\overline{EOC}$ signal.

T / F

⟶

SAQ 3.3f (cont.)

(*iv*) Digital signals are restricted to one of two levels whereas analogue signals can vary continuously between 0 and 5 V.

T / F

(*v*) The use of an ADC converter is the only way of reading an analogue signal by computer.

T / F

Response

(*i*) This statement is completely wrong. The quantisation error is to do with how many steps are used by the ADC is representing the voltage range (from say 0 to 10 V). An 8 bit ADC would divide this into $2^8 - 1 = 255$ steps so that each step involves a change in voltage of $10/255 = 0.039$ V. On the other hand a 16-bit ADC would divide the 10 V range into $2^{16} - 1 = 65535$ steps with intervals of 0.00015 V. The quantisation error is nothing to do with the probability of failure of any of the bits of the ADC and exists even for a perfectly working ADC.

(*ii*) This is also false. To obtain the complete signal in the computer, including full details of the noise, we would have to sample at least twice the rate of the highest frequency component. On the other hand, if the noise was a small component and random, we could accumulate data at a lower frequency and expect an error to be introduced because of the noise. The error could be reduced by taking repeated measurements at the lower sampling rates and averaging the results. If random, the noise would average out but the analytical signal (hopefully non random!) would not.

(*iii*) This is true provided that the ADC is fast enough to guarantee the data will be valid when they are read by the computer. To be on the safe side, it is good practice to monitor for the EOC signal (or data valid, whichever is provided by the ADC).

(*iv*) The second half of this statement is false. Analogue signals are not restricted to the range 0 to 5 V. For example the analogue output sent to a recorder may be of the order of 100 mV. This would probably need amplifying before being fed to the ADC to obtain a compatible voltage range.

(*v*) Again this is not true. Earlier in this section brief mention was made of the voltage to frequency converter which provided a digital waveform whose frequency was proportional to the original analogue voltage. The digital waveform could be sampled at a single input pin on a digital interface.

Another example is when an instrument such as a digital pH meter, or a digital voltmeter, provides multi-byte datum values to be read by the computer. These instruments involve analogue to digital conversion but the user does not communicate directly with them. Control signals are usually available for 'run/hold' or 'freeze' the display to allow the computer to read a sequence of bytes without the datum changing. Readings from such instruments may be read into the computer in several ways including BCD, as described earlier, or using RS232 or IEEE 488 communications to be discussed in the next section.

****

SAQ 3.4a

In the context of parallel communication between computers under handshake control, indicate whether or not each statement is true (T) or false (F).

(*i*) If the sending computer halts, the receiver will wait until the appropriate handshake signals have been received and so give the appearance of halting too.

T / F

⟶

SAQ 3.4a (cont.)

(*ii*) If the receiving computer shuts down, the sender will continue to pass data which will then be lost.

T / F

(*iii*) The ASCII character for '1' is identical with the binary representation of 1.

T / F

(*iv*) Integers in the range 0 to 255 can be passed from one computer to another as a single byte. Other integers or decimal numbers cannot be sent from one computer to another using parallel communication.

T / F

(*v*) Parallel communication is normally used on short distances.

T / F

Response

(*i*) This is true. The receiver monitors the 'data available' signal before reading the byte present at the port used for parallel transmission. Usually signals are chosen to be active low,and the default state is high. So if the power to the sender was temporarily interrupted, the program in the sender would be lost and the DAV line would remain high and the receiving machine would wait for ever.

(*ii*) This statement is false. Because of the two-way communication involved in handshaking, if one machine shuts down the other will apparently 'hang-up', but in fact it would be waiting for the completion of a handshake exchange.

(*iii*) This is false also. Binary 1 is 00000001 in terms of 8 bits. On the other hand, the ASCII code for 1 is 0110001 (or decimal 49). You must always correctly differentiate between the genuine binary representation of a digit and the ASCII code. With the former you can carry out arithmetic operations and obtain correct answers, but not so with ASCII code.

(*iv*) Integers and decimal numbers in general are stored in a computer as a sequence of bytes. In some machines 5 bytes are used to represent integers and 6 for decimal numbers. The precise format may vary from one computer to another which may cause problems in transmitting the sequence of bytes which represents an integer value or a decimal number. An alternative approach is to transmit a number (integer or decimal) as a sequence of digits each represented by its ASCII code. Since this code is accepted as standard from one machine to another problems of internal representation are avoided. Of course, there is the extra complication that the sending machine must assemble the string of ASCII characters to be sent and the receiver must re-convert the sequence of digits into a number which is stored in correct internal format.

(*v*) This is true because on the one hand the cabling is expensive (eight lines for data, two for handshaking and a common ground) and on the other hand only low voltages are used ($\simeq$5 V or $\simeq$0 V) for the digital signals. In a noisy electrical environment the signal may become degraded leading to data transmission errors. For high rates of data transfer the integrity of the signals may be affected by 'reflections' of the electronic pulses along the data lines. All of these problems are reduced if the cable lengths for parallel communications are kept short (say 1 metre or less).

****

SAQ 3.4b

Which one of the following serial data packets corresponds to the transmission of the ASCII character 'A' using 300 baud even parity and two stop bits?

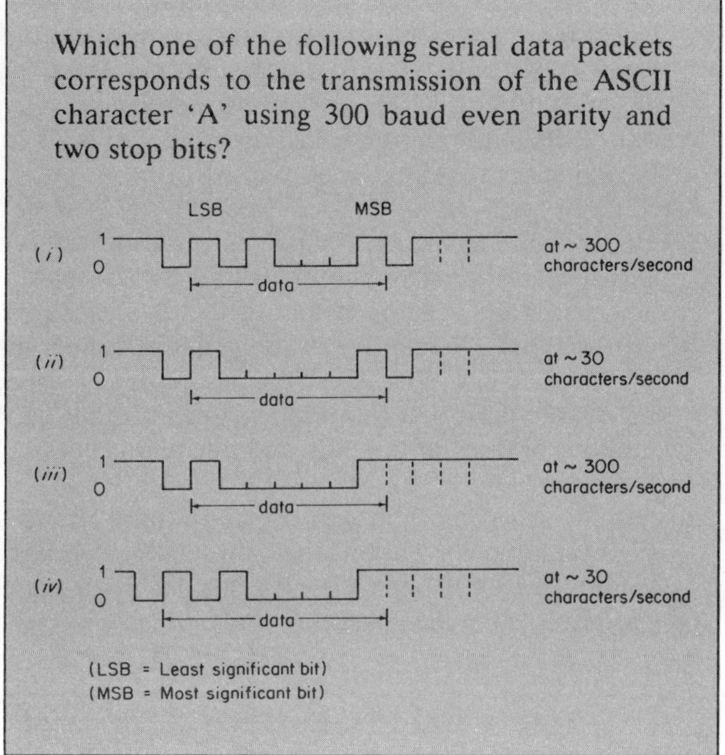

Response

(*i*) No, this one is wrong on a number of counts. Firstly the baud rate is the number of bits transmitted per second. Since each character takes 11 bits (7 for the character, 2 stop, 1 parity and 1 start bit) to transmit at about 300 characters per second we would have a baud rate of about 3300, which is a non-standard rate and does not correspond to the required 300 baud. Secondly the datum transmitted is 1000101 = 69 or the ASCII code for 'E' (remember the most significant bit in the serial data is on the right whereas conventionally we write binary numbers with the most significant bit on the left). Finally, we wanted even parity, but adding up the number of ones in the data plus parity bits gives 3. This is odd parity rather than even.

(*ii*) Yes this is the correct one. The datum is of even parity because there are 2 ones in the data plus parity bits. The baud rate of 300, with 11 bits transmitted per serial packet means about 300/11 $\approx$ 27 characters per second, or roughly 30/second. Finally the datum transmitted corresponds to 1000001 = 65 which is the ASCII code for 'A'.

(*iii*) No, this one is wrong. Firstly the question refers to 300 baud which corresponds to about 30 characters/second not 300 characters/second. See the response to (*i*) above for more detail on this point. Secondly, although the datum transmitted corresponds to the character 'A', the parity is wrong. For even parity, the parity bit should have been set to 0, so that the total number of ones in the data and parity bits is even.

(*iv*) No, this is wrong. The transmission rate of about 30 characters per second is correct but the character sent is wrong. It corresponds to 1000101 = 69 or 'E'. For this incorrect character the parity bit is correctly set to 1.

****

SAQ 3.4c

Write a stepwise design to read a sequence of characters received by a UART and assemble all the characters together in the string variable X$.

Assume that the UART status register can be read from port 201 and each byte that is received can be read from port 202. Concentrate on the problem of reading in the characters one at a time and assume that the UART has been properly configured already for you. ⟶

SAQ 3.4c (cont.)

The string of characters will be terminated by a carriage return (ASCII code 13), and bit 7 of the status register will be set to 1 if a character has been received since the last read of the data port 202.

Response

(Assume UART already configured for use).

1. Initialise string variable X$ = " ".

2. Read status byte from port 201.

3. Mask off all but bit 7 of the status byte and check if bit 7 is set to 1 or not.

4. If bit 7 is set to 0, repeat from step 2.

5. Read the byte received and strip off the parity bit by ANDing with the byte 01111111. Put the result in Z.

6. If Z = 13 a carriage return was received therefore go to 8.

7. Add the ASCII character for the byte received (stored in Z) to X$(Z), then repeat from step 2 to get next character.

8. Exit, as complete string of characters is stored in X$.

****

SAQ 3.4d

(*i*) What is the minimum number of connecting wires needed to establish RS232 communication between two computers? What problems can arise if this minimum linkage is used?

(*ii*) Why should RS232 signal lines never be connected directly to electronic circuits which use TTL voltage levels?

(*iii*) Which of the following are not standard baud rates?

(1) 300, (2) 400, (3) 600,

(4) 800, (5) 1200, (6) 1600,

(7) 2400, (8) 4800, (9) 9600

Response

(*i*) Three wires are needed to establish serial communication at the minimum level. These are the transmit, receive and common ground lines. A problem which can arise is that the data being sent may arrive too quickly for the receiving software to cope. An over-run error would be produced in the receiving UART and data would be lost. This is particularly likely to happen when the receiving software is written in BASIC and so to avoid it one should use low baud rates (300 baud or less).

(*ii*) The reason is that RS232 signal lines use a different voltage range. For TTL the range is about 2.8 to 5 volts for a logic 1, and between 0 and about 0.8 for a logic 0. In contrast, for RS232 communication, a logic 1 is represented by between -3 V and -15 V, and a logic 0by between $+3$ V and $+15$ V. Since the voltage levels are so different damage to TTL circuitry would occur if you connected them directly.

(*iii*) The non-standard ones are (2) 400, (4) 800 and (6) 1600.

The standard ones are related by multiples of two: 300, 600, 1200, 2400, 4800, 9600. (110 and 19,200 band are also standard values).

****

SAQ 3.4e

Designate each of the following statements as either true (T) or false (F).

(*i*) An IEEE-488 instrumentation interface is correctly described as bit-parallel and byte-serial.

T / F

(*ii*) A big advantage of the IEEE-488 system is that any number and variety of instruments and devices can be connected to the controlling computer as long as each has the appropriate interface.

T / F

(*iii*) Two instruments connected through IEEE-488 interfaces to the same controller can have the same primary and secondary addresses provided they are different types of instrument.

T / F

(*iv*) Although the IEEE-488 system is a high-speed instrumentation interface, rapid data capture is never possible when the application program is written in BASIC.

T / F

Response

(*i*) This is true. The data bus for IEEE-488 interface has 8 lines to transmit in parallel 8 bits of data. However, commands and data usually take the form of a sequence of ASCII characters and these are transmitted one at a time. In this sense the system is bit-parallel and byte-serial.

(*ii*) No, this statement is false as there is a strict limit to the number of devices which can be connected at anyone time. Including the controller the limit is 15.

(*iii*) No, that is false. The primary and secondary addresses (taken together) of a device must be distinct. It is possible to include a number of interfaces together under the same primary address by giving each interface a distinct secondary address. Then, if you inadvertently have two devices with the same secondary address, you will obtain a bus error signal.

(*iv*) The some extent this depends on the implementation of the driver software for the IEEE 488 interface. A good quality system will allow you to define arrays into which data can be loaded directly from the interface without executing BASIC statements. Thus high speed data capture or data output becomes possible even using BASIC programs. To conclude the statement given is false because there are situations where it can be done.

****

SAQ 4.1a

The determination of Na^+ by the method of standard additions requires the following solutions to be dispensed into the measuring vessel under computer control.

(A) the sample,
(B) standard Na^+ solution,
(C) rinse solution to prepare the measuring vessel for the next sample.

For each of the above solutions (A), (B) and (C) choose the most appropriate method of dispensing the solution, bearing in mind both the accuracy required and cost.

(*i*) A peristaltic pump operating at 240 V AC.

(*ii*) A 24 volt stepper motor driving a threaded rod which in turn drives a syringe.

(*iii*) Weigh the measuring vessel as a means of monitoring the amount dispensed by gas displacement of the solution from its storage bottle.

(*iv*) A syringe capable of delivering a single shot, equal to the capacity of the syringe, at one stroke.

Response

(*A*) We shall need enough sample solution to cover the electrode system. Assuming electrodes of standard size, and a small measuring vessel, about 30 cm^3 of liquid should suffice.

If we used a computer driven syringe to deliver the sample we should easily be able to deliver a sample volume accurate to less than 0.1 cm^3 or about 0.3% which is certainly accurate enough for our purpose. An alternative but less accurate method would involve a peristaltic pump switched on and off under computer control. The sources of error here are likely to be the fluctuations in pumping rate and the quantisation of the delivery into the measuring vessel by the drop size of the liquid leaving the delivery tube. These factors will depend on the quality of the peristaltic pump, the condition of the tubing used with the pump, and the diameter of the exit of the delivery tube. Although the accuracy achieved will depend on the apparatus used, it should be possible to dispense 30 cm^3 of sample solution to an accuracy of 1 or 2% which should be adequate for the present application. Furthermore a peristaltic pump operated on a time basis is technically easier to implement under computer control than say a motor driven syringe and is also cheaper. In spite of these advantages, in this case study we opt for the computer driven syringe on the grounds of finer control of the volume dispensed.

The disadvantage with the gravimetric method of option (*iii*) is the cost of the balance. To obtain the digital resolution required, and recognising the need for a computer interface, we would probably incur a cost equivalent to that of the pX meter.

The main problem with option (*iv*) is the accuracy with which the liquid is metered. For a particular sample it is true that the volume delivered by single shot dispensation could be adjusted to give the correct volume accurately. However, a different sample may need a different volume and it would be useful to be able to change the sample volume under computer control. If a fixed volume is acceptable then this method is a good choice. In this case study we opted for a variable sample volume delivered by a motor driven syringe.

(*B*) The standard additions method requires the accurate addition of small amounts of standard solution to the sample contained in the measuring vessel. Typically an addition may correspond to 1 cm^3 of standard added to 30 cm^3 of sample. Clearly we need to achieve high accuracy in these additions. This criterion certainly excludes option (*iv*) (single shot dispensing) unless a standard volume of addition can be accepted. It would be a pity to constrain the automated operation of the equipment in this way, and so option (*iv*) is excluded from this case study as a means of dispensing standard solution. Option (*i*) which uses a peristaltic pump is unsatisfactory for low volumes even though the possibility of varying the size of the addition exists. Option (*iii*) using a balance is too expensive and so for this case study we choose option (*ii*), the motor driven syringes.

(*C*) There is nothing critical about the rinse solution which needs to be dispensed. The simplest method is probably to use a peristaltic pump (option (*i*)) although option (*iv*) could also be used. We select option (*i*) for this case study.

****

SAQ 4.1b

Which one of the following experimental arrangements would you use to empty the measuring vessel between measurements on different samples?

Computer-controlled valve

Waste

(*i*) In-line liquid control valve with gravity feed. ⟶

SAQ 4.1b (cont.)

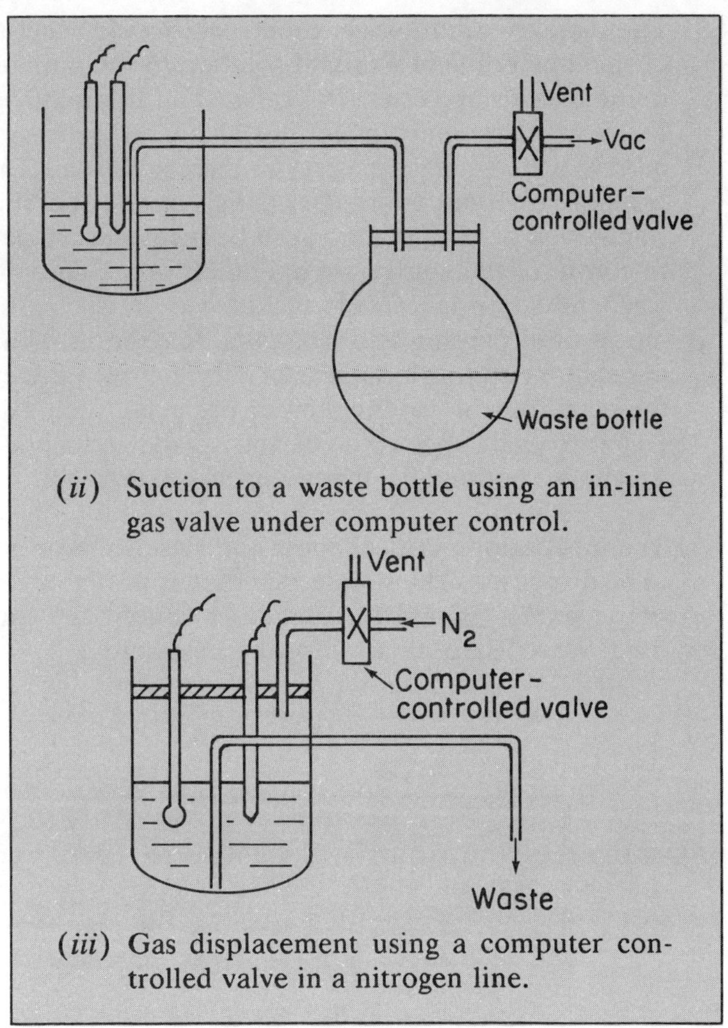

(*ii*) Suction to a waste bottle using an in-line gas valve under computer control.

(*iii*) Gas displacement using a computer controlled valve in a nitrogen line.

Response

(*i*) This is perhaps the most obvious one to choose. There are, however, some experimental problems with it.

The first is that when air is entrained in the waste pipe below the computer controlled valve it is possible to obtain erratic draining. Sometimes when the valve is activated, and when a column of liquid still remains in the waste pipe, draining takes place smoothly. But how long do we keep the valve open? The obvious answer is long enough to drain all the liquid as determined by trial and error. To be on the safe side we might keep the valve open too long. This inevitably introduces a slug of air into the waste pipe, which then causes unpredictable draining when the next sample is to be drained. One way round this is fit some sort of sensor (either optical or conductometric) above the drain tube's valve. The computer could monitor this and switch off the valve before air is trapped in the waste pipe. This however is another complication which makes this choice unattractive. Another drawback with this particular design is that the liquid is in direct contract with the valve. To avoid problems of corrosion we would probably opt for an expensive stainless steel valve or an even more expensive one based on PTFE.

(*ii*) This is a successful design and is the one adopted in this case study. The computer-controlled valve is not in contact with the liquid and so an inexpensive gas valve can be used. The action is very positive in the sense that as soon as the valve opens the liquid is removed from the measuring vessel down to the last drop if the vessel has a curved base and the glass drain tube is carefully fitted to reach the lowest point.

(*iii*) This option is similar to option (*ii*) except we blow out the liquid rather than apply suction.

Again an in line gas valve can be used (which is an inexpensive approach) and virtually all the liquid can be removed by carefully tailoring the entrance to the drain tube to meet the lowest point of the base of the measuring vessel.

However the big disadvantage is that we need a sealed measuring vessel which is difficult to arrange. Furthermore we would need a computer controlled vent to allow us to add liquid without forcing the existing contents of the vessel up the waste pipe where at worst it could be lost and at best it would not be accessible to any stirring action we may employ to ensure mixing after each addition. For these reasons this approach is not adopted.

****

SAQ 4.2a

(*i*) Is the following statement true?

Electromechanical relays can only be used for mains voltage switching.

(*ii*) A solid-state relay can be switched on by supplying a 5 V signal with a current loading of 8 mA. Which of the following connections to a single pin of a computer output port, capable of providing 1 mA at 5 V, will allow computer control of the relay?

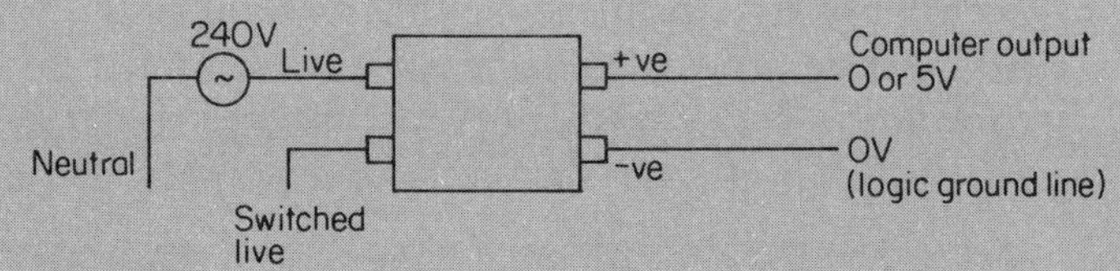

(*a*) Direct connection

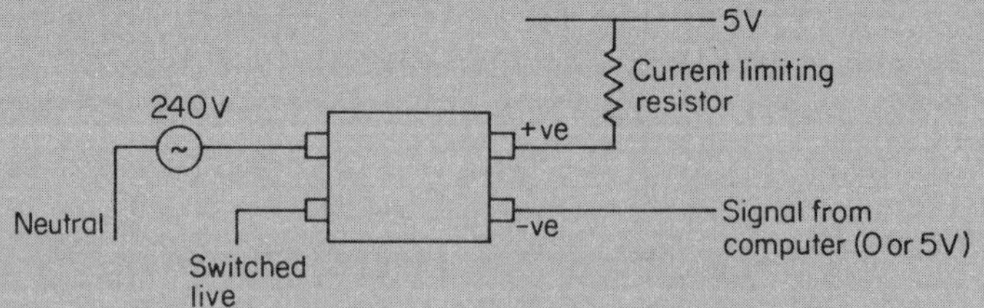

(*b*) Use of external power to provide the current to drive the input side of the solid-state relay.

Response

(*i*) No, the statement is not true. Electromechanical relays are available to cover a wide range of voltages and currents. Most relays can cope with quite high voltages on their output side but their current carrying capacity varies over the range. For low currents we have the so-called 'reed relays' which take only a small current to operate them (of the order of mA) and can conduct about 0.5 A typically. At the other end of the scale we have heavy duty relays capable of conducting 10 A at 30 V DC for example.

(*ii*) The direct connection in (*a*) may not work because the 5 V signal supplied by the computer interface to switch on the relay can only supply 1 mA (the technical jargon is 'to source' 1 mA). The connection shown in (*b*) is an improvement because as a general rule a computer interface can accept (or 'sink') more current than it can source. When the computer supplies a 5 V signal, in this case, the relay is switched *off* because the voltage is the same at both input pins to the relay. (A difference of at least 3 V with the correct polarity is normally required to switch on the relay). On the other hand when the computer supplies a 0 V signal the relay is switched *on* because we have a difference of 5 V across low voltage inputs to the relay. The current flows from the 5 V rail, through the current limiting resistor, passes to the computer interface pin, which is at 0 V and will remain so provided it can 'sink' the current supplied. Since an interface can usually sink more current than it supplies, this arrangement (*b*) is more likely to succeed than (*a*).

Finally note that (*b*) uses inverse logic compared with (*a*). To switch on the relay (*a*) requires the computer to output a digital '1' whereas with (*b*) a digital '0' would be needed.

****

SAQ 4.2b

The following questions all refer to a stepper motor with excitation coils A, B and C, and 24 steps per revolution with a clockwise rotation corresponding to the excitation sequence A,B,C etc.

(*i*) Which one of the following sequences would provide one quarter of a turn anti-clockwise from a point where coil C is energised?

(*a*) ABC ABC ABC ABC
(*b*) CBA CBA CBA
(*c*) BAC BAC
(*d*) CAB CAB

(*ii*) What time delay is required between changes in coil excitation to give a clockwise rotation of 2 revolutions per minute?

(*iii*) What problem could arise if the time delay between excitation of stepper motor coils, in a rotation sequence, is too short?

Response

(*i*) For 24 steps per revolution we need 6 steps to produce a quarter turn. The next coil to be energised after C would be B for an anti-clockwise turn. Bearing these points in mind:

(*a*) is incorrect because there are too many steps and the rotation is clockwise;
(*b*) this is an anti-clockwise rotation but coil B should have been the first one in the sequence;
(*c*) this is correct;
(*d*) this would produce a clockwise rotation starting from coil C.

(*ii*) Two revolutions per minute, with 24 steps per revolution, equals 48 steps in 60 seconds.

The delay time would therefore be $60/48 = 1.25$ s

(*iii*) Excitation pulses would be missed with the result that the motor would not rotate as far as it should.

****

SAQ 4.2c

Which one of the following arrangements would generate an electrical signal at X which would be equivalent to a digital value '0' at the limit of travel of the syringe barrel (not the piston) when the syringe is full.

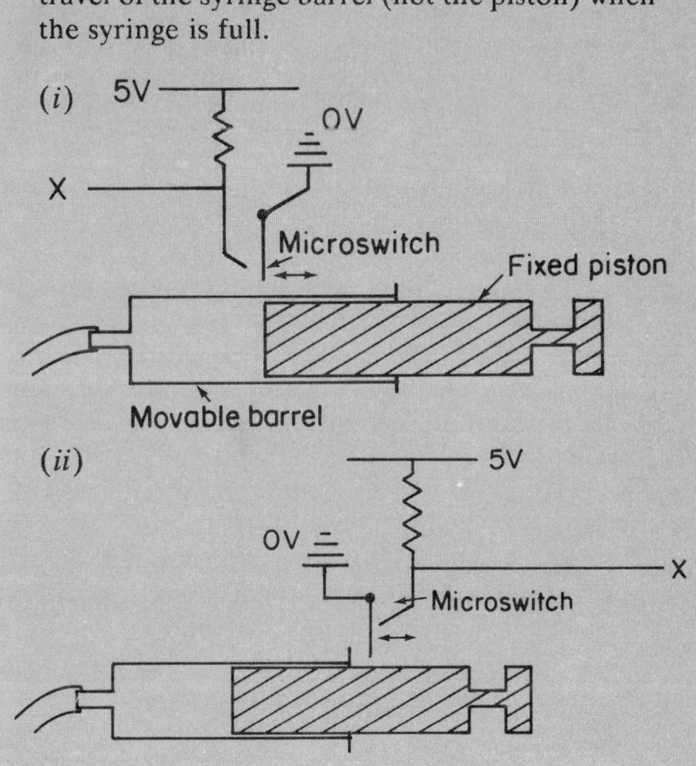

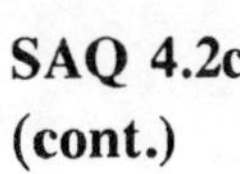

SAQ 4.2c (cont.)

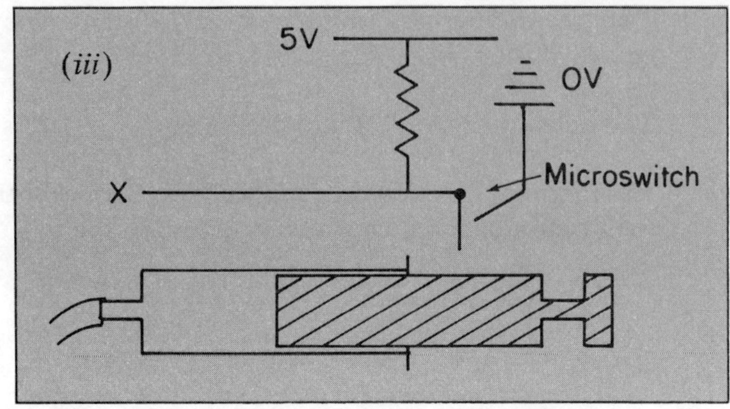

Response

(*i*) This is the correct answer because as the syringe is filled the barrel moves to the left. Eventually the lip of the barrel catches the microswitch and makes a connection with the 0 V line. Before the end of travel is reached point X is connected directly to the 5 V line. Thus before the end of travel is reached the digital value at X is '1' and when the syringe is full the signal changes to '0'.

(*ii*) The microswitch here is mounted on the wrong side of the lip of the syringe barrel. It would not be activated when fitting the syringe. On the other hand it could be used to indicate that the syringe is empty because as liquid is dispensed the barrel moves from left to right. In this case a full or partially empty syringe would produce a digital value for X equal to '0' and when the syringe is empty X would indicate a digital '1'.

(*iii*) Once again the microswitch is positioned wrongly as discussed in (*ii*) above. In this case it could be used to indicate whether or not the syringe is empty. A full or partially empty syringe would correspond to a digital value of X = 1 and an empty one would correspond to X = 0.

****

SAQ 4.2d

If the pX meter gives a reading of −128 mV, which one of the following sets (A,B,C,D) of signals at the output pins of the meter (see Fig. 4.2g) is correct?

Pin number at pX meter	*Digital values* A	B	C	D
1	1	0	0	0
2	0	0	0	0
3	0	0	0	0
4	0	1	1	1
5	0	0	0	0
6	0	1	1	1
7	1	0	0	0
8	0	0	0	0
9	0	1	1	1
10	0	0	0	0
11	0	0	0	0
12	1	0	0	0
13	0	0	0	0
14	0	0	0	1
15	1	0	1	1
16	1	1	1	1

Response

(*a*) This is incorrect for a number of reasons. The first digit should be 8 (recall the datum value was −128) which is 1000 in binary. This is certainly the pattern for pins 1 to 4 *but* it is the wrong way round. Pin 4 is the most significant bit, so Pin 4 should have the value '1', the others being '0'. The same error occurs with the other digits. Finally, as pin 14 is low, the datum is invalid anyway!

(*b*) This is also wrong. The BCD information is correct but in this case the datum is incorrect and the datum corresponds to +128 mV (rather than −128 mV) since pin 15 is low.

(*c*) The only thing wrong with this option is pin 14 is low which means that the data ready signal is not set.

(*d*) This option is the correct one.

****

SAQ 4.3a

Write the dispensing, refilling and initial charging subroutines for syringe 2 starting at line number 5000.

Response

The motor of syringe 2 is controlled by writing numbers 8, 16 and 32 in sequence to Port D by means of variable M2%. The position of the syringe is monitored by reading PC (4 = empty, 8 = full). Otherwise the subroutines are very similar to that for syringe 1:

```
5000 REM ** DISPENSE SYRINGE 2 *****************
5010 REM ** N2% STEPS AT A TIME
5020 FOR I%=1 TO N2%
5030       REM ** CHECK & REFILL IF EMPTY
5040       X%=PEEK(PC)
5050       IF (X% AND 4)=4 THEN GOSUB 5200
5060       M2%=2*M2%
5070       IF M2%>32 THEN M2%=8
5080       POKE PD,M2%
5090       REM ** PAUSE
5100       FOR X%=1 TO Q%:NEXT X%
5110       NEXT I%
5120 RETURN
```

(NB Make M2% = 32 on initialisation.)

```
5200 REM ** REFILL SYRINGE 2 ***************
5210 PRINT "SYRINGE 2 REFILLING"
5220 M2%=M2%/2
5230 IF M2%<8 THEN M2%=32
5240 POKE PD,M2%
5250 REM ** PAUSE
5260 FOR X%=1 TO Q%:NEXT X%
5270 REM ** CHECK & REPEAT UNTIL FULL
5280 X%=PEEK(PC)
5290 IF (X% AND 8)<>8 THEN 5220
5300 PRINT "SYRINGE 2 FULL"
5310 RETURN

5600 REM ** CHARGING SYRINGE 2 ***************
5610 PRINT "EMPTYING SYRINGE 2 BEFORE CHARGING"
5620 M2%=2*M2%
5630 IF M2%>32 THEN M1%=8
5640 POKE PD,M2%
5650 REM ** PAUSE
5660 FOR X%=1 TO Q%:NEXT X%
5670 REM ** CHECK & REPEAT UNTIL EMPTY
5680 X%=PEEK(PC)
5690 IF (X% AND 4)<>4 THEN 5620
5700 GOSUB 5200:REM REFILL
5710 RETURN
```

****

SAQ 4.3b	Write in correct order, but without line numbers, the statements necessary to obtain an emf value from BCD data input through Ports A, B and C.

Response

Here we simply pull together the statements discussed in the text. We must first have a loop which tests pin 6 of Port B until it is set:

```
IF (PEEK(PB) AND 64)=0 THEN (loop back)
```

Then the digits are read:

Digits 1 and 2 from PA:

```
P%=PEEK(PA)
Z=(P% AND 15)
Y=(P% AND 240)/16
```

Digit 3 from PB:

```
P%=PEEK(PB)
X=(P% AND 15)
```

The emf value is calculated from X, Y and Z and assigned to variable V:

```
V=100*X+10*Y+Z
```

The sign of V must be negative if pin PB5 is set, ie

```
IF (PEEK(PB) AND 32)=32 THEN V=-V
```

While digit 4 is not needed in the standard addition program, it may be required when the meter operates in the pH mode. This digit can have the value 0 or 1 only, corresponding to tens of pH. It is found by testing bit 7 of Port B:

```
P%=PEEK(PB)

IF (P% AND 128)=128 THEN W=1 ELSE W=0
```

SAQ 4.3c

> Outline a procedure and write a program to determine the factor S of the equation
>
> $E = B + S*\log[Na^+]$
>
> if the syringes of the standard addition apparatus contain two different standard solutions.

Response

The essential operations are:

Drain (if necessary and rinse the vessel.
Dispense one standard solution (concentration C1).
Measure emf (E1).
Drain and rinse the vessel.
Dispense the second standard (C2).
Measure emf (E2).
Calculate S.

The emf difference depends on the concentration ratio:

$$E2 - E1 = S * \log(C2/C1)$$

The ratio should be chosen to give a reasonable change of emf. (eg If C2/C1 = 10 the change is 55–60 mV but if C2/C1 = 2 it is 16–18 mV).

A suitable program uses the initialisation procedure as before up to line 1220. The remainder of the initialisation and the main program (lines 1230 to 2290) are replaced by the new lines:

```
1230 REM ** DETERMINATION OF CONSTANT *********
1240 INPUT "CONCENTRATION OF SOLUTION 1 ?", C1
1250 INPUT "HOW MANY STEPS OF SOLUTION 1 ?", N1%
1260 PRINT
```

```
1270 INPUT "CONCENTRATION OF SOLUTION 2 ?", C2
1280 INPUT "HOW MANY STEPS OF SOLUTION 2 ?", N2%
1290 PRINT
1300 PRINT "DETERMINATION COMMENCING"
1310 REM ** RINSE VESSEL
1320 GOSUB 6200:REM DRAIN
1330 GOSUB 6000:REM ADD SUPPORT
1340 GOSUB 6200:REM DRAIN
1350 REM ** TAKE STANDARD 1 & MEASURE
1360 GOSUB 4000:REM ADD STANDARD 1
1370 GOSUB 6400:REM MIX
1380 GOSUB 3000:REM MEASURE
1390 E1=E:REM EMF FOR C1
1400 REM ** REPEAT _ DRAIN & RINSE
1410 GOSUB 6200
1420 GOSUB 6000
1430 GOSUB 6200
1440 REM ** NOW STANDARD 2
1450 GOSUB 5000
1460 GOSUB 6400
1470 GOSUB 3000
1480 E2=E:REM EMF FOR C2
1490 REM ** CALCULATE FACTOR
1500 S=(E2-E1)/LOG(C2/C1)
1510 PRINT "FACTOR S IS ";INT(10*S+0.5)/10
1520 END
```

Instead of using two different solutions a procedure could be devised to determine the factor from measurements made with one standard and the supporting electrolyte in varying proportions.

****

SAQ 4.4a

> Write a subroutine starting at line 3400 for adding alkali when the pH is too low.

Response

The number of motor steps is either 25 or 5 in this case. These numbers are chosen to suit the particular situation (concentration of alkali, composition of liquid).

```
3400 REM ** PH TOO LOW ***********************
3410 REM ** DECIDE ADDITIONS (N2%)
3420 IF P<6 THEN N2%=25 ELSE N2%=5
3430 GOSUB 5000:REM ADD ALKALI
3440 GOSUB 3000:REM MEASURE pH
3450 REM ** PH HIGH ENOUGH?
3460 IF P<6.8 THEN 3420
3470 RETURN
```

****

SAQ 4.4b

Write statements which keep an operator informed of the pH and the current operation. (eg the VDU displays the message 'pH=7.8 Adding acid 2 steps at a time'.)

Response

The statements are probably best located in the subroutines for adjusting the pH., eg:

```
3422 PRINT "pH= "; INT(100*P+0.5)/100

3424 PRINT "Adding alkali "; N2%; "steps at a time"
```

The purpose in rounding the pH to 2 decimal places is not so much to make sure that the second decimal is correct but simply to avoid an output with eight or ten figures. Many computers can take care of the rounding process through print formatting ('print using') statements.

It would probably be desirable to arrange for this information to be displayed always at the same position on the VDU. To do this a TAB should be included in the statement and perhaps also a statement for clearing part of the screen.

SAQ 5.1a

An analogue–digital converter needs 50 microseconds to complete a conversion and store the result.

How much computer time per item is required by the time-average method if the signal-to-noise ratio is to be enhanced by a factor of four over that for a single measurement?

Would ensemble averaging effect any saving in time?

Response

Since the factor is proportional to the square root of the number of measurements a factor of 4 requires that 16 replicate measurements be made. At 50 microseconds per measurement the time necessary is 800 microseconds or 0.8 milliseconds.

So long as the measurements are being controlled by a BASIC program this kind of procedure should work all right but difficulties may arise if a machine code program is used because the program may try to take readings too quickly. It is essential that enough time is allowed for conversion to be completed and the result stored before the next conversion is started. This can be a problem even in BASIC programs when the conversion time is long and it may be necessary to introduce a delay between readings.

To achieve the same effect by ensemble averaging would require 16 runs to be made. Since 50 microseconds must be allowed for each reading no time could be saved.

****

SAQ 5.1b

Smooth the data given below by the 5-point moving average method but give double weight to the central point.

Data (as in example program)

4	4	7	8	8	12	18	21	22	31	36	40
50	57	55	69	73	75	81	91	97	99	97	95
94	93	95	87	85	82	78	67	58	56	47	45
42	35	26	24	23	15	9	5	6	5	2	

Response

The only change from the program of the example is in the use of five points and in doubling the weighting given to the central point.

```
100 REM ** 5–POINT SMOOTHING. CENTRAL WEIGHTING
110 REM ** DIMENSION ARRAYS
120 DIM R(50),S(50),W(5)
130 REM ** GET RAW DATA INTO ARRAY
140 READ N%
150 FOR I%=1 TO N%
160     READ R(I%)
170     NEXT I%
180 REM ** INITIALISE WORKING ARRAY
190 FOR I%=1 TO 4
200     W(I%)=R(I%)
210     NEXT I%
```

```
220 REM ** NOTE W(5) STILL TO COME
230 REM ** MAIN LOOP FOR NEW ARRAY
240 REM ** LOSE 2 ITEMS AT EACH END
250 FOR I%=3 TO N%-2
260     REM ** BRING IN W(5) & AVERAGE
270     W(5)=R(I%+1)
280     S=W(1)+W(2)+2*W(3)+W(4)+W(5)
290     S(I%)=S/6
300     REM ** 5-POINT AVERAGE BUT DIVIDE BY 6
310     REM ** NOW MOVE ALONG
320     FOR J%=1 TO 4
330         W(J%)=W(J%+1)
340         NEXT J%
350     NEXT I%
360 REM ** NEW ARRAY NOW COMPLETE
370 REM ** NOW SEE OLD AND NEW (YOU PLOT)
380 FOR I%=3 TO N%-2
390     R(I%)=INT(R(I%)+0.5)
400     S(I%)=INT(S(I%)+0.5)
410     PRINT R(I%),S(I%)
420     NEXT I%
430 END
500 REM ** DATA STATEMENTS
510 DATA 47
520 DATA 4,4,7,8,8,12,18,21,22,31,36
530 DATA 40,50,57,55,69,73,75,81,91,97
540 DATA 99,97,95,94,93,95,87,85,82,78
550 DATA 67,58,56,47,45,42,35,26,24,23
560 DATA 15,9,5,6,5,2
```

Output from this program is tabulated below in a different format. (The program can be made to format output by adjusting the field size and modifying line 410). The new points have been rounded to the nearest integer though plotting them without rounding is more satisfactory:

Original:	(4)	(4)	7	8	8	12	18	21
Smoothed:	–	–	6	7	9	10	13	16
Original:	22	31	36	40	50	57	55	69
Smoothed:	21	25	30	36	42	48	55	60
Original:	73	75	81	91	97	99	97	95
Smoothed:	66	71	77	83	89	94	96	96
Original:	94	93	95	87	85	82	78	67
Smoothed:	95	95	93	91	88	85	80	75
Original:	58	56	47	45	42	35	26	24
Smoothed:	68	61	55	49	45	40	35	29
Original:	23	15	9	5	6	(5)	(2)	
Smoothed:	25	20	15	11	8	–	–	

****

SAQ 5.1c

Write and run a program to smooth the following data by the Savitsky–Golay method for a 7-point cubic smooth. If possible plot the raw and smoothed points.

Data (47 points as before):

4	4	7	8	8	12	18	21	22	31	36	40
50	57	55	69	73	75	81	91	97	99	97	95
94	93	95	87	85	82	78	67	58	56	47	45
42	35	26	24	23	15	9	5	6	5	2	

Response

For 7-point cubic smoothing the convoluting integer set is:

$-2 \ 3 \ 6 \ 7 \ 6 \ 3 \ -2$

and the normalising factor is 21.

The program is similar to that used for the previous SAQ, the main difference being in the convoluting integers and normalising factor.

```
100 REM ** 7-POINT SMOOTHING
110 REM ** DIMENSION ARRAYS
120 DIM R(50),S(50),W(7)
130 REM ** GET RAW DATA INTO ARRAY
140 READ N%
150 FOR I%=1 TO N%
160     READ R(I%)
170     NEXT I%
180 REM ** INITIALISE WORKING ARRAY
190 FOR I%=1 TO 6
200     W(I%)=R(I%)
210     NEXT I%
220 REM ** NOTE W(7) STILL TO COME
230 REM ** MAIN LOOP FOR NEW ARRAY
240 REM ** LOSE 3 AT EACH END
250 FOR I%=4 TO N%-3
260     REM ** BRING IN W(7) & AVERAGE
270     W(7)=R(I%+3)
280     REM ** SUM IN TWO STAGES (LONG!)
290     S=-2*(W(1)+W(7))+3*(W(2)+W(6))
300     S=S+6*(W(3)+W(5))+7*W(4)
310     S(I%)=S/21
320     REM ** MOVE ALONG
330     FOR J%=1 TO 6
340         W(J%)=W(J%+1)
350         NEXT J%
360     NEXT I%
370 REM ** NEW ARRAY NOW COMPLETE
380 REM ** SEE OLD AND NEW (PLOT POINTS)
```

```
390     FOR I%=4 TO N%-3
400     R(I%)=INT(R(I%)+0.5)
410     S(I%)=INT(S(I%)+0.5)
420     PRINT R(I%),S(I%)
430     NEXT I%
440 END
500 REM ** DATA STATEMENTS
510 DATA 47
520 DATA 4,4,7,8,8,12,18,21,22,31,36
530 DATA 40,50,57,55,69,73,75,81,91,97
540 DATA 99,97,95,94,93,95,87,85,82,78
550 DATA 67,58,56,47,45,42,35,26,24,23
560 DATA 15,9,5,6,5,2
```

The original and smoothed data are tabulated below:

Original:	(4)	(4)	(7)	8	8	12	18	21
Smoothed:	–	–	–	7	10	13	16	20
Original:	22	31	36	40	50	57	55	69
Smoothed:	25	29	35	43	48	54	61	66
Original:	73	75	81	91	97	99	97	95
Smoothed:	71	77	83	90	96	98	98	95
Original:	94	93	95	87	85	82	78	67
Smoothed:	95	94	92	89	86	81	75	68
Original:	58	56	47	45	42	35	26	24
Smoothed:	60	53	49	45	40	34	29	24
Original:	23	15	9	5	(6)	(5)	(2)	
Smoothed:	20	15	10	6	–	–	–	

You might care to follow up this SAQ by running the program using a greater number of points for smoothing. You will then notice how detail is lost and heights (amplitudes, Y values) reduced as more points are included in the averaging process. In practice the analyst must decide how much smoothing is required in any particular case.

****

SAQ 5.2a

(*i*) Run the example program but alter the test IF F% = 2 AND G% = 1 ... (line 390) to IF F%<>G% ... and explain the result.

(*ii*) Draw a flow diagram for the segment of program which tests for the presence of a peak (lines 360–480).

Response

(*i*) The original version tests to see if the amplitudes have changed from *rising* to *not rising*. If this change has occurred then K% is increased by one to fix P%(K%) at the highest point since K% was last changed.

By altering the test to IF F%<>G% the variable K% will also increase by one when there is a change from falling to rising and so valleys as well as peaks will be detected.

If you are interested you might alter some of the data points and try to detect shoulders by similar methods. We shall not pursue this as we are about to deal with another method of examining a series of peaks.

(*ii*) The variables involved in testing are listed below:

F% value 1 if current height is greater than previous, otherwise value 2

G% value of F% from previous point

I% index of current point

S(I%) height of current point (smoothed)

H height of previous point

K% index of peaks

P%(K%) number of highest point so far found in peak K%

We enter the testing routine with a smoothed height S(I%)...

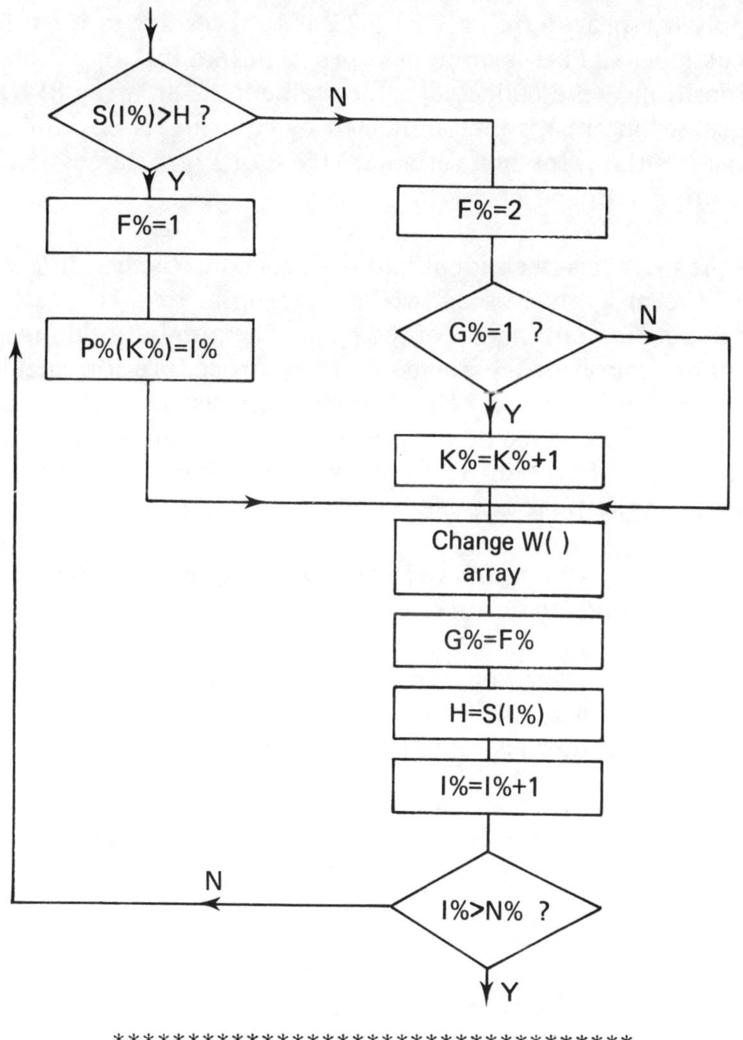

****

SAQ 5.2b

Suggest how the position of a selected peak of a spectrum like that of Fig. 5.2a might be located precisely.

Response

To apply the procedure used in 5.2.2 it is necessary to identify the particular peak. This is done by a program like that of 5.2.1 where peak positions were collected as the elements of an array P%(K%). The peak of interest is then indicated by entering its position at the keyboard as the value of a variable, M%. For Fig. 5.2a possible M% values are 24, 46 and 56.

Once the peak has been identified the procedure is the same as that part of the program in 5.2.2 which, starting at line 410, calculates the true position and height of the peak. The number of points used in the application of the Savitsky–Golay procedure for calculating the precise position and height must be chosen in the light of the number of points which define the peak. For example, no more than 7 points would be used for the peak at position 46 because of the proximity of the peak at position 56.

It should perhaps be remarked that the procedure outlined is computational rather than spectroscopic. The parameters found for spectral peaks may be the correct answers to a mathematical exercise but their interpretation in terms of chemical structure and concentration must take account of the influence of adjacent peaks, especially when there is more than one component.

****

SAQ 5.2c

The simplest and most direct method of producing a first derivative is to plot the differences between successive points (ie plot Y(I%)–Y(I%−1) against I%). Use the data of the example program to do this. Compare the plot obtained by applying this simple method to the raw data with that obtained after smoothing the data.

Response

An outline of the main points of a possible program is given below. Your program should give similar results though it may differ in detail:

The initialisation procedures are the same as in the example program and the smoothing may be done by any of the methods discussed. Storing the raw and smoothed points in arrays makes subsequent programming easier so let us suppose that they are stored in R() and S(). If a 5-point smoothing formula has been employed the calculation and plotting of the first derivative curves then involves a simple loop like:

```
FOR I%=4 to N%-2
R=R(I%)-R(I%-1)
D=S(I%)-S(I%-1)
(Plot R and D against I% after scaling)
NEXT I%
```

The loop starts with point 4 so that the first difference is R(4) − R(3). The first two points are ignored because they are lost in the smoothed array.

You may find that a straightforward procedure like this is adequate in many cases but when second and higher derivatives or more accurate calculations are required or considerable smoothing is needed the least-squares procedure is superior. One advantage of the convolution method is that smoothing and differentiation are done in the same operation.

****

SAQ 5.3a

The simple method of adding ordinates may be regarded as an approximate form of the trapezoidal rule which takes the sum of amplitudes and multiplies the sum by the strip width (no multiplying or dividing by 2).

Apply this approximation to the data in the text example but read the data into an array before carrying out the calculation.

Response

To perform the calculation it is not strictly necessary to use an array. If calculations are to be done later, however, an array is very useful.

```
100 REM ** SIMPLEST INTEGRATION
110 REM ** DIMENSION ARRAY FOR DATA
120 DIM Y(50)
130 REM ** FIRST READ N% AND H
140 READ N%
150 READ H
160 REM ** READ INTO ARRAY
170 FOR I%=1 TO N%
180     READ Y(I%)
190     NEXT I%
200 REM ** NOW ADD - START WITH S=ZERO
210 S=0
220 FOR I%=1 TO N%
230     S=S+Y(I%)
240     NEXT I%
```

```
250 REM ** MULTIPLY BY H FOR AREA
260 A=H*S
270 PRINT "AREA="; A
280 END
300 REM ** DATA STATEMENTS
310 DATA 31,10
320 DATA 0,1,2,5,8,12,18,28,42,65
330 DATA 100,141,172,190,199,199
340 DATA 190,174,151,126,101,76,53
350 DATA 35,21,12,6,3,2,1,0
```

****

SAQ 5.3b

The 27 numbers below represent signal amplitudes measured at regular intervals of 5 units. Determine the area of the peak by Simpson's Rule, taking the base line as a straight line between the amplitudes at points 3 and 25.

Data:

58 54 52 52 53 55 60 65 72
80 88 99 110 123 132 135 133 116
95 74 54 40 33 29 26 26 27

Response

The program below has been written to allow the limits of integration (points 3 and 25) to be entered at the keyboard after the graph has been displayed. An alternative procedure would be to include the limits in the data but if this is done you do not have the opportunity of inspecting the peak before deciding where to draw the base line.

```
100 REM ** SIMPSON'S RULE. SELECT BASE
110 REM ** N% (ODD) & H FROM DATA
120 DIM Y(50)
130 READ N%,H
140 REM ** READ Y VALUES INTO ARRAY
150 FOR I%=1 TO N%
160     READ Y(I%)
170     NEXT I%
180 REM ** (DISPLAY GRAPH ON SCREEN)
190 PRINT "SELECT 2 POINTS (BOTH ODD OR BOTH EVEN)"
200 PRINT "TO DEFINE BASELINE AND LIMITS FOR AREA";
210 INPUT X1%,X2%
220 REM ** ADD FIRST AND LAST TO GIVE S
230 S=Y(X1%)+Y(X2%)
240 REM ** ADD TERMS TO BE MULT'D BY 4
250 T=0
260 FOR I%=X1%+1 TO X2%-1 STEP 2
270     T=T+Y(I%)
280     NEXT I%
290 REM ** MULTIPLY BY 4 & ADD TO S
300 S=S+4*T
310 REM ** ADD TERMS TO BE MULT'D BY 2
320 T=0
330 FOR I%=X1%+2 TO X2%-2 STEP 2
340     T=T+Y(I%)
350     NEXT I%
360 REM ** MULTIPLY BY 2 & ADD TO S
370 S=S+2*T
380 REM ** SUM NOW COMPLETE
390 REM ** CALCULATE SURPLUS AREA
400 E=(X2%-X1%)*(Y(X2%)-(Y(X2%)-Y(X1%))/2)
410 PRINT "AREA ="; H*S/3-E
420 END
500 REM ** DATA STATEMENTS
510 REM ** NUMBER OF POINTS & WIDTH
520 DATA 27,5
530 REM ** AMPLITUDES
540 DATA 58,54,52,52,53,55,60,65,72,80
550 DATA 88,99,110,123,132,135,133,116
560 DATA 95,74,54,40,33,29,26,26,27
```

Using points 3 and 25 to define the base line the answer obtained was 4393 units. This comes from 8683 for the area between the curve and the X axis less 4290 for the 'surplus' area below the base line.

****

SAQ 5.4a

Suggest how the location and integration program might be modified (*i*) to yield peak times relative to that of a selected peak, (*ii*) to take account of a base line which is not zero but which may change in a predictable manner.

Response

(*i*) The relative times cannot be calculated until all peaks have been processed and so the peak positions must be stored in a array or in memory locations. We therefore modify the subroutine which commences at line 600 and store the position or time of peak J% in an array:

```
622 P%(J%)=I%-1
```

Of course, array P%() must be dimensioned. After the percentage areas have been printed the times are calculated relative to the selected peak and the results printed.

(*ii*) In a very simple case the base line has a fixed amplitude other than zero. If this amplitude is A0 it is only necessary to subtract this value from every amplitude read, eg:

```
272 A=A%-A0
322 A=A%-A0
```

You will note that integer variables must be abandoned if A0 does not have an integral value. If no other change is made the points are plotted as if the base line were of zero amplitude.

One stage more complex is a base line which starts at zero and increases linearly. The equation for such a line is a simple function of I% and so we might write:

```
322 B=M*I%
324 A=A-B
```

Provided M has been given a value the base (B) is worked out for each amplitude. Thus, if we make M=0.024 at the initialisation stage the base is 2.4 when I% is 100. Obviously, the base line can be made as variable as we like by controlling the dependence of B on I%. This dependence could be through a mathematical relationship or by setting up an array of base line values from a blank run.

SAQ 6.1a

The program specification requires that the computer accumulates readings to obtain a baseline value before and after measuring each standard and unknown solution. With reference to these measurements, indicate whether or not each of these statements is true (T) or false (F).

(*i*) A drifting baseline is fully compensated for by taking the mean baseline before and after each measurement of sample or standard solution.

(T / F)

(*ii*) The output to the recorder, which is used as an input signal to the computer, should be adjusted so that a zero reading is obtained by the computer when solvent is sprayed into the flame.

(T / F)

SAQ 6.1a (cont.)

(*iii*) The digital readings, corresponding to the analogue voltage in the millivolt range, should be scaled to give a proper absorbance values before using them to compute analytical results.

(T / F)

Response

(*i*) This statement is false. The simple averaging of the baseline before and after measuring a sample or standard does take some account of baseline drift but it cannot fully compensate for it, even if the drift is strictly linear. The point is illustrated by the diagrams (*a*) and (*b*) below.

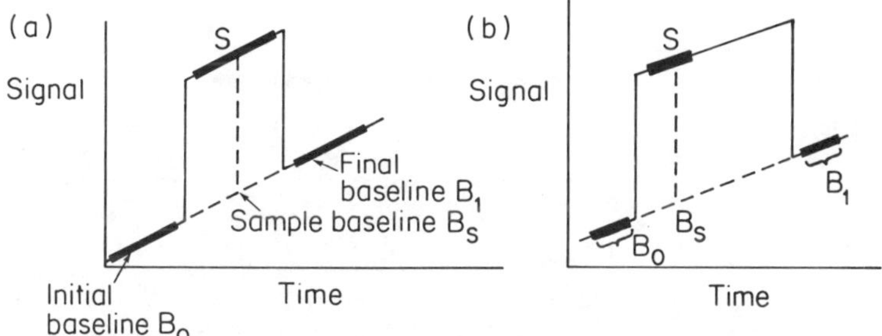

In both diagrams, the time period when the computer is accumulating data is indicated by the thickened line. In case (*a*), B_O is the mean reading when spraying solvent before measuring a sample or standard. The mean value *S* corresponds to reading for the sample or standard, and B_1 is the mean baseline measurement following the peak. The estimate of the baseline immediately below the centre of the peak, B_s, is given quite well as the mean of B_O and B_1. The reason this works is that the data are accumulated the whole time the sample or standard is sprayed. The situation is markedly different in case (*b*). Here we spray sample or standard for much greater time period than that devoted to accumulating data. When we eventually come back to spraying solvent extra drift has occurred than in case (*a*) and the mean of B_O and B_1 is not equal to B_s.

(*ii*) This is false also. Assuming the output is proportional to absorbance, we can simply subtract the baseline value from that of the sample or standard. There is therefore no need to attempt to alter the gain to achieve a zero reading when spraying solvent.

(*iii*) This statement is false too. It may be better to see readings on the screen as true absorbances, but as far as the analytical result is concerned it is not necessary. The important point is that the reading displayed should be proportional to absorbance, and the constant of proportionality should hold for all of the calibration standards and any unknown solution which is analysed, as well as measurements on the baseline.

****

SAQ 6.1b

The readings, Y, obtained from the AA spectrophotometer, produced the following calibration curve over the range 0 to 5 ppm.

$$Y = 10 * X - 0.2 * X^2$$

(*i*) Calculate the values of X for a reading of Y = 40, by use of the standard solution, given below, for a quadratic equation expressed in the form: ⟶

SAQ 6.1b (cont.)

$$AX^2 + B * X + C = 0.$$

$$X = \frac{-B \pm \sqrt{B^2 - 4 * A * C}}{2 * A}$$

(general solution)

Of the two values of X obtained from the solution of the quadratic equation, which one is physically reasonable?

(*ii*) Under what circumstances is it impossible to calculate a value for X from the general solution given in (*i*) above?

Response

(*i*) Substituting Y = 40 gives the equation:

$$40 = 10 * X - 0.2 * X^2$$

Re-arranging this to correspond to the general form gives the expression:

$$0.2 * X^2 - 10 * X + 40 = 0$$

or A = 0.2, B = −10, C = 40.

Therefore for X we have,

$$X = \frac{-(-10) \pm \sqrt{(-10)^2 - 4 * (0.2) * (40)}}{2 * (0.2)}$$

$$X = \frac{10 \pm \sqrt{68}}{0.4} = \frac{10 \pm 8.25}{0.4}$$

$X = 4.38$ ppm or 45.6 ppm

The lower value is physically resonable because it is within the range of the calibration curve.

(*ii*) First the equation is invalid when A = O, since we need to divide by 2 * A to obtain X. If A were O the result would be infinite and a computational error (digital overflow) would occur as the computer tried to calculate a number great than it can store.

Another problem arises when $4 * A * C > B^2$. Then to obtain a value for X we would have to take the square root of a negative number and this would not give rise to a physically meaningful result. This situation can arise if for some reason the value of C becomes too large. To appreciate what this means in terms of the original measurement, recall that C is effectively equal to the reading Y. (As $Y = 10 * X - 0.2 * X^2$ becomes $0.2 * X^2 - 10 * X + Y = 0$ in the standard form, it follows that the constant term C = Y). But how could we obtain a reading Y which was beyond the range of the calibration curve? An obvious possibility is if we measure an unknown solution which is too concentrated. Another could arise from an instrumental error due to something obstructing the light beam through the flame. A possibility which is particularly important in this case study is when we wish to compute the concentration X for a reading of Y + 5%. It could happen that the computation is valid for the reading Y, but invalid, for the above reasons, when we artificially increase the reading by 5%. The program has to guard against generating a computational error by always checking that B^2 is greater than or equal to $4 * A * C$, and if it is not the calculation for X is avoided and a message output to the operator.

****

SAQ 6.2a

The flow chart in Fig. 6.2a has individual blocks identified by letters in the top right–hand corner. In answering the questions below, refer to these block identification letters.

The user of the program starts his run by adding a new element to disc. He then continues by processing 4 calibration standards followed by 2 unknown solutions. By reference to the flow chart in Fig. 6.2a, which one of the following sequences correctly describes the user's route through the system.

(*i*) A,B,E,B,C,F,G,B,E,J,K,L,M,N,O,L,M, N,O,L,M,N,O,L,M,N,O,L,M,N,O,P,Q,R, S,T,B,exit

(*ii*) A,B,C,F,G,B,E,J,K,L,M,N,O,P,Q,R,S,T, Q,R,S,T,B,exit

(*iii*) A,B,C,F,G,B,E,J,K,L,M,N,O,L,M,N, O,L,M,N,O,P,Q,R,S,T,Q,R,S,T,B,exit

Response

(*i*) This option is wrong for two reasons. First the sequence A,B,E,B,C implies that the operator initially tried to retrieve information from the disc for the element in question. This may be realistic if he or she thought the information was on file, but the question did specify the addition of a new element to the disc data file. A more substantial reason for rejecting this response is that only one sample is analysed. The sequence Q,R,S,T is that for analysis of an unknown, and as such it should be traversed twice in this case.

(*ii*) This route is wrong because only one standard would be processed and only one sample analysed. In fact no calibration would be possible using only one standard since we need at least three measurements to determine the three unknown coefficients in the equation which describes the calibration curve.

(*iii*) This is the correct choice. The user enters the system via the option to add new data to disc and then proceeds to process 4 standards and 2 unknowns.

****

SAQ 6.2b

How would you modify the program design given in Fig. 6.2a to prevent the user from trying to produce a calibration curve based on less than 3 standard solutions?

Response

Immediately following the block labelled 'O', the number of standards should be checked. If the standard counter, ST is less than 3 then execution should continue from the beginning of block L. If 3 or more standards have already been run then the question 'Any more standards?' can be put as in the original design.

****

SAQ 6.2c

Suppose the the operator made a mistake and put the same standard solution in twice (with the correct concentration in each case), would this lead to any computational problems for the program as designed?

Response

Yes, this could cause a problem. In the extreme example where 3 standards were input but each had the same concentration, the computer would be unable to calculate the coefficients a, b, c in the equation for the calibration curve. The equations would become ill–conditioned and no meaningful results could be obtained. Mathematical techniques are available for checking that the equations used are not ill–conditioned but a discussion of them is beyond the scope of this introduction. A well-designed program would check that the equations are not ill–conditioned before attempting the calculation. The interested reader is referred to one of the statistical texts mentioned in the study guide.

****

SAQ 6.3a

Would the following section of program fail if line 30 were omitted?

```
10 X=OPENIN"DFILE"
20 INPUT#X,NN:REM read the number of
                    chemical elements on file
30 IF NN=0 THEN 70
40 FOR I = 1 TO NN
50     INPUT#X, EL$(I),WV$(I),SW$(I),
                    FO$(1),RA$(I),LC$(I)
60     NEXT I
70 CLOSE#X
```

Response

Yes, under certain circumstances the program would not work properly. The problem arises when NN = 0, which corresponds to no elements on file, and is due to the fact that a FOR–NEXT loop is always executed at least once even if the first value of the looping variable (I = 1 here) is greater than the upper limit (NN = 0). Thus line 50 will be executed with I = 1 and an attempt will be made to read non-existent data from the file. This will generate an error typically reported as EOF (end of file) at line 50. Whenever you write a program involving FOR-NEXT loops it is good practice to check the value of the upper limit of the loop against the initial value of the looping variable to ensure invalid loops are not executed.

****

SAQ 6.4a

Which of the following binary patterns corresponds to DVM reading of +89.4 mV?

	Port A	Port B	Port C
(*i*)	01001001	00001000	10000000
(*ii*)	10000000	01001001	11000000
(*iii*)	00001000	10010100	10000000

Response

A reading of +89.4 mV means digit 1 is 0, digit 2 is 8, digit 3 is 9, and digit 4 is 4 as can be seen from Fig. 6.4a. As digits 1 and two make up the byte read from Port A, the binary pattern for Port A is therefore 0000 1000 (corresponding to decimal 0 and 8 respectively). Similarly for Port B, which holds digits 3 and 4 we have the binary pattern 1001 0100 (or decimal 9 and 4 respectively). Port C should provide a data valid signal on bit 7 and a zero at bit 6 indicates a positive reading. The correct answer is therefore:

Port A	Port B	Port C
00001000	10010100	10000000

which corresponds to option C. The other two options have the various digits assigned to the wrong bits in the Port which normally holds them and in one case, option B, bit 6 indicates a negative reading.

****

SAQ 6.5a

> In obtaining a response to the opening menu, would the computer accept lower case rather than capital letters?

Response

The program as written would not accept lower case letters in place of capitals. Each keyboard character has an internationally accepted code, called the ASCII code. The codes for A,B,C,D are 65,66,67,68 respectively. In contrast the codes for a,b,c,d are 97,98,99,100. Line 1180 obtains the ASCII code for the character in X$ and line 1190 checks the range. One way of allowing upper and lower case characters would be to change line 1190 and add a new line 1192 as follows.

```
1190 IF (X<65 OR X>68) THEN GOTO 1192 ELSE 1200

1192 IF (X<97 OR X>100) THEN GOTO 1170 ELSE X=X-32
```

If the letter is a capital (A to E) execution passes to line 1200 as before, but if it is lower case line 1192 is executed. If the letter is not a to e the input is rejected. If it is acceptable we subtract 32 from the ASCII value to produce an upper case ASCII code (65,66,67,68,69).

Units of Measurement

For historic reasons a number of different units of measurement have evolved to express quantity of the same thing. In the 1960s, many international scientific bodies recommended the standardisation of names and symbols and the adoption universally of a coherent set of units—the SI units (Système Internationale d'Unités)—based on the definition of five basic units: metre (m); kilogram (kg); second (s); ampere (A); mole (mol); and candela (cd).

The earlier literature references and some of the older text books, naturally use the older units. Even now many practicing scientists have not adopted the SI unit as their working unit. It is therefore necessary to know of the older units and be able to interconvert with SI units.

In this series of texts SI units are used as standard practice. However in areas of activity where their use has not become general practice, eg biologically based laboratories, the earlier defined units are used. This is explained in the study guide to each unit.

Table 1 shows some symbols and abbreviations commonly used in analytical chemistry. Table 2 shows some of the alternative methods for expressing the values of physical quantities and the relationship to the value in SI units.

More details and definition of other units may be found in the *Manual of Symbols and Terminology for Physicochemical Quantities and Units*, Whiffen, 1979, Pergamon Press.

Table 1 *Symbols and Abbreviations Commonly used in Analytical Chemistry*

Symbol	Meaning
Å	Angstrom
$A_r(X)$	relative atomic mass of X
A	ampere
E or U	energy
G	Gibbs free energy (function)
H	enthalpy
J	joule
K	kelvin (273.15 + t °C)
K	equilibrium constant (with subscripts p, c, therm etc.)
K_a,K_b	acid and base ionisation constants
$M_r(X)$	relative molecular mass of X
N	newton (SI unit of force)
P	total pressure
s	standard deviation
T	temperature/K
V	volume
V	volt ($J\ A^{-1}\ s^{-1}$)
a, a(A)	activity, activity of A
c	concentration/ mol dm^{-3}
e	electron
g	gramme
i	current
s	second
t	temperature / °C
bp	boiling point
fp	freezing point
mp	melting point
$\approx$	approximately equal to
$<$	less than
$>$	greater than
e, exp(x)	exponential of x
ln x	natural logarithm of x; ln x = 2.303 log x
log x	common logarithm of x to base 10

Table 2 *Alternative Methods of Expressing Various Physical Quantities*

1. **Mass (SI unit : kg)**

$$\text{g} = 10^{-3}\ \text{kg}$$
$$\text{mg} = 10^{-3}\ \text{g} = 10^{-6}\ \text{kg}$$
$$\mu\text{g} = 10^{-6}\ \text{g} = 10^{-9}\ \text{kg}$$

2. **Length (SI unit : m)**

$$\text{cm} = 10^{-2}\ \text{m}$$
$$\text{Å} = 10^{-10}\ \text{m}$$
$$\text{nm} = 10^{-9}\ \text{m} = 10\text{Å}$$
$$\text{pm} = 10^{-12}\ \text{m} = 10^{-2}\ \text{Å}$$

3. **Volume (SI unit : m^3)**

$$\text{l} = \text{dm}^3 = 10^{-3}\ \text{m}^3$$
$$\text{ml} = \text{cm}^3 = 10^{-6}\ \text{m}^3$$
$$\mu\text{l} = 10^{-3}\ \text{cm}^3$$

4. **Concentration (SI units : mol m^{-3})**

$$\text{M} = \text{mol l}^{-1} = \text{mol dm}^{-3} = 10^3\ \text{mol m}^{-3}$$
$$\text{mg l}^{-1} = \mu\text{g cm}^{-3} = \text{ppm} = 10^{-3}\ \text{g dm}^{-3}$$
$$\mu\text{g g}^{-1} = \text{ppm} = 10^{-6}\ \text{g g}^{-1}$$
$$\text{ng cm}^{-3} = 10^{-6}\ \text{g dm}^{-3}$$
$$\text{ng dm}^{-3} = \text{pg cm}^{-3}$$
$$\text{pg g}^{-1} = \text{ppb} = 10^{-12}\ \text{g g}^{-1}$$
$$\text{mg\%} = 10^{-2}\ \text{g dm}^{-3}$$
$$\mu\text{g\%} = 10^{-5}\ \text{g dm}^{-3}$$

5. **Pressure (SI unit : N m^{-2} = kg m^{-1} s^{-2})**

$$\text{Pa} = \text{Nm}^{-2}$$
$$\text{atmos} = 101\ 325\ \text{N m}^{-2}$$
$$\text{bar} = 10^5\ \text{N m}^{-2}$$
$$\text{torr} = \text{mmHg} = 133.322\ \text{N m}^{-2}$$

6. **Energy (SI unit : J = kg m^2 s^{-2})**

$$\text{cal} = 4.184\ \text{J}$$
$$\text{erg} = 10^{-7}\ \text{J}$$
$$\text{eV} = 1.602 \times 10^{-19}\ \text{J}$$

Table 3 *Prefixes for SI Units*

Fraction	Prefix	Symbol
10^{-1}	deci	d
10^{-2}	centi	c
10^{-3}	milli	m
10^{-6}	micro	μ
10^{-9}	nano	n
10^{-12}	pico	p
10^{-15}	femto	f
10^{-18}	atto	a

Multiple	Prefix	Symbol
10	deka	da
10^{2}	hecto	h
10^{3}	kilo	k
10^{6}	mega	M
10^{9}	giga	G
10^{12}	tera	T
10^{15}	peta	P
10^{18}	exa	E

Table 4 *Recommended Values of Physical Constants*

Physical constant	Symbol	Value
acceleration due to gravity	g	9.81 m s^{-2}
Avogadro constant	N_A	$6.022\ 05 \times 10^{23}$ mol^{-1}
Boltzmann constant	k	$1.380\ 66 \times 10^{-23}$ J K^{-1}
charge to mass ratio	e/m	$1.758\ 796 \times 10^{11}$ C kg^{-1}
electronic charge	e	$1.602\ 19 \times 10^{-19}$ C
Faraday constant	F	$9.648\ 46 \times 10^{4}$ C mol^{-1}
gas constant	R	8.314 J K^{-1} mol^{-1}
'ice-point' temperature	T_{ice}	273.150 K exactly
molar volume of ideal gas (stp)	V_m	$2.241\ 38 \times 10^{-2}$ m^{3} mol^{-1}
permittivity of a vacuum	ϵ_o	$8.854\ 188 \times 10^{-12}$ kg^{-1} m^{-3} s^{4} A^{2} (F m^{-1})
Planck constant	h	$6.626\ 2 \times 10^{-34}$ J s
standard atmosphere pressure	p	101 325 N m^{-2} exactly
atomic mass unit	m_u	$1.660\ 566 \times 10^{-27}$ kg
speed of light in a vacuum	c	$2.997\ 925 \times 10^{8}$ m s^{-1}